PSYCHIATRY, PSYCHOANALYSIS, AND THE NEW BIOLOGY OF MIND

PSYCHIATRY, PSYCHOANALYSIS, AND THE NEW BIOLOGY OF MIND

By

Eric R. Kandel, M.D.

With commentaries by

Arnold M. Cooper, M.D.
Steven E. Hyman, M.D.
Thomas R. Insel, M.D.
Donald F. Klein, M.D.
Joseph LeDoux, Ph.D.
Eric J. Nestler, M.D., Ph.D.
John M. Oldham, M.D.
Judith L. Rapoport, M.D.
Charles F. Zorumski, M.D.

American
Psychiatric
Publishing, Inc.

Washington, DC
London, England

Copyright © 2005 American Psychiatric Publishing, Inc.
ALL RIGHTS RESERVED

Manufactured in the United States of America on acid-free paper
09 08 07 06 05   5 4 3 2 1
First Edition

Typeset in Adobe's Berkeley and Eurostile

American Psychiatric Publishing, Inc.
1000 Wilson Boulevard
Arlington, VA 22209-3901
www.appi.org

Library of Congress Cataloging-in-Publication Data
Kandel, Eric R.
   Psychiatry, psychoanalysis, and the new biology of mind / by Eric R. Kandel ; with commentaries by Arnold M. Cooper ... [et al.].—1st ed.
       p. ; cm.
   Includes bibliographical references and index.
   ISBN 1-58562-199-4 (alk. paper)
   1. Psychiatry. 2. Psychoanalysis. 3. Neurobiology. 4. Molecular biology.
   [DNLM: 1. Psychoanalysis—Collected works. 2. Mental Processes—physiology—Collected Works. 3. Molecular Biology—methods—Collected Works. 4. Psychiatry—methods—Collected Works. WM 460 K16p 2005] I. Title.
   RC435.2.K36 2005
   616.89—dc22   2004029916

British Library Cataloguing in Publication Data
A CIP record is available from the British Library.

For
Paul and Minouche,
who have taught me much
about the
biology of mind

CONTENTS

CONTRIBUTORS

Thomas D. Albright, Ph.D.
Professor, The Salk Institute for Biological Studies, San Diego, California; Investigator, Howard Hughes Medical Institute, Chevy Chase, Maryland

Arnold M. Cooper, M.D.
Stephen P. Tobin and Dr. Arnold M. Cooper Professor Emeritus in Consultation-Liaison Psychiatry, Weill Cornell Medical College; Training and Supervising Analyst, Columbia University Center for Psychoanalytic Training and Research, New York, New York

Steven E. Hyman, M.D.
Provost, Harvard University; Professor of Neurobiology, Harvard Medical School, Cambridge, Massachusetts

Thomas R. Insel, M.D.
Director, National Institute of Mental Health, Bethesda, Maryland

Thomas M. Jessell, Ph.D.
Investigator, Howard Hughes Medical Institute, Chevy Chase, Maryland; Professor of Biochemistry and Molecular Biophysics, Columbia University Medical Center, New York, New York

Donald F. Klein, M.D.
Professor of Psychiatry, Columbia University College of Physicians and Surgeons, New York, New York

Joseph LeDoux, Ph.D.
Henry and Lucy Moses Professor of Science, Center for Neural Science, New York University, New York, New York

Eric J. Nestler, M.D., Ph.D.
Lou and Ellen McGinley Distinguished Professor and Chairman, Department of Psychiatry; Professor, Center for Basic Neuroscience, The University of Texas Southwestern Medical Center, Dallas, Texas

John M. Oldham, M.D.
Chairman, Department of Psychiatry and Behavioral Sciences, Medical University of South Carolina, Charleston, South Carolina

Herbert Pardes, M.D.
President and Chief Executive Officer, New York-Presbyterian Hospital, New York, New York

Michael I. Posner, Ph.D.
Professor Emeritus, University of Oregon, Eugene, Oregon; Adjunct Professor, Department of Psychiatry, Weill Medical College of Cornell University, New York, New York

Judith L. Rapoport, M.D.
Chief, Child Psychiatry Branch, National Institutes of Health, Bethesda, Maryland

Charles F. Zorumski, M.D.
Samuel B. Guze Professor and Head of Psychiatry, Professor of Neurology, Washington University School of Medicine, St. Louis, Missouri

FOREWORD

Eric Kandel is the first American psychiatrist ever to have won the Nobel Prize in physiology or medicine and only the second psychiatrist to have done so in the prize's 102-year history. Much of Kandel's work was done at Columbia University College of Physicians and Surgeons, where for the past 20 years he has been University Professor, the institution's highest academic rank. He is one of only 11 scholars at Columbia who have been awarded that distinction. In addition, he is a professor in the departments of psychiatry, physiology, and biochemistry.

I first met Eric in the late 1960s, when my family and I were vacationing in Wellfleet, Massachusetts, a small village on Cape Cod where Eric and his wife, Denise, have their summer home. Over the ensuing years, I came to know him well, both as a colleague and as a friend. While I was head of the National Institute of Mental Health, I repeatedly sought his counsel. By the time I arrived at Columbia as chairman of the Department of Psychiatry, I was not surprised to find that Eric was one of the mainstays of the academic community and that his advice was valued at all levels of the university.

Eric Kandel received his undergraduate degree at Harvard College and his medical training at New York University School of Medicine. He took postdoctoral training at the National Institutes of Health from 1957 to 1960 and then a residency in psychiatry at the Massachusetts Mental Health Center, which is affiliated with Harvard Medical School. In 1962, he took a fellowship year abroad at the Institut Marey in Paris with Ladislav Tauc, where he began work on the marine snail *Aplysia*.

Eric arrived at Columbia in 1974 as the founding director of the Center for Neurobiology and Behavior. In 1984, he, Richard Axel, and James Schwartz were asked by the Howard Hughes Medical Institute to form a Howard Hughes Medical Institute in Neural Science at Columbia. They, in turn, recruited Tom Jessell and Steven Siegelbaum. These steps helped the

Center for Neurobiology and Behavior at Columbia become what is arguably the leading research group in the world in brain science. As director of the center, Eric organized the neural science course for medical students at Columbia, a course that combines the basic biology of brain and behavior with an introduction to clinical neurology and psychiatry. With James Schwartz and later with Tom Jessell, Eric edited *Principles of Neural Science,* now generally recognized as the standard textbook in the field.

Eric's research has focused on the molecular basis of synaptic plasticity in the central nervous system and on the relationship of this plasticity to cognitive functions. By approaching simple forms of learning in *Aplysia* on the cellular and molecular levels, and then combining this approach with molecular genetics and extending it to forms of learning in the mouse, Eric opened up the study of long-term synaptic plasticity and its relationship to learning and memory storage.

His discoveries are many. He provided the first direct evidence that learning leads to changes in synaptic strength at specific synapses and that memory is associated with the persistence of those changes. Moreover, Eric's success at linking molecular biology and behavior has been extraordinarily influential. He has shown that long-term memory differs from short-term memory in requiring the activation of a cascade of genes and that this genetic program leads to the growth of new synaptic connections.

Eric's work—and Eric himself—has been a source of pride for psychiatry. He has fostered the careers of a number of important young investigators and has thereby stimulated greatly the flow of people and ideas into the molecular study of behavior.

For those achievements, Eric was elected a member of the National Academy of Sciences in 1974 and went on to receive honorary degrees from 12 academic institutions. He has been honored with the Albert Lasker Basic Medical Research Prize, the National Medal of Science awarded to him by President Reagan, the Gairdner International Award for Outstanding Achievement in Medical Science, the Harvey Prize in Medicine awarded by the Technion in Israel, the Wolf Prize of Israel, the Bristol-Myers Squibb Award for Distinguished Achievement in Neuroscience Research, and the Heineken Prize in Medicine from Holland, in addition to the Nobel Prize in physiology or medicine.

The eight previously published articles assembled here cover a range of issues in neurobiology, psychiatry, and psychoanalysis and have influenced the progress of psychiatry over the last 3 decades. Eric's early interest in psychoanalysis and his clinical training have given him a deep appreciation of the implications of his work for psychology. In turn, his continued interest in psychiatry has had a great impact on the overall direction of his work. Central to Eric's thinking has been the idea that a fuller understanding of the biological processes of learning and memory would illuminate our understanding of behavior and of its disorders, an aspiration shared by many leaders of psychiatry, including Sigmund Freud.

The outstanding essays collected in this volume will prove invaluable reading for everyone interested in these areas of research, and they will be treasured as a resource for our thinking about the future.

Herbert Pardes, M.D.

INTRODUCTION

Consistent with my long-standing aspiration to become a psychoanalyst, I trained in psychiatry at the Massachusetts Mental Health Center of Harvard Medical School in the early 1960s and completed a residency there. Then I changed direction. I decided not to obtain psychoanalytic training or even to have a clinical practice. Rather, I spent the next 40 years doing biological research—developing a reductionist approach to learning and memory, first in snails and then in mice.

As a result of the radical change in my career path, I am often asked, "What did you gain from your psychiatric training? Was it profitable for your career as a neural scientist?" I am always surprised by such questions because it is clear to me that my training in psychiatry and my interest in psychoanalysis are central to my thinking; they have provided me with a perspective on behavior that has influenced almost every aspect of my work. Had I skipped residency training and gone to France earlier to spend time in a molecular biology laboratory—say, with Jacques Monod and François Jacob in Paris—I might have worked on the molecular biology of the brain at a slightly earlier point in my career. But the overarching ideas that have influenced my work and fueled my interest in conscious and unconscious memory derive from a perspective on the mind that psychiatry and psychoanalysis opened up for me.

The essays in this volume reflect that influence. They also reveal neurobiology's influence on my view of psychiatry and psychoanalysis; namely, my hope that molecular biology will provide a fresh perspective on the study of behavior and that the ensuing insights will lead to a new science of the mind, one that is grounded in the rigorous empirical framework of molecular biology yet incorporates the humanistic concepts of psychoanalysis.

When I look back on these essays, some written more than 30 years ago, I appreciate even more that one of the great privileges of an academic career is being allowed to pursue different interests at different times in one's life. At various points in my career, I have grappled with psychoanalysis and psychiatry, with cellular neurobiology, and, since 1980, with molecular biology. I have benefited greatly from the freedom afforded me by academic life and have learned both from the nuanced analytic thinking of psychoanalysis and psychiatry and from the rigorous methods of modern biology. The transitions in my career were not accidental; they reflected the evolution of my thinking about how memory storage might be studied most effectively.

My interest in memory dates back to early childhood. I was born in Vienna in November 1929, 8 years before Austria embraced Hitler. In April 1939, my brother and I emigrated to the United States, followed by our parents a few months later. As an undergraduate at Harvard College, I tried to understand my European past. I took honors in an area called History and Literature and wrote my dissertation on the attitudes toward National Socialism of three German writers. Each of the writers I examined—Carl Zuckmayer, Hans Carossa, and Ernst Junger—represented a different position along the spectrum of intellectual responses to National Socialism. Zuckmayer, a courageous liberal and lifelong critic of National Socialism, left Germany early and went first to Austria and then to the United States. Hans Carossa, a physician-poet, took a neutral position and remained in Germany. He and other more passive intellectual opponents of Hitler undertook a so-called inner emigration: physically, they remained in Germany, but their spirits lay elsewhere. Ernst Junger, a dashing German military officer in World War I, extolled the virtues of the warrior and was an intellectual precursor of the Nazis.

Although I was fascinated with the subject of my thesis—it was my first large-scale, independent intellectual accomplishment—I did not continue the study of modern European intellectual history. While at Harvard, I befriended several students who had been born in Vienna and whose parents were psychoanalysts from the Freud circle. Through Anna Kris (later Anna Kris Wolf), I became interested in psychoanalysis. She introduced me to her parents, Ernst and Marianne Kris, two gifted and remarkable people who were extremely enthusiastic about the future of psychoanalysis and encouraged me to think about a career in their field.

In the 1950s, psychoanalysis seemed to have a future filled with promise. It had developed a family of new insights into the mind and unconscious mental processes, as well as into the complex factors that motivate human behavior. I found the possibility of becoming a psychoanalyst exciting. To pursue this career in the early 1950s, one needed to go to medical school, so I enrolled at New York University School of Medicine in 1956.

Throughout my 4 years of medical school, I stayed with the plan of becoming a psychoanalysis-oriented psychiatrist. In my senior year, believing that a psychoanalyst should have some insight into the biology of brain function, I took an elective period in a neurophysiology laboratory at Columbia University. This belief was a minority view within the psychoanalytic community, but it certainly was not unique. Two psychoanalysts whom I had met, Lawrence Kubie and Mortimer Ostow, had been trained in neurology. Both of them appreciated the importance of Wilder Penfield's finding that stimulating the surfaces of the temporal lobe elicits memory-like experience. Both Kubie and Ostow wrote articles for the *Psychoanalytic Quarterly* indicating that some of the ideas of psychoanalysis might be validated by direct exploration of the human brain. To me, the idea that one could delineate key psychoanalytic ideas about mental structure in biological terms was simply fascinating.

My elective period at Columbia gave me the chance to work with two outstanding neurophysiologists, Dominick Purpura and Harry Grundfest. The experience influenced me profoundly. It also led to a very desirable postdoctoral position at the National Institutes of Health (NIH), a position I was given in large part, I suspect, because of Grundfest's nomination. Upon finishing my medical internship, I went to the National Institute of Mental Health at the NIH as a research associate and member of the Public Health Service. I entered the NIH an enthusiastic but inexperienced research scientist and came out 3 years later a competent neurophysiologist.

I was extremely lucky. During my second year at the NIH, Alden Spencer (a fellow research associate) and I obtained the first intracellular recording from the hippocampus, a part of the mammalian brain studied by Wilder Penfield and Brenda Milner concerned with memory. Alden and I were euphoric. Two young, relatively inexperienced scientists had succeeded in opening up a new area of research. We obtained our initial results in the fall of 1958. I had already made a commitment to start my residency in psychiatry in July 1959, so I wrote to Dr. Jack Ewalt, director of the Massachusetts Mental Health Center and professor of psychiatry at Harvard Medical School, telling him of my situation and asking if it were possible to have a 1-year extension. He replied immediately that I should stay as long as necessary. That third year proved crucial not only for my collaborative work with Alden on the hippocampus, but also for my maturation as a scientist.

Encouraged by this cordial beginning, I visited Dr. Ewalt upon my arrival at the Massachusetts Mental Health Center on July 1, 1960. I asked him if it might be possible to have some space and modest resources to set up a laboratory. He looked at me in astonishment then pointed to the pile of résumés from the 22 other residents who were about to begin their training. "Who do you think you are?" he bellowed. "What makes you think that you are better than any one of these?"

I was completely taken aback, both by the content of his remarks and by the tone. In all my years as an undergraduate at Harvard and a medical student at NYU, none of my professors had ever talked to me like that. I assured him that I had no illusions about how my clinical skills compared to those of my peers, but that I did have 3 years of research experience that I did not want to lie dormant.

Dr. Ewalt told me to go to the wards and take care of patients. I left his office depressed and confused, and I briefly entertained the idea of switching to the residency program at the Boston Veterans Administration. Jerry Lettvin, a neurobiologist and friend to whom I described the conversation, urged me to take the position at the Veterans Administration, stating, "Working at the Massachusetts Mental Health Center is like swimming in a whirlpool."

Nevertheless, because of the excellent reputation of the psychiatry residency program, I decided to swallow my pride and stay at the center. It proved a wise decision: a few days later, I went across the street to the physiology department of Harvard Medical School and discussed my situation with Elwood Henneman, a senior neurophysiologist on the faculty, who immediately offered me space in his laboratory. Several weeks later, Dr. Ewalt approached me and said, "I gather from my colleagues at the medical school, Steven Kuffler and Elwood Henneman, that you are a good person to invest in. What do you need? How can I help you?" He then made available all the resources necessary to continue research in Henneman's laboratory throughout my 2 years of residency training.

In those days, residents worked from 8:30 A.M. to 5:00 P.M. with only a rare duty on evenings and weekends. As a result, I was able to carry out a fairly interesting and somewhat original series of experiments on hypothalamic neuroendocrine cells in my spare time. In those experiments, I found that neuroendocrine cells have all the electrophysiologic characteristics of conventional nerve cells. It was the first direct evidence that cells in the brain that look glandular and secrete hormones also act like nerve cells in their signaling capability.

From that time onward, Dr. Ewalt became a strong supporter of my research, and he went out of his way to help me develop as a scientist. While I was still a second-year resident, he appointed me to a permanent civil service position in the state of Massachusetts, thereby giving me the opportunity to stay indefinitely. Three years later, in 1965, when I was beginning to think about moving to New York, the highly influential and charismatic psychoanalyst Grete Bibring stepped down as head of the Department of Psychiatry at Beth Israel Hospital. Dr. Ewalt and Howard Hiatt, the head of the search committee and chairman of the Department of Medicine at Beth Israel, offered me the position.

Even though I decided to pass up that opportunity in favor of a research career, I have remained deeply indebted to Dr. Ewalt for the support he gave me and other research-oriented people. He believed that basic science would be important to the future of psychiatry, an attitude that was at odds with the predominant view at the Massachusetts Mental Health Center. Elvin Semrad, the head of clinical services, and most of our supervisors were heavily oriented toward psychoanalytic theory and practice. Few of them thought in biological terms, few were familiar with psychopharmacology, and most discouraged us from reading the psychiatric or even the psychoanalytic literature because they thought we should learn from our patients and not from our books.

I hasten to add that despite the narrowness and lack of inquisitiveness of many of the clinical teaching faculty, we nevertheless learned a great deal as residents. We had excellent supervision for individual psychotherapy. Moreover, we learned a great deal from one another. I was fortunate to have in my residency class several people with extraordinary intellectual gifts: Judith L. Rapoport, Joseph Schildkraut, Paul Wender, Alan Hobson, Paul Sapier, Tony Kris, Ernest Hartmann, George Vaillant, and Dan Buie all emerged in later years as leaders of American psychiatry. We influenced each other, and to some degree we influenced our faculty.

The residents organized a discussion group on descriptive psychiatry that met monthly. Tony Kris invited Mark Altschule, an outstanding Harvard internist who was interested in schizophrenia, to lead the discussion. We took turns presenting an original essay that we had prepared for the occasion. I remember, in particular, Vaillant presenting an outstanding paper on good prognosis and bad prognosis in schizophrenia. Prior to our arrival, the Massachusetts Mental Health Center had almost never invited outside speakers to address the residents or the faculty. This was a reflection of the vaunted self-confidence of Harvard and Boston at large, which is best represented by the canard of the Boston matron who, when asked about her travels, responded, "Why should I travel? I'm already there."

Kris, Schildkraut, and I therefore also initiated academic grand-rounds, which brought important people with new views to the hospital to address the staff. While at NIH, I had been spellbound by a lecture by Seymour Kety in which he reviewed the contributions of genetics to schizophrenia. I therefore thought we might kick off our lecture series with that topic. In 1961, I could not find a single psychiatrist in all of Boston who knew anything about the subject. Somehow, I found out that Ernst Mayr, the great evolutionary biologist at Harvard, was a friend of the late Franz Kallmann, a pioneer in the genetics of schizophrenia. Mayr generously agreed to come and give us two (splendid) lectures on the genetics of mental illness.

I had entered medical school with a strong conviction about the promising future of psychoanalysis. Now, I found myself questioning psychoanalysis. I was not alone: my view was shared by Paul Wender, Alan Hobson, Judith Livant Rapoport, Tony Kris, and Ernest Hartmann. Of this group, Tony, Ernest, and I were probably the most sympathetic to psychoanalysis.

At NIH, I had gotten off to a good start in my studies of the hippocampus, which has a role in the process of storing memories of facts and events. However, the function and neural circuitry of the hippocampus are very complex, and I gradually realized that if the power of modern biology were to be applied to the study of memory, the effort would have to begin with the simplest examples of learning and memory—that is, with a radical reductionist approach. I therefore decided to study the marine snail *Aplysia*. I had decided that after 2 years at the Massachusetts Mental Health Center, I would spend a year in Paris working on *Aplysia*. Fortunately, this proved to be a productive and immensely enjoyable career choice.

I returned to Harvard in 1963 as an instructor in the Department of Psychiatry. I carried out full-time research on *Aplysia* and spent a few hours a week supervising residents in their psychotherapeutic work. As I look back, I am amazed to see how far psychiatry has progressed in the last 40 years. Even as recently as 1980, several psychoanalysts held the extreme view that biology is irrelevant to psychoanalysis. Two anecdotes illustrate this view: one relating to my career decision, the other to the relationship between psychiatry and biology.

In 1965, I made what was probably the most difficult career decision of my life. Despite the fact that I was a good therapist and enjoyed working with patients, I decided I would not apply for training at the Boston Psychoanalytic Institute as I had planned (and as many of the other residents at the center in fact did). Rather, I would devote myself to full-time research. In an upbeat frame of mind, having put this decision behind us, my wife and I took a brief holiday. We accepted Henry Nunberg's invitation to spend a few days at his parents' summer home in Yorktown Heights, New York. Henry was a very good friend, and Denise and I knew his parents moderately well. His father, Herman Nunberg, was an outstanding psychoanalyst and an influential teacher whose textbook I much admired for its clarity. He also had a broad, albeit dogmatic, interest in many aspects of psychiatry. At our first dinner together, after I had enthusiastically outlined my new career plans, Herman Nunberg looked at me in amazement and muttered, "It sounds to me as if your analysis was not fully successful; you seem never really to have quite resolved your transference."

I found that comment both humorous and irrelevant—and reminiscent of Elvin Semrad's failure to understand that a psychiatrist's interest in brain research need not imply a rejection of psychoanalysis. On reading an earlier version of this introduction, my longtime friend and colleague Donald Klein, a hard-nosed academic psychiatrist, commented, "Another issue that you may, or may not, want to deal with is whether you have undergone a fruitful personal analysis. Those who have been through an analysis, and feel it was useful, often maintain a positive attitude toward analytic ideas in general. Of course, bad outcomes breed negative evaluation." I certainly think that my analytic experience has been useful to me, and there is no question that this positive attitude contributes to my insistence (Klein would call it my delusional optimism) that biology can transform psychoanalysis into a scientifically grounded discipline. But the point I would emphasize here is that if Herman Nunberg were alive today, it is almost inconceivable that he would pass the same judgment on a psychiatrist who moved into brain science.

In 1986, I attended a symposium in New Haven in honor of Morton Reiser's retirement as chairman of the Department of Psychiatry at Yale University. Mort invited me and several other colleagues to give talks. One of the invitees was Mort's close associate Marshall Edelson, a philosopher of the mind and professor of psychiatry at Yale. In his lecture, Edelson recapitulated some of the arguments he had developed in his book *Hypothesis and Evidence in Psychoanalysis*. He argued that the effort to connect psychoanalytic theory to a neurobiologic foundation, or to try to develop ideas about how different mental processes are mediated by different systems in the brain, should be resisted as an expression of logical confusion. Mind and body must be dealt with separately. He went on to say that scientists might eventually conclude that the distinction between mind and body is not merely a temporary methodologic stumbling block stemming from the inadequacy of our current ways of thought; rather, it is a logical and conceptual necessity that no future developments will ever mitigate.

When my turn came, I gave a paper on learning and memory in the snail. I pointed out that all mental processes, from the most routine to the most sublime, emanate from the brain and that all mental illness, irrespective of symptomatology, must be associated with distinctive alterations in the brain. Edelson rose during the discussion and said that while he agreed that psychotic illnesses were disorders of brain function, the disorders that Freud described and that are seen in practice by psychoanalysts, such as obsessive-compulsive neurosis and anxiety states, could not be explained on the basis of brain functioning.

Edelson's views and Herman Nunberg's more personal judgment represent idiosyncratic extremes, but they were representative of the thinking of a surprisingly large number of psychoanalysts not so many years ago. The insularity of these views hindered psychoanalysis from growing during the recent golden age of biology, a hindrance that I hope will soon disappear.

Marianne Goldberger has made an interesting point regarding these issues: it was not that Nunberg, or perhaps even Edelson, thought that mind and brain were separate; it was rather that they did not know how to join them. Since the 1980s, the way in which mind and brain should be joined has become clearer, and consequently psychiatry has emerged in a new role. It has become not only a beneficiary of modern biological thought but also a stimulus to that thought. In the last few years, I have seen significant interest in the biology of the mind, even within the psychoanalytic community. The next step is to incorporate components of a psychoanalytic perspective into the modern biology of the mind and to create a unified view, from mind to molecules that will be intellectually inspiring to psychiatrists and therapeutically satisfying to patients. Who would predict that this new biology of mind could prove to be not only central to psychiatry and psychoanalysis but also of interest to the whole academic enterprise? For there is every reason to believe that the biology of the mind will be the central pursuit of modern scholarship in the twenty-first century much as the biology of the gene was the central pursuit during the last half of the twentieth century.

This volume includes eight published essays, arranged thematically rather than chronologically, and an afterword. In each case they are preceded by an introductory commentary written respectively by Judith L. Rapoport, Tom Insel, Arnold Cooper, Donald Klein, Joseph LeDoux, Eric Nestler, Steve Hyman, Charles Zorumski, and John Oldham. I am grateful to each of them for their scholarly essays and to Herbert Pardes, my longtime colleague and friend, for his introductory comments.

The first essay, "Psychotherapy and the Single Synapse: The Impact of Psychiatric Thought on Neurobiologic Research," addresses the issues raised by Edelson. It is based on the first annual Elvin Semrad Memorial Lecture, which I gave at the Harvard Club in Boston on June 9, 1978. Semrad was the clinical role model at the Massachusetts Mental Health Center. An extraordinarily charismatic person and a brilliant interviewer of patients, he made an indelible impression both on our patients and on us. Patients could remember several years later a single encounter with Semrad, even though they might have completely forgotten anything that happened to them in therapy with us. Semrad had a magical influence on many residents because of his poetic insights into patients and their diseases. He strongly encouraged us to sit with psychotic patients, listen to them carefully, and care about them. He discouraged reading and research because he believed that reading interfered with our ability to listen to and learn about patients directly from them. One of his famous epigrammatic remarks was "There are those who care about people and those who care about research." Semrad was less concerned about a stronger intellectual fabric for psychiatry or advancing knowledge toward that goal than about developing therapists who could empathize more deeply with patients.

I respected Semrad and learned from him, but we disagreed profoundly about the role of research and the function of a training program in psychiatry. In particular, I was disappointed at his failure to see that better patient care in psychiatry cried out for new knowledge through more focused research and that it was the job of Harvard Medical School to provide an environment in which such knowledge could grow. The essay is an attempt to explain what neurobiologic research could mean for psychotherapy and how a unified psychoanalytic and biological perspective could influence one's work in the laboratory. The theme of this essay is that insofar as psychotherapy works, it works at the same level—that of neural circuits and synapses—as drugs do, a point of view that the field is only now beginning to explore.

"A New Intellectual Framework for Psychiatry" is an extended version of a talk I gave in 1997 on the occasion of the hundredth anniversary of the New York State Psychiatric Institute. The thrust of my argument is that the future of psychiatry is deeply rooted in its past and in its connection with biology and that the training of residents in psychiatry and neurology should begin on common ground. Much as internal medicine is the basic training for residents who become cardiologists or nephrologists, so too the biology of the brain should be the common focus of first-year neurology and psychiatry residents. "A New Intellectual Framework for Psychiatry" was published in the *American Journal of Psychiatry,* and it stimulated the largest number of letters the journal had received in recent years in response to a single article—not all of them positive.

"Biology and the Future of Psychoanalysis: A New Intellectual Framework for Psychiatry Revisited" was written in response to those letters, many of which focused on the relevance of biology to the future of psychoanalysis. I outlined two alternative futures for psychoanalysis: In one, psychoanalysis evolves as a hermeneutic discipline focused on elaborating Freud's powerful set of intuitive insights into the mind, without attempting to document its key conclusions. In the other, psychoanalysis becomes a source of rich ideas that can be tested experimentally—that is, an experimentally based science of the mind. The latter course requires psychoanalysis to collaborate with other experimental sciences. I find it encouraging that the ability to detect functional changes in the brain after psychotherapy has opened up a new, objective way of evaluating the effects of psychotherapy on individual patients.

"From Metapsychology to Molecular Biology: Explorations Into the Nature of Anxiety" is an expanded version of the John Flynn Memorial Lecture presented to the Department of Psychiatry at Yale University School of Medicine, which I gave in a modified form in response to an invitation from Donald Klein at the American College for Neuropsychopharmacology. Published in 1983, the essay argues that psychiatry is badly in need of animal models of psychiatric disorders. Learned fear is a prime example: one can use animals ranging in evolutionary complexity from snails to monkeys to study learned fear because fear is a universal behavior and is conserved in evolution. I also developed the idea that molecules and cellular mechanisms for learning and memory might represent a molecular alphabet that could be combined in various ways to produce a range of adaptive and maladaptive behaviors. This idea turned out to be useful in some of my later work.

"Neurobiology and Molecular Biology: The Second Encounter" is based on a summary I gave at the Cold Spring Harbor Symposium of 1983 at the invitation of James Watson, then director of the Cold Spring Harbor Laboratory. I first met Watson in the early 1970s, while I was on the faculty at NYU. He had become interested in neural science and thought that the Cold Spring Harbor Laboratory should give a summer course in this area. He asked me to help organize one based on *Aplysia*. I was fortunate to recruit JacSue Kehoe and Philippe Ascher, two outstanding young French neurobiologists working on *Aplysia,* to lead the experimental portion of the course. Two parallel lecture series were also organized for the course, one by Jack Byrne, Larry Squire, Kier Pearson, and me and the other by John Nicholls. The courses were very successful, sparking a long-term interest in neurobiology at Cold Spring Harbor. Watson punctuated this interest with a symposium in 1975 on "The Synapse" and again with the 1983 symposium, which was historic. Neuroscientists had been aware of the growth of molecular biology for some time, and many outstanding molecular biologists—including Francis Crick, Seymour Benzer, Sidney Brenner, and James Watson—had already moved into neurobiology. But it was only with the emergence of recombinant DNA that molecular biology began to have an explosive impact on the field of neurobiology. The 1983 symposium signaled the beginning of extraordinary activity in neuroscience.

"Neural Science: A Century of Progress and the Mysteries That Remain," published in *Cell* in 2000, is a collaborative effort with Tom Albright, Tom Jessell, and Michael Posner. It was written at the request of the editors of *Cell* to review the achievements of brain biology in the twentieth century. In some ways, this essay is an update of the report given at Cold Spring Harbor in 1983. By 2000, the range and ambition of neuroscience was broader, extending from genes to mental processes. The four of us outlined the emergence of the new science of the mind: a great unification into one intellectual framework of behavioral psychology, cognitive psychology, neuroscience, and molecular biology.

"The Molecular Biology of Memory Storage: A Dialogue Between Genes and Synapses" is the lecture I gave at the Karolinska Institute when I was awarded the Nobel Prize in physiology or medicine in 2000. I shared the prize with Arvid Carlsson and Paul Greengard. Carlsson, who had discovered that dopamine is a key modulatory transmitter in the brain, suggested that reduced dopaminergic transmission is critical to Parkinson's disease, whereas enhanced transmission contributes to schizophrenia. Greengard had discovered that dopamine acts on a receptor that increases the amount of the second messenger cAMP and that cAMP activates a specific kinase, the cAMP-dependent protein kinase. This kinase phosphorylates a variety of substrate proteins in the cell to initiate synaptic actions. Greengard went on to describe that a variety of transmitters act through second messengers. I was recognized for finding that learning depends on changes in synaptic strength. A transient alteration in the strength of synaptic connections gives rise to short-term memory, while the growth of new synaptic connections extends the memory. Learning recruits modulatory neurotransmitters that act on receptors to increase cAMP, which activates the cAMP-dependent protein kinase and leads to the increased synaptic strength needed for short-term memory. In long-term memory, the cAMP-dependent protein kinase moves into the nucleus of the cell and prompts the activation of genes that lead to the growth of new synaptic connections.

"Genes, Brains, and Self-Understanding: Biology's Aspirations for a New Humanism" is an abbreviated version of the commencement address I gave to the graduating class of 2001 at Columbia University College of Physicians and Surgeons. In it I explore the implications of the mapping of the human genome for medicine in general and for psychiatry and mental health in particular. I also indicate that the new science of the mind and the human genome studies have social implications that will be important for the future of medicine.

I conclude with a brief essay, "Afterword: Psychotherapy and the Single Synapse Revisited," in which I suggest that the time is ripe for psychiatry to take a major step forward. Psychiatry has been revitalized by effective new drugs, and it is being revolutionized by molecular biology, genetics, and neuroimaging. Now we need to use the power of biology and cognitive psychology to take up the task of healing the many mentally ill persons who do not benefit from drug therapy. We need to put psychotherapy on a scientific basis and to explore its biological consequences, using imaging and other empirical means of evaluation. In this way, we may be able to explore which form of psychotherapy is most effective for different categories of patients.

All of these essays were written during the 30 years in which I have been a member of Columbia University College of Physicians and Surgeons and the Department of Psychiatry. I was recruited to Columbia in 1974 as the founding director of the Center for Neurobiology and Behavior, which ultimately included Alden Spencer, James H. Schwartz, Irving Kupfermann, Richard Axel (who went on to win the Nobel Prize in physiology or medicine in 2004), Tom Jessell, John Koester, Steven Siegelbaum, Rene Hen, Lorne Role, Michael Shelanski, Samuel Schacher, Jack Martin, Claude Ghez,

Mickey Goldberg, and Daniel Saltzman, among many others. The center contains one of the most remarkable neural science groups in the world. Many of the ideas discussed in these essays evolved out of interactions with my colleagues, and it has been one of the great joys and privileges of my career to have matured scientifically in this heady intellectual environment.

In 1984, Richard Axel, James H. Schwartz, and I were invited by Donald Fredrickson, president of the Howard Hughes Medical Institute, to develop a Howard Hughes Medical Institute program in neuroscience at Columbia. This allowed us to recruit Thomas Jessell from Harvard and to keep Steve Siegelbaum at Columbia. I was appointed senior investigator. The leadership of the Howard Hughes Medical Institute has consistently encouraged Hughes investigators to have a long-term perspective on their work so they will tackle challenging problems. Research on the molecular biology of learning and memory certainly meets both of those criteria, especially for someone who came to this problem from psychoanalysis!

Eric R. Kandel, M.D.

"PSYCHOTHERAPY AND THE SINGLE SYNAPSE"

Judith L. Rapoport, M.D.

"Psychotherapy and the Single Synapse: The Impact of Psychiatric Thought on Neurobiologic Research," written in 1979, presents a lucid and remarkably timely review of advances in understanding the impact of experience on biological structure and function. From research on early adverse rearing experiences possibly mediated by sensory deprivation, to data from Spitz's early clinical observations, to Harlow's primate rearing manipulations, to Hubel and Wiesel's elegant experiments on visual deprivation, Eric Kandel shows how crucial the right experience at the right time can be for normal psychologic and neurobiologic development.

Eric Kandel's own work was based on the habituation and sensitization training of *Aplysia californica,* taking advantage of its limited wiring system that made relevant neural cells easily identifiable. Using electrophysiological single cell recording, Kandel and his students showed long-term changes in the form of decreased sensory motor excitation occurring during habituation. In sensitization, a parallel and opposite change took place, documenting that presynaptic facilitation now heightened the transmitter release following exposure to a noxious stimulus.

Kandel has since extended this work remarkably and in doing so has enhanced our understanding of behavioral change at a molecular level with widespread implications for learning and memory. There has indeed been a great shift in the field of psychiatry: academic psychiatric centers are research focused and various combinations of brain imaging, molecular genetics, and epidemiology provide the mainstay of most medical school–based clinical psychiatric research (for excellent examples, see Caspi et al. 2003 and Hariri 2002). To accelerate this progress, it would be relatively simple to merge research facilities for neurology and psychiatry departments at many medical centers.

The 1979 message of this paper, highlighting the duality of our mental health practitioners, still stands today. I too recall the duality of our Massachusetts Mental Health Care residency experience with Kandel, although I differ with his recollection in that it never seemed leisurely! We were indeed focused on our personal clinical experience, and those of a hoard of supervisors so numerous that I sped or ran to assorted private offices to dissect my clinical exchanges. Unlike my Swarthmore College undergraduate education, where critique of primary sources was the core teaching vehicle, the sole focus at Massachusetts Mental Health Center was one's own person as the therapeutic tool. As such, it was a combined support group and initiation ceremony. There was a monthly seminar with Ives Hendricks in his home on Beacon Hill; his clinical teaching was superb and eccentric, and the food was superb! The Hendricks seminar experience was balanced by Sam Horenstein's Saturday morning neurology rounds. Sam usually contented himself with documenting subtle neurologic signs in our psychiatric patients. With limited tools, both seminars were essentially descriptive. The real winds of change came from the growing use of antipsychotics and antidepressants. Even though we knew we were in a new age, senior nursing staff still showed us where the hot packs and cold packs were, in case we ever were to need them. Thanks to neuroleptics, we didn't.

Now back to the future. The practice of mental health therapy today shows an even greater duality, and a troubled one. It would be impossible to merge treatment staffs of psychiatry and neurology facilities. The dramatic changes in psychopharmacology of the 50s and 60s have provided medications that are now part of the general medical armamentarium and are (statistically) less likely to be prescribed by psychiatrists. Neuroscience has influenced the treatment of neurological disorders such as Alzheimer's disease, but major new medical treatments for this severe psychiatric illness have been slow to arrive, and we still do not understand the mechanism of important older drugs such as lithium or the basis for the unique efficacy of clozapine. The daily work of a psychopharmacologist in managed care is almost a caricature of our 1979 therapeutic duality; the heavy caseload in today's medication clinics prohibits all but the most superficial interpersonal experience. (This fuels the current crisis in psychiatric training, as such jobs provide little career satisfaction.)

In fairness, we already have some clinical evidence for therapy-induced change in brain circuitry. The best documented therapy in our field is behavior therapy, which is also where these neurobiological changes have been shown (Schwartz et al. 1996). By extension, one might easily envision behavioral treatments of anxiety disorders, monitored by fMRI "office checks" on amygdala activation.

In contrast, more psychodynamic therapy/psychoanalysis is now practiced primarily by psychologists who have no neuroscience training. There have been a few advances in evaluating the efficacy of focused, interpersonally oriented treatment (Weissman et al. 1979). The attempt to develop a unified biological basis for understanding and furthering the psychiatric treatment has not been hampered by a lack of willingness of neuroscientists to take on the biological strata for complex social behaviors (Insel and Young 2001). Here too, one might envision neurobiological aspects of "bonding" applied in future therapies. But in 2004, most would agree that we still have a long way to go in reconciling the relationship between biology and psychiatry, and many would debate whether to even go there. Certainly across all of medicine, even for conditions for which a specific biological cause is known, *short of an absolute cure,* there remains a crucial need for caring clinicians who can help the patient to understand and deal with his or her illness.

References

Caspi A, Sugden K, Moffitt T, et al: Influence of life stress on depression: moderation by a polymorphism in the 5-HTT gene. Science 301:386–389, 2003

Hariri A, Mattay VS, Tessitore A, et al: Serotonin transporter genetic variation and the response of the human amygdala. Science 297:400–403, 2002

Insel T, Young LJ: The neurobiology of attachment. Nat Rev Neurosci 2:129–136, 2001

Kandel E: Psychotherapy and the single synapse. N Engl J Med 301:1028–1037, 1979

Schwartz JM, Stoessel PW, Baxter LR, et al: Systematic changes in cerebral glucose metabolic rate after successful behavior modification treatment of obsessive-compulsive disorder. Arch Gen Psychiatry 53:109–113, 1996

Weissman M, Prousoff B, Dimascio A, et al: The efficacy of drugs and psychotherapy in the treatment of acute depressive episodes. Am J Psychiatry 136:555–558, 1979

CHAPTER 1

PSYCHOTHERAPY AND THE SINGLE SYNAPSE

The Impact of Psychiatric Thought on Neurobiologic Research

Eric R. Kandel, M.D.

This article was originally published in the *New England Journal of Medicine,* Volume 301, Number 19, 1979, pp. 1028–1037.

From the Division of Neurobiology and Behavior, Departments of Physiology and Psychiatry, Columbia University, College of Physicians and Surgeons, and the New York State Psychiatric Institute.

Supported by a Research Scientist Award (MH-18558) and by grants (MH-26212 and NS-12744) from the National Institutes of Health.

Based on the first annual Elvin V. Semrad Memorial Lecture, given at the Harvard Club of Boston, June 9, 1978. Dr. Semrad was born in 1909, in Abie, Nebraska. Educated in Nebraska, he trained in psychiatry at the Boston Psychopathic Hospital, at the McLean Hospital, and at the Boston Psychopathic Institute. In 1952 he was appointed to the faculty of Harvard Medical School and became clinical director of the Massachusetts Mental Health Center. Dr. Semrad had recently retired from his position as professor of clinical psychiatry at Harvard Medical School when he died suddenly on October 7, 1976.

The title of this lecture is at best premature and more likely absurd, but I have adopted it for two reasons. In the first place, I want to emphasize the continuing tension within psychiatry between biologic and psychologic explanations of behavior. Secondly, I want to consider the simplistic but perhaps useful idea that the ultimate level of resolution for understanding how psychotherapeutic intervention works is identical with the level at which we are currently seeking to understand how psychopharmacologic intervention works—the level of individual nerve cells and their synaptic connections.

I will discuss the second issue later. First, I should like to consider the tension within psychiatry. Although this tension is long-standing and almost universal, I first encountered it in 1960, when I entered psychiatric residency training at the Massachusetts Mental Health Center. In looking about I was struck by the fact that our residency cohort, a very congenial and intelligent group, was nonetheless split in a fundamental way on one basic issue: the degree to which we accepted the current psychoanalytic view of the mind as providing the adequate conceptual framework for future work in psychiatry. On this issue we were divided into two groups: the hard-nosed and the soft-nosed.

The hard-nosed residents, many of whom were attracted to the humane and existential aspects of the analytic perspective, thought that the psychoanalytic view of the mind was slightly vague, difficult to verify (or discredit), and therefore limited in its powers. The hard-nosed yearned for more substantial knowledge and were drawn to new ways of thought. In particular, many were drawn to biology. By contrast, most soft-nosed residents had little direct interest in the biology of the brain, which they thought had promised much to psychiatry but delivered little. The soft-nosed saw the future of psychiatry not simply in the development of a better body of knowledge but in the development of better therapists—therapists qualified to provide more effective treatment to very disturbed patients. Needless to say, this distinction is drawn too boldly. Many residents then held, and probably still hold, aspects of both views. But the distinction does draw attention to a fundamental tension, a difference in worldview that existed in the psychiatric world around us as well as in ourselves. I think that most of us at that time simply failed to appreciate two aspects of the relation between biology and psychiatry: we failed to appreciate that the conflicted relation between biology and psychiatry is not unique but is characteristic of the interaction between closely related fields of science, and we did not know that in other fields of science this relation has often aided the advancement of knowledge. People (and noses) may fall by the wayside, but the related scientific disciplines usually profit and move on.

As pointed out by a number of students of science, most recently by the biologist E. O. Wilson (1977), there exists for most parent disciplines in science an antidiscipline. The antidiscipline generates creative tension within the parent discipline by challenging the precision of its methods and its claims. For example, for my own parent discipline, cellular neurobiology, there stands at a more fundamental level the antidiscipline of molecular biology, and for molecular biology there stands at a more fundamental level structural (physical) chemistry. In this context it is clear that neurobiology is the new antidiscipline for which psychology in general and psychiatry in particular are the parent disciplines.

I say "new" antidiscipline because as knowledge advances and scientific disciplines change, so do the disciplines impinging on them. In the period from 1920 to 1960, psychiatry derived its main intellectual impetus from psychoanalysis. During this phase, its most powerful antidisciplines were philosophy and the social sciences (Hook 1959). Since 1960, psychiatry has begun (again) to derive its main intellectual challenge from biology, with the result that neurobiology has been thrust into the position of the new antidiscipline for psychiatry. Modern neurobiology had its first impact on psychiatry when it provided insights into the actions of psychotherapeutic drugs. But most of us believe that this is only the beginning and that in the near future, neurobiology will address a matter of more general and fundamental importance: the biology of human mental processes. When it comes to mental function, however, biologists are badly in need of guidance. It is here that psychiatry, as guide and tutor of its antidiscipline, can make a particularly valuable contribution to neurobiology. Psychology and psychiatry can illuminate and define for biology the mental functions that need to be studied if we are to have a meaningful and sophisticated understanding of the biology of the human mind.

Given the potential power of neurobiology and the vision of psychiatry, we may well ask why this type of complementarity was not viable before. The answer to that question is surprisingly simple. The relevant branches of biology—ethology and neurobiology—were, until recently, simply not mature enough, either technically or philosophically, to address higher-order problems related to mental processes. On the appropriate level of resolution, the cellular level, neurobiology has only recently become capable of accomplishing for psychology and psychiatry what other antidisciplines have traditionally accomplished for their parent disciplines—to expand and enlighten the discipline by providing a new level of mechanistic understanding.

I hasten to emphasize that I do not mean "to displace." As Wilson has pointed out, an antidiscipline is usually narrower in scope than its parent discipline. The antidiscipline can succeed in revitalizing and reorienting the parent discipline. It forces a new set of approaches, new methodologies and new insights, but it does not provide a broader, more coherent framework; it does not produce richer paradigms. Although neurobiology can provide key insights into the human mind, psychology and psychoanalysis are potentially deeper in content. The hard-nosed propositions of neurobiology, although scientifically more satisfying, have considerably less existential meaning than do the soft-nosed propositions of psychiatry. If neurobiology is at all equal to the task, the sciences of the mind are likely to absorb the relevant techniques and ideas generated by neurobiology and, having absorbed them, move on.

This very dichotomy of antidiscipline and parent discipline indicates how the two disciplines can most fruitfully interact. In this interaction, psychiatry has a double role. On the one hand, it must seek answers to questions on its own level—questions related to the diagnosis and treatment of mental disorders. On the other hand, psychiatry must pose the questions that its antidiscipline need answer. One of the powers of psychology and psychiatry, it seems to me, lies in their perspective and, most of all, in their paradigms, their specific views of certain interrelated variables.

I would like to consider the synergistic interaction between psychiatry and biology by describing two paradigms that psychology and psychiatry have defined for neurobiology and that are now being addressed on the cellular level: the effects on later development of certain types of social and sensory deprivation in early life, and the mechanisms of learning.

These two classes of studies are paradigmatic in several senses. In purely behavioral terms, the studies represent examples of the sorts of issues that behavioral science in general and psychiatry in particular must summarize and call to the attention of neurobiology. In addition, the studies are interesting from a methodologic point of view because they illustrate how behavioral models must be simplified and redefined so that they can be effectively tackled on progressively more mechanistic levels.

Deprivation in Early Childhood

Experiments ranging from complex ones in the human infant to simple ones in laboratory animals have documented the existence of a set of critical stages for normal psychologic development. During these stages, the subject must interact with a normal social and perceptual environment if development is to proceed normally. Unless animals and human beings are raised for the first year (or longer) in what the psychoanalyst Heinz Hartmann (1958) first called "an average expectable environment," later social and sensory development is disrupted, sometimes disastrously.

Before formal studies on maternal deprivation were performed, a few anecdotal examples of social isolation were collected by anthropologists and clinicians. From time to time, children had been discovered living in an attic or a cellar, with minimal social contact, perhaps spending only a few minutes a day with a caretaker, a nurse, or a parent. Children so deprived in early childhood are often later found to be speechless and lacking in social responsiveness. It is difficult, however, to analyze exactly what went wrong with these children. One often does not know whether the child was severely retarded mentally from the beginning. In addition, one does not know the nature or degree of social isolation. But further information on isolation has been gained from studies of children reared in public institutions.

In a classic series of studies, the psychoanalyst René Spitz compared the development of infants raised in a foundling home for abandoned children with the development of infants raised in a nursing home attached to a women's prison (Spitz 1945, 1946; Spitz and Wolf 1947). Both institutions were reasonably clean and provided adequate food and medical care. The babies in the nursing home were all cared for by their mothers. Because they were in prison and away from their families, the mothers tended to pour affection onto their infants in the time allotted each day. By contrast, in the foundling home, the infants were cared for by nurses, each of whom was responsible for seven infants. As a result, the children in the foundling home had much less contact with other human beings than did those in the nursing home. The two institutions also differed in another respect. In the nursing home the cribs were open, and the infants could readily watch the activity in the ward. They could see other babies playing and observe the mothers and staff going about their business. In the foundling home, the bars of the cribs were covered with sheets that prevented the infants from seeing outside and thus dramatically reduced the sensory environment. In short, the children in the foundling home lived under conditions of sensory, as well as social, deprivation.

Spitz followed a group of infants at the two institutions from birth through their early years. At the end of the first 4 months of life, the children in the foundling home scored better than those in the nursing home on a number of developmental indices. This difference suggested to Spitz that genetic factors did not favor the infants in the nursing home. However, 8 months later, at the end of the first year, the children in the foundling home had fallen far below those in the nursing home, and syndromes developed that Spitz, like Eckstein-Schlossmann (1926) before him, called "hospitalism" (now often called "anaclitic depression"). The children were withdrawn, they showed little curiosity or gaiety, and they were highly susceptible to infection. In the second and third years of life, when the children in the nursing home were walking and talking like family-reared children, the children in the foundling home were retarded in their development and showed slowed reactions to external stimuli. Only two of 26 children in the foundling home were able to walk, only these two spoke out at all, and even they could say only a few words. Normal children at this age are fairly agile, speak hundreds of words, and can construct sentences (Bloom 1970).

Although Spitz's studies have been criticized for their methodologic weakness (Pinneau 1955), several aspects of the studies have been confirmed (Bowlby 1975; Dennis 1960; Engel and Reichsman 1956; Provence and Lipton 1962). For example, in a study of an orphanage in Teheran where social and sensory stimulation were minimal, Dennis (1960) found that 60% of the 2-year-olds were not capable of sitting up unassisted, and 85% of the 4-year-olds were not yet walking on their own. The studies of Spitz thus stand as a landmark; they define a paradigm that has since been studied repeatedly and profitably.

The next step was to develop an animal model of infant social isolation. This step was taken accidentally by Margaret and Harry Harlow, two psychologists working at the University of Wisconsin. In an attempt to raise a stock of sturdy and disease-free monkeys for experimental work, the Harlows separated the infant monkeys from their mothers a few hours after birth, to feed them a special formula and rear them with special hygienic precautions. The newborn monkeys were fed daily by remote control and observed through one-way mirrors. Monkeys reared in isolation for a year proved to be seriously impaired socially and psychologically. When returned to the monkey colony, an isolated monkey did not play with other monkeys, and its grooming and other social interactions were minimal. When attacked, the monkey did not defend itself. Much of its activity was self-directed and consisted of self-clasping, self-mouthing, and self-mutilating acts, such as chewing on its fingers and toes. It also tended to crouch in a corner and rock back and forth in a manner reminiscent of autistic children. When these monkeys reached sexual maturity they did not mate, and several mature females that were artificially inseminated ignored their offspring. This profound social and psychologic damage resulted from only 6 months of total isolation during the first years. Comparable periods of isolation in later life had little effect on social behavior. These findings suggest that in monkeys, as in human beings, there is a critical period for social development (Harlow 1958; Harlow et al. 1965; Suomi and Harlow 1975).

The Harlows next sought to determine what ingredients had to be introduced into the isolation experience to prevent the development of the isolation syndrome. They found that giving the isolated monkey a surrogate mother, a cloth-covered wooden dummy, elicited clinging behavior in the isolate but was insufficient to allow the emergence of normal social behavior. Social development occurred normally only if, in addition to a surrogate mother, the isolated monkey had contact, for a few hours each day, with a peer who spent the rest of its day in the monkey colony. Recently, Suomi and Harlow (1975) found that the syndrome can sometimes be fully reversed by certain monkey psychotherapists—monkeys with certain specific characterologic traits. However, unlike the traits that Dr. Semrad nurtured in his residents, the characteristics of a successful monkey psychotherapist include an obstinate and truculent pursuit, an unmitigated insistence on continued interaction with the socially withdrawn monkey, until the isolate responds, after 6 months of "therapy," with an apparent flight into health—almost, as it were, out of desperation.

Even restricted sensory deprivation has dire consequences, again, initially revealed through clinical studies. In 1932, von Senden summarized the literature on children born with congenital cataracts that were removed much later in life. The cataracts deprived these children of patterned visual experience but allowed them to see diffuse light. Tested after removal of the cataracts in the teenage years or later, they could not discriminate patterns well. They learned readily to recognize color but had only a limited ability to discriminate forms. Some required months to distinguish a square from a circle. Some never learned to recognize people whom they saw daily (Wertheimer 1951).

Similar results were later obtained in monkeys by Austin Riesen and his colleagues, who reared newborn chimpanzees in the dark: by 3–4 months of age, the normal chimpanzee readily learns to discriminate among visual stimuli and between friends and strangers (Riesen 1958). The infant chimpanzee recognizes and welcomes its caretaker but shows fear and avoidance of strangers. A chimpanzee reared in the dark for over a year and then restored to a normal environment does not learn readily to recognize and avoid objects and cannot discriminate vertical from horizontal lines. Only after weeks of living in a normal environment does the animal learn to distinguish friend from foe. These abnormal responses are not due simply to the absence of sensory stimulation early in life but are due to the absence of patterns of stimulation. A chimpanzee brought up with sensory stimulation in the form of an unbroken field of light, produced by enclosing the head in a translucent dome of plastic that permits normal intensity of stimulation without the contours of the normal visual environment, is just as blind as the animal reared in darkness. Thus, the development of normal perception—that is, the capacity to distinguish between objects in the visual world—requires exposure to patterned visual stimulation early in infancy.

How is this accomplished? Can we begin to relate the interaction between the perceptual environment and the brain during the critical period to the function of individual nerve cells? In an imaginative series of studies in newborn kittens and monkeys, Hubel and Wiesel examined the effects of visual deprivation on cellular responses in the primary visual (striate) cortex (Hubel 1967; Hubel and Wiesel 1977; Hubel et al. 1977; Wiesel and Hubel 1963). They found that a normal adult monkey has good binocular interaction. Most cells in the cortex respond to an appropriate stimulus presented to either the left or right eye; only a small proportion respond exclusively to one eye or the other (Figure 1–1 and Figure 1–2). However, if a monkey is raised from birth to 3 months with one eyelid sutured closed, the animal will be permanently blind in that eye. Electrical recordings made from single nerve cells in the striate cortex after removal of the occluding sutures show that the affected eye has lost its ability to control cortical neurons. Only a very few cells can be driven from the deprived eye. Similar visual deprivation in an adult has no effect on vision.

Hubel and Wiesel next found that visual deprivation in newborn monkeys profoundly alters the organization of the ocular-dominance columns. Normally, the fibers from the lateral geniculate nucleus for each eye end in separate and alternating areas of the cortex, giving rise to equal-sized columns dominated alternately by one or the other eye (Figure 1–3A). The radioautographic data of Hubel and Wiesel show that after deprivation, the columns receiving input from the normal eye are much widened at the expense of those receiving input from the deprived eye (Figure 1–3B). As indicated in Figure 1–3, these changes may occur because the geniculate cells that receive input from the closed eye regress and lose their connections with cortical cells, whereas the geniculate cells that receive input from the opened eye sprout and connect to cortical cells previously occupied by input from the other eye.

These studies have provided direct evidence that sensory deprivation early in life can alter the structure of the cerebral cortex. What I find particularly interesting is that Hubel and Wiesel had physiologic evidence for the effect of sensory deprivation in 1965. Using standard techniques, they failed at that time to find any evidence of structural changes in the cortex. Only in 1970, with the development of new radioautographic labeling techniques for mapping connections among neurons (Cowan et al. 1972), were they able to demonstrate the disturbance anatomically. Thus, in a larger sense, their studies make us realize that we are just beginning to explore the structural organization of the brain and the alterations that may be caused by experience and by disease. It is no wonder that an understanding of the biologic basis of most forms of mental illness has been beyond our reach until now.

It will be interesting, in the future, to see whether social deprivation of the sort studied by Harlow leads to deterioration or distortion of connections in other areas of the brain.

Learning in the Adult

The effect of patterning of environmental experience on brain function is, of course, not limited to early development. Sensory and social stimuli con-

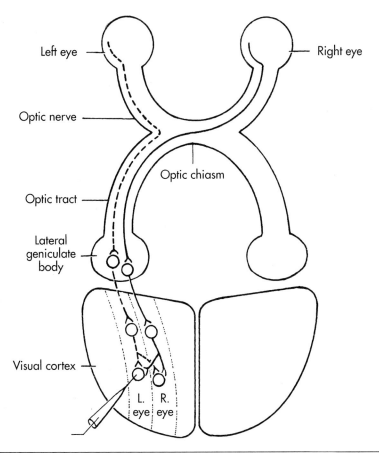

FIGURE 1–1. Diagram of the retinal geniculocortical pathway in higher mammals, showing that the input from the two eyes is segregated until integration is achieved by neurons in the visual cortex.

The left half of each retina is indicated by a solid line for the right eye and a dotted line for the left eye. The axons of cells in the lateral geniculate body form synaptic connections with neurons in the striate (visual) cortex. These cortical neurons are organized into separate columns and receive input from only one eye, but their axons are sent to adjacent columns as well as along their own column. This effect creates a mixing of inputs and allows most cells in the cortex to receive input from both eyes.

stantly impinge on the brain and produce consequences of varying intensity and duration. The most clear-cut and best understood of these consequences is learning. Learning is defined as a prolonged or even relatively permanent change in behavior that results from repeated exposure to a pattern of stimulation (Thorpe 1956). I use learning as my second example of the effects of patterning because I believe that the mechanisms of learning represent a key

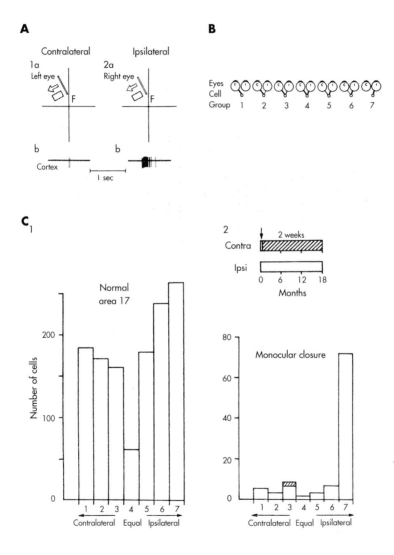

FIGURE 1-2. Binocular interaction and plasticity in the monkey's visual cortex *(opposite page)*.

Part A shows a receptive field of a typical neuron in the visual (striate) cortex as mapped from the left eye (1a) and from the right eye (2a). The neuron responds with a train of action potentials to a diagonal bar of light moving to the left. Each diagram shows the visual field as seen by one eye. Although the two are superimposed, they are drawn separately here for clarity. The fields in the two eyes are similar in orientation, position, shape, and size and respond to the same form of stimulus, in this case, a moving bar. The cell responds more effectively when the stimulus is presented to the ipsilateral eye (A2b) than to the contralateral eye (A1b). F denotes the location of the foveal region in the visual field.

On the basis of the responses illustrated in A, Hubel and Wiesel divided the response properties of cortical neurons into the seven ocular-dominance groups in B. If a cell (small circles) in the visual cortex is influenced only by the contralateral eye (c), it falls into Group 1. If it receives input only from the ipsilateral eye (i), it falls into Group 7. For the intermediate groups, one eye may influence the cell much more than the other (Groups 2 and 6), or the differences may be slight (Groups 3 and 5). According to these criteria, the cell in A would fall into Group 6.

Part C shows ocular-dominance histograms in normal and monocular monkeys. The histogram in C1 is based on 1,256 cells recorded from area 17 in normal adult and juvenile monkeys. The cells in layer 4 were excluded. The histogram in C2 was obtained from one monkey in which the right eye was closed for 2 weeks to 18 months, and recordings were made from the left hemisphere. The shadings in the histogram indicate cells with abnormal responses.

Source. Adapted from Hubel 1967, Hubel and Wiesel 1977, Hubel et al. 1977, and Wiesel and Hubel 1963.

---

to an understanding of character development and of the amelioration of characterologic disorders produced by psychotherapeutic intervention.

The ability to learn from experience is certainly the most remarkable aspect of human behavior. We are in many ways the embodiment of what we have learned. In man as well as other animals, most forms of behavior involve some aspects of learning and memory. Moreover, many psychologic and emotional problems are thought to be learned—that is, they are thought to result, at least in part, from experience. And insofar as psychotherapeutic intervention is successful in treating mental disorders, it presumably succeeds by creating an experience that allows people to change.

As in studies of social and sensory deprivation, the major questions in biologic studies of behavior and learning were first posed 70 years ago, but the ability to answer them was gained only recently. Here, as in investigations of the critical developmental period, this ability came with progressively simpler experimental systems. The most consistent progress has

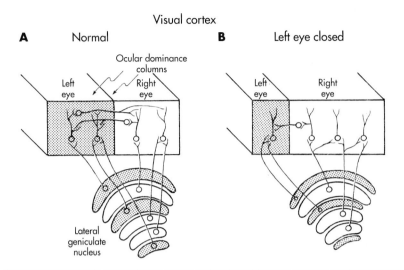

FIGURE 1–3.   Diagram showing changes in the dimensions of the cortical columns for eye preference after closure of the left eye.

After deprivation, the columns receiving input from the normal (right) eye are widened at the expense of those receiving input from the deprived (left) eye.

resulted from studies of two simple forms of nonassociative learning: habituation and sensitization. Each of these forms is evident in human beings but can also be explored effectively in a variety of simple animal models. I will first consider habituation.

Habituation

Habituation, perhaps the simplest form of learning, is a decrease in a behavioral response resulting from repeated presentation of the initiating stimulus. A common example is the habituation of an "orienting response" to a new stimulus. When a novel stimulus such as a loud noise is presented for the first time, one's attention is immediately drawn to it, and one's heart rate and respiratory rate increase. If the same noise is repeated, one rapidly learns to recognize the sound and one's attention and bodily responses gradually diminish (that is why one can become accustomed to working in a noisy office). In this sense, habituation is learning to recognize and to ignore stimuli that have lost novelty or meaning. Besides being important in its own right, habituation is frequently involved in more complex learning, which includes not only acquiring new responses but also eliminating incorrect responses.

The first approach to an animal model of habituation was made by Sherrington in 1906. In the course of studying the behavior underlying posture and locomotion, he observed that habituation of certain reflex forms of behavior, such as the flexion withdrawal of a limb to stimulation of the skin, occurred with repeated stimulation and that recovery occurred only after many seconds of rest. With characteristic prescience, Sherrington suggested that the habituation of the withdrawal reflex was due to a functional decrease in the effectiveness of the set of synapses through which the motor neurons for the behavior were repeatedly activated. This problem was subsequently reinvestigated by Spencer, Thompson, and Neilson, who found close parallels between habituation of the spinal reflexes in the cat and habituation of more complex behavioral responses in man (Spencer et al. 1966). Moreover, by recording intracellularly from motor neurons, Spencer and his colleagues began the modern study of habituation. They found, as Sherrington had suggested, that the depression of the behavior was due to a decrease in the synaptic convergence onto the motor cells. However, the central synaptic pathways of the flexion-withdrawal reflex in the cat are complex, involving many as yet unspecified connections through interneurons. As a result, further analysis of habituation has required still simpler systems in which the behavioral response can be reduced to one or a series of monosynaptic connections.

My colleagues and I have extended the analyses of habituation and sensitization in studies of the marine snail *Aplysia californica*. This animal has a defensive withdrawal reflex of its respiratory organ, the gill, which is similar to the defensive reflexes of mammals, and habituation of this reflex shows all the features that characterize habituation in vertebrates, including man (Kandel 1976; Pinsker et al. 1970). Moreover, the wiring diagram of this behavior is remarkably simple, consisting of 6 identified motor neurons that mediate the behavior and a group of 24 sensory neurons that connect directly onto the motor neurons. There are also several interneurons that receive input from the sensory neurons and converge on the motor neurons (Figure 1–4). Activity in a sensory neuron leads to release of a chemical transmitter substance that interacts with the receptors on the external membrane of the motor cell and reduces its membrane potential. If the membrane potential is reduced sufficiently, the motor cell will fire an action potential. The synaptically produced reduction in membrane potential is therefore called an excitatory synaptic potential (Eccles 1964). In response to the first stimulus, the sensory neurons produce large excitatory synaptic potentials in the motor cells, causing these cells to discharge rapidly and produce a brisk withdrawal. With habituation training, the synaptic potential in the motor cell gradually becomes smaller; it produces fewer spikes, and the behavior is reduced. Finally, the synaptic potential becomes very small, at which point no behavior is produced. After a single training session involving 10 stimuli, the memory for this event (as evidenced by a reduced synaptic potential and behavior) is short, persisting for only minutes or hours. However, after four repeated training sessions spaced over consecutive days, the memory for habituation is prolonged, persisting for more than 3 weeks.

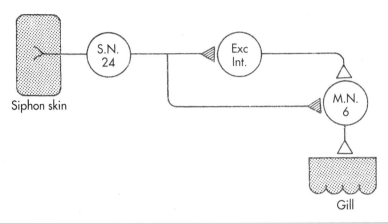

FIGURE 1–4. Diagram indicating the neural circuit for the gill-withdrawal reflex in *Aplysia californica*.

Of about 24 mechanoreceptor sensory neurons that innervate the siphon skin, only one is shown for the purpose of simplification. In this reflex, the site of the plasticity (hatched triangles) underlying habituation is at the terminals of the sensory neurons on the central target cells—the interneurons and the motor neurons.

The critical change underlying short-term habituation occurs at the excitatory chemical synapses that the sensory neurons make on the motor neurons. With repeated stimulation, these synapses become less effective functionally because they release progressively less transmitter. Transmitter release depends on the influx of calcium into the terminals with each action potential. Analyses of the mechanisms that produce habituation indicate that the reduced output of neurotransmitter and the resultant depression of synaptic transmission is caused by a prolonged decrease in calcium influx (Klein and Kandel 1978).

What are the limits of this plasticity? How much can the effectiveness of a given synapse change, and how long can such a change endure? Can long-term habituation produce a complete and prolonged inactivation of a previously functioning synapse? In an effort to answer these questions, the connections between the sensory neurons and a given motor neuron were compared in control animals and animals examined after the acquisition of long-term habituation (Castellucci et al. 1978). In the control animals, 90% of sensory neurons produced detectable connections to the major motor cells (Figure 1–5). By contrast, after long-term habituation, only 30% of the sensory neurons produced detectable connections onto the motor cell, and this effect lasted for over a week; these connections were only partially restored at 3 weeks. Thus, fully functioning synaptic connections were inactivated for over a week as a result of a simple learning experience—several brief sessions of habituation training of 10 trials each.

Thus, whereas short-term habituation involves a transient decrease in synaptic efficacy, long-term habituation leads to prolonged and profound functional inactivation of a previously existing connection. These data provide direct evidence that long-term change in synaptic efficacy can underlie a specific instance of long-term memory. Moreover, at a critical synapse such as this one, relatively few stimuli produce long-term synaptic depression.

Sensitization

Sensitization, the opposite of habituation, is the process whereby an animal learns to increase a given reflex response as a result of a noxious or novel stimulus. Thus, sensitization requires the animal to attend to stimuli that potentially produce painful or dangerous consequences. Like habituation, sensitization can last from minutes to days and weeks, depending on the pattern of stimulation (Pinsker et al. 1973). In this discussion, I will focus on the short-term form.

At the cellular level, sensitization also involves altered transmission at the synapses made by the sensory neurons on their central target cells. Specifically, sensitization involves a mechanism called presynaptic facilitation, whereby the neurons mediating sensitization end on the terminals of the sensory neurons and enhance their ability to release transmitter (Figure 1–6). Thus, the same synaptic locus is regulated in opposite ways by opposing forms of learning: it is depressed by habituation and enhanced by sensitization. The transmitter released by the neurons that mediate presynaptic facilitation (which is thought to be serotonin) acts on the terminals of the sensory neurons to increase the level of cyclic AMP (cAMP). Cyclic AMP, in turn, acts (perhaps through phosphorylation of a membrane channel) to increase calcium influx and thereby enhance transmitter release (Brunelli et al. 1976; Cedar and Schwartz 1972; Cedar et al. 1972; Hawkins et al. 1976; Klein and Kandel 1978; Schwartz et al. 1971) (Figure 1–7).

How effective a restoring force is sensitization? Can it restore the completely inactivated synaptic connections produced by long-term habituation? We have found that study sensitization not only reversed the depressed behavior but restored the effectiveness of synapses that had been functionally disconnected and would have remained so for over a week (Carew et al. 1979) (Figure 1–6B).

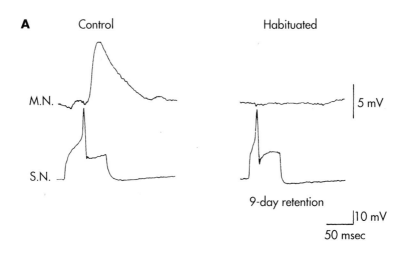

A

Control   Habituated

M.N.

S.N.

9-day retention

5 mV

10 mV
50 msec

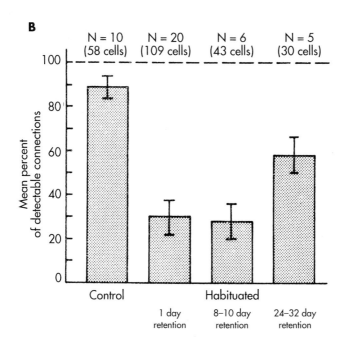

B

N = 10   N = 20   N = 6   N = 5
(58 cells) (109 cells) (43 cells) (30 cells)

Mean percent
of detectable connections

Control   Habituated

1 day   8–10 day   24–32 day
retention   retention   retention

FIGURE 1–5.   Long-term habituation *(opposite page)*.

In A, a synaptic connection between a sensory neuron (S.N.) and the motor neuron (M.N.) L7 is compared in control (untrained) animals and in animals that have been subjected to long-term habituation training. In control animals, the synaptic connections produce a large excitatory synaptic potential. The synaptic connection in habituated animals is undetectable. The sensory neuron was depolarized intracellularly to trigger a single action potential and evoke a synaptic potential in the gill motor neuron L7.

In B, the mean percentage of detectable connections is shown in control and habituated animals tested at three intervals after long-term habituation training. The error bars indicate the S.E.M.
Source.   Adapted from Castellucci VF, Carew TJ, Kandel ER: "Cellular Analysis of Long-Term Habituation of the Gill-Withdrawal Reflex of *Aplysia californica.*" *Science* 202:1306–1308, 1978. Used with permission.

Thus, in these simple instances, learning does not involve a dramatic anatomic rearrangement in the nervous system. No nerve cells or even synapses are created or destroyed. Rather, learning of habituation and sensitization changes the functional effectiveness of previously existing chemical synaptic connections and, in these instances, does so simply by modulating calcium influx in the presynaptic terminals. Thus, a new dimension is introduced in thinking about the brain. These complex pathways, which are genetically determined, appear to be interrupted not by disease but by experience, and they can also be restored by experience.

Implications for the Classification and Understanding of Psychiatric Disorders

The finding that dramatic and enduring alterations in the effectiveness of connections result from sensory deprivation and learning leads to a new way of viewing the relation between social and biologic processes in the generation of behavior. There is a tendency in psychiatry to think that biologic determinants of behavior act on a different "level of the mind" than do social and functional determinants. For example, it is still customary to classify psychiatric illnesses into two major categories: organic and functional. The organic mental illnesses include the dementias and the toxic psychoses; the functional illnesses include the various depressive syndromes, the schizophrenias, and the neuroses. This distinction stems from studies in the nineteenth century, when neuropathologists examined the brains of patients at autopsy and found a disturbance in brain architecture in some diseases and a lack of disturbance in others. The diseases that produced clear (gross) ev-

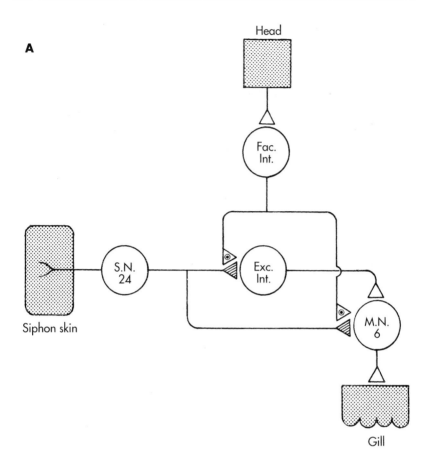

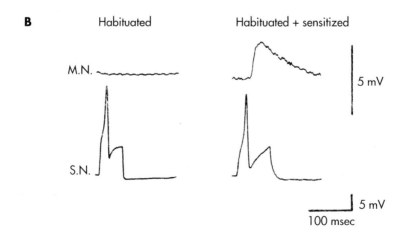

FIGURE 1–6. Scheme of circuit for presynaptic facilitation (A) and restoration of synaptic transmission and behavior by a sensitizing stimulus after long-term habituation (B) *(opposite page)*.

In A, stimuli to the head activate neurons that excite facilitative interneurons (Fac. Int.). The facilitating cells, in turn, end on the synaptic terminals of the sensory neurons (S.N.), where they modulate transmitter release. *Exc. Int.* denotes excitatory interneurons, and *M.N.* motor neuron.

In B, a typical undetectable excitatory synaptic potential from a habituated animal and a typical detectable excitatory postsynaptic potential from a sensitized animal are shown.

Source. Adapted from Carew T, Castellucci VF, Kandel ER: "Sensitization in *Aplysia*: Restoration of Transmission in Synapses Inactivated by Long-Term Habituation." *Science* 205:417–419, 1979. Used with permission.

idence of brain lesions were called organic, and those that lacked these features were called functional. Studies of the critical developmental period and of learning have shown that this distinction is artificial. Sensory deprivation and learning have profound biologic consequences, causing effective disruption of synaptic connections under some circumstances and reactivation of connections under others. Instead of distinguishing between mental disorders along biologic and nonbiologic lines, it might be more appropriate to ask, in each type of mental illness, to what degree is this biologic process determined by genetic and developmental factors, to what degree is it due to infectious or toxic agents, and to what degree is it socially determined? In each case, even in the most socially determined neurotic illness, the end result is biologic. Ultimately, all psychologic disturbances reflect specific alterations in neuronal and synaptic function. And insofar as psychotherapy works, it works by acting on brain functions, not on single synapses, but on synapses nevertheless. Clearly, a shift is needed from a neuropathology also based only on structure to one based on function.

An Overview

Cellular studies of the critical stages of development and of learning have shown that genetic and developmental processes determine the connections between neurons; what they leave unspecified is the strength of the connections. It is this factor—the long-term efficacy of synaptic connections—that is played on by environmental effects such as learning. What learning accomplishes in the instances so far studied is to alter the effectiveness of preexisting pathways, thereby leading to the expression of new patterns of behavior. As a result, when I speak to someone and he or she listens to me, we not only make eye contact and voice contact but the action of the neu-

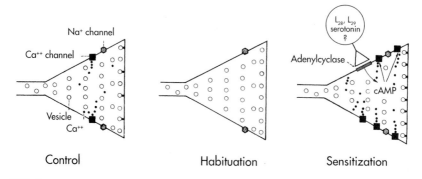

FIGURE 1–7. Model of short-term habituation and sensitization.

ronal machinery in my brain is having a direct and, I hope, long-lasting effect on the neuronal machinery in his or her brain, and vice versa. Indeed, I would argue that it is only insofar as our words produce changes in each other's brains that psychotherapeutic intervention produces changes in patients' minds. From this perspective, the biologic and psychologic approaches are joined. I would hope that the deep-seated dualism that once caused psychiatry and neurobiology to split into hard-nosed and soft-nosed attitudes will prove to be only a transient interlude in the history of psychiatry. Certainly, in their day, Meynert, Wagner-Jauregg, and Freud had little difficulty in appreciating philosophically what my residency cohort and I lost sight of and what we can now again assert, perhaps with slightly more sophistication: what we conceive of as our mind is an expression of the functioning of our brain.

References

Bloom L: Language Development: Form and Function in Emerging Grammars. Cambridge, MA, MIT Press, 1970

Bowlby J: Attachment theory, separation anxiety, and mourning, in American Handbook of Psychiatry, 2nd edition, Vol 6. Edited by Arieti S, Hamburg DA, Brodie H. New York, Basic Books, 1975, pp 292–309

Brunelli M, Castellucci V, Kandel ER: Synaptic facilitation and behavioral sensitization in Aplysia: possible role of serotonin and cyclic AMP. Science 194:1178–1181, 1976

Carew T, Castellucci VF, Kandel ER: Sensitization in Aplysia: restoration of transmission in synapses inactivated by long-term habituation. Science 205:417–419, 1979

Castellucci VF, Carew TJ, Kandel ER: Cellular analysis of long-term habituation of the gill-withdrawal reflex of Aplysia californica. Science 202:1306–1308, 1978

Cedar H, Schwartz JH: Cyclic adenosine monophosphate in the nervous system of Aplysia californica, II: effect of serotonin and dopamine. J Gen Physiol 60:570–587, 1972

Cedar H, Kandel ER, Schwartz JH: Cyclic adenosine monophosphate in the nervous system of Aplysia californica, I: increased synthesis in response to synaptic stimulation. J Gen Physiol 60:558–569, 1972

Cowan WM, Gottlieb DI, Hendrickson AE, et al: The autoradiographic demonstration of axonal connections in the central nervous system. Brain Res 37:21–51, 1972

Dennis W: Causes of retardation among institutional children: Iran. J Genet Psychol 96:47–59, 1960

Eccles JC: The Physiology of Synapses. Berlin, Springer-Verlag, 1964

Eckstein-Schlossmann E: Zur Frage des Hospitalismus im Säuglingsanstalten. Z Kinderheilk 42:31–38, 1926

Engel GL, Reichsman F: Spontaneous and experimentally induced depression in an infant with gastric fistula: a contribution to the problem of depression. J Am Psychoanal Assoc 4:428–452, 1956

Harlow HF: The nature of love. Am Psychol 13:673–685, 1958

Harlow HF, Dodsworth RO, Harlow MK: Total social isolation in monkeys. Proc Natl Acad Sci USA 54:90–97, 1965

Hartmann H: Ego Psychology and the Problem of Adaptation. New York, International University Press, 1958

Hawkins R, Castellucci V, Kandel ER: Identification of individual neurons mediating the heterosynaptic facilitation underlying behavioral sensitization in Aplysia. Abstracts Soc Neurosci 2:235, 1976

Hook S (ed): Psychoanalysis: Scientific Method and Philosophy. New York, New York University Press, 1959

Hubel DH: Effects of distortion of sensory input on the visual system of kittens. Physiologist 10:17–45, 1967

Hubel DH, Wiesel TN: Ferrier lecture: functional architecture of macaque monkey visual cortex. Proc R Soc Lond B Biol Sci 198:1–59, 1977

Hubel DH, Wiesel TN, LeVay S: Plasticity of ocular dominance columns in monkey striate cortex. Philos Trans R Soc Lond B Biol Sci 278:377–409, 1977

Kandel ER: Cellular Basis of Behavior: An Introduction to Behavioral Neurobiology. San Francisco, CA, WH Freeman, 1976

Klein M, Kandel ER: Presynaptic modulation of voltage-dependent Ca^{2+} current: mechanism for behavioral sensitization in Aplysia californica. Proc Natl Acad Sci USA 75:3512–3516, 1978

Pinneau SR: The infantile disorders of hospitalism and anaclitic depression. Psychol Bull 52:429–451, 1955

Pinsker HM, Kandel ER, Castellucci V, et al: Analysis of habituation and dishabituation in Aplysia, in Biochemistry of Simple Neuronal Models. Vol 2, Edited by Costa E. New York, Raven Press, 1971, pp 351–373

Pinsker HM, Hening WA, Carew TJ, et al: Long-term sensitization of a defensive withdrawal reflex in Aplysia. Science 182:1039–1042, 1973

Provence S, Lipton RC: Infants in Institutions: A Comparison of Their Development With Family Reared Infants During the First Year of Life. New York, International Universities Press, 1962

Riesen AH: Plasticity of behavior: psychological aspects, in Biological and Biochemical Bases of Behavior. Edited by Harlow HF, Woolsey CN. Madison, University of Wisconsin Press, 1958, pp 425–450

Schwartz JH, Castellucci VF, Kandel ER: Functioning of identified neurons and synapses in the abdominal ganglion of Aplysia in the absence of protein synthesis. J Neurophysiol 34:939–953, 1971

Sherrington CS: The Integrative Action of the Nervous System. New Haven, CT, Yale University Press, 1906

Spencer WA, Thompson RF, Neilson DR Jr: Response decrement of the flexion reflex in the acute spinal cat and transient restoration by strong stimuli. J Neurophysiol 29:221–239, 1966

Spitz RA: Hospitalism: an inquiry into the genesis of psychiatric conditions in early childhood. Psychoanal Study Child 1:53–74, 1945

Spitz RA: Hospitalism: a follow-up report on an investigation described in Volume 1, 1945. Psychoanal Study Child 2:113–117, 1946

Spitz RA, Wolf K: Anaclitic depression: an inquiry into the genesis of psychiatric conditions in early childhood, II. Psychoanal Study Child 2:313–342, 1947

Suomi SJ, Harlow HF: The role and reason of peer relationships in rhesus monkeys, in Friendship and Peer Relations. Edited by Lewis M, Rosenblum LA. New York, Wiley, 1975, pp 153–186

Thorpe WH: Learning and Instinct in Animals. Cambridge, MA, Harvard University Press, 1956

von Senden M: Raum- und Gestalt-auffassung bei Operierten blindgeborenen vor und nach der Operation. Leipzig, JA Barth, 1932

Wertheimer M: Hebb and Senden on the role of learning in perception. Am J Psychol 64:133–137, 1951

Wiesel N, Hubel DH: Single-cell responses in striate cortex of kittens deprived of vision in one eye. J Neurophysiol 26:1003–1007, 1963

Wilson EO: Biology and the social sciences. Daedalus 2:127–140, 1977

"A NEW INTELLECTUAL FRAMEWORK FOR PSYCHIATRY"

Thomas R. Insel, M.D.

In "A New Intellectual Framework for Psychiatry," Eric Kandel aims to integrate psychiatry with the biological insights of 1998, specifically addressing the relationship of cognition and behavior to brain processes (Kandel 1998). He notes the need to enhance psychiatric training with neuroscientific expertise and describes the importance of biology for a comprehensive understanding of mental processes. Kandel provides five principles that frame this understanding, some of which may have seemed provocative in 1998: 1) all mental processes are neural, 2) genes and their protein products determine neural connections, 3) experience alters gene expression, 4) learning changes neural connections, and 5) psychotherapy changes gene expression. He concludes this thoughtful paper with a description of "unconscious" processing in patients with hippocampal lesions, noting that neuroscience might provide a new framework for psychoanalysis as well as psychiatry in general.

In the 7 years since Kandel's paper, biology has been transformed by several landmark events and discoveries, rendering Kandel's call for integration even more important. The most historic event occurred in 2003 when the Human Genome Project published the full sequence of the human genome, mapping 30,000 genes across nearly 3 billion bases of DNA. The human sequence not only provides an unprecedented opportunity to study how our species differs from its mammalian relatives, it also demonstrates the remarkable sequence similarity across humans, with 99.9% homology between individuals. A current project, the International Haplotype Mapping Project, is working to describe the nature of human variation, identifying where the 0.1% of difference between individuals emerges across the 3 billion bases of DNA (The International HapMap Consortium 2003). With the advent of new technologies for high-throughput sequencing, projects that in 1998 required tens of thousands of hours (such as the sequencing of a new microbe) now are routinely completed by a single postdoctoral fellow in a day.

The past 7 years can also be considered an era of biological pluralism, sometimes noted as the era of systems biology. Decades of studying a single gene or a single neurotransmitter have given way to techniques that permit the measurement of thousands of RNAs or proteins simultaneously. Recall that the entire body of scientific literature in this field prior to 1998 focused on roughly 1% of the genome. Indeed, the few neurotransmitters, receptors, and transporters studied in neuroscience totaled perhaps 30 amines and proteins, products of less than 0.1% of the genome. We now suspect that 20,000 genes are expressed in the brain, with as many as 6,000 expressed exclusively in the brain. Not surprisingly, in the past 6 years, much of biology has moved into a discovery phase, exploring which genes are expressed in the brain, where and when they are expressed, and how they respond to experience. Neuroanatomic maps of cytoarchitecture can now be redrawn based on molecular fingerprints of individual cells and brain nuclei (Zirlinger et al. 2001). There is no doubt that, as Kandel stated in 1998, 1) genes and proteins determine neural connections and 2) experience, including psychotherapy, alters gene expression. The molecular players and the cellular rules by which neural systems develop and experience alters gene expression are just being revealed. One thing is already clear: serotonin and dopamine will be only two of hundreds of important factors that future psychiatrists will need to know about.

Systems neuroscience has also advanced beyond the study of single electrodes and single brain regions to the widespread use of multielectrode arrays and various new imaging techniques to visualize multiple brain regions simultaneously. The simplistic (and even the complex) network diagrams of hierarchical organization in the brain have given way to dynamic models of neural activity, involving abundant recursive connections between brain regions and subtle temporal and state changes that have been hypothesized to underlie mental function (Abbott 2001). While there is no question that, as Kandel stated, "all mental processes are neural," we are now beginning to understand how neural activity measured in ensembles of cells or in field potentials of millions of cells binds information together to create memory, attention, or consciousness (Reynolds and Desimone 2003).

While molecular, cellular, and systems neuroscience have advanced so rapidly over the past 7 years, has psychiatry embraced or ignored this progress? Anyone reading the *American Journal of Psychiatry* during this time will recognize the abundant findings of psychiatric genetics and the increasing impact of neuroimaging. The human genome map, the haplotype map, and rapid genotyping are already beginning to revolutionize our approach to psychiatric genetics, allowing gene findings from linkage studies and high-throughput studies of variations in candidate genes associated with psychiatric illness. While almost no one expects that genetics will discover a Mendelian "cause" for any of the major mental illnesses, the discovery of variations associated with vulnerability should reveal the architecture for each of these illnesses that predisposes for risk, just as we have seen for hypertension and other genetically complex medical disorders. Similarly, the profile of gene expression in schizophrenia and bipolar disorder can be investigated by interrogating thousands of genes in select brain areas (Middleton et al. 2002).

Neuroimaging of regional function, in vivo neurochemistry, and connectivity have allowed psychiatric researchers to peer inside the "black box" of the brain. In this research area, part of the integration with neuroscience that Kandel hoped for in 1998 has arrived, although thus far cognitive scientists, not psychiatric patients, have been the chief beneficiaries. Studies with fMRI have provided remarkable insights into how the brain parses language, recognizes faces, and encodes emotion. Recent studies have described the neurobiology of repression (Anderson et al. 2004), romantic love (Bartels and Zeki 2000), and the unconscious (Henson 2003). But the technology, remarkable as it is, remains correlational with an unclear relationship to the millisecond world of neural function. PET studies of receptors and transporters may be more easily interpreted, but the field lacks many of the radioligands needed. And Kandel's call for studies measuring changes in regional activity with psychotherapy or psychopharmacological treatment remains largely unanswered (note, however, Goldapple et al. 2004).

While research in psychiatry has begun to embrace the power of molecular, cellular, and systems neuroscience, this scientific excitement has not yet influenced clinical practice by refining diagnosis or informing treatment. Furthermore, these advances have been conspicuously ignored by training programs. Most psychiatry residency programs remain focused on psychodynamic psychotherapy or applied psychopharmacology with little exposure to the revolutions occurring in neurobiology or cognitive science. While many of America's best colleges have developed departments or majors in neuroscience, medical schools continue to divide the mind from the brain, forcing students to choose between psychiatry and neurology. Judging from recent recruiting statistics, both psychiatry and neurology are stalled, in spite of the enormous interest in neuroscience from students entering medical school. The intellectual framework Kandel foresees for psychiatry may ultimately require that both psychiatry and neurology are reframed as clinical neuroscience disciplines. Patients with mental disorders—autism, Tourette's syndrome, schizophrenia, and bipolar disorder—have brain illnesses. And patients with Parkinson's disease, Alzheimer's disease, and most neurologic disorders have mental symptoms as core features of their illness.

As Kandel predicted, neuroscience can bring intellectual excitement back to psychiatry. This is even more evident and more necessary in 2005 than in 1998. The neurobiological tools are available to study the most mysterious aspects of mental life, including unconscious processes, emotion, and drives. Genetics may bring validity, not just reliability, to psychiatric diagnosis helping clinicians to understand the many subtypes of psychosis and ultimately predicting which treatment is most suitable for each patient. Just as important, psychiatry can bring to biology and to the rest of medicine an appreciation for the complexity of mental life. But for Kandel's vision to be realized, psychiatry will need to embrace neuroscience, not just as a research tool but as the scientific basis for clinical training and everyday practice. Martin Luther King Jr. once remarked that the sad part of the Rip van Winkle story was not that he awoke to a world that did not recognize him, but that he slept through the American Revolution (King 1959). Let us hope, along with Eric Kandel, that modern psychiatry does not sleep through one of the most exciting periods of progress in understanding the biological basis of cognition and behavior.

References

Abbott LF: The timing game. Nat Neurosci 4:115–116, 2001
Anderson MC, Ochsner KN, Kuhl B, et al: Neural systems underlying the suppression of unwanted memories. Science 303:232–235, 2004
Bartels A, Zeki S: The neural basis of romantic love. Neuroreport 11:3829–3834, 2000
Goldapple K, Segal Z, Garson C, et al: Modulation of cortical-limbic pathways in major depression. Arch Gen Psychiatry 61:34–41, 2004
Henson RN: Neuroimaging studies of priming. Prog Neurobiol 70:53–81, 2003
The International HapMap Consortium: The International HapMap Project. Nature 426:789–796, 2003

Kandel ER: A new intellectual framework for psychiatry. Am J Psychiatry 155:457–469, 1998

King ML Jr: "Remaining awake through a great revolution." Morehouse College Commencement, June 2, 1959. Available at: http://www.stanford.edu/group/King/liberation_curriculum (document 590602–005). Accessed February 15, 2004.

Middleton FA, Mirnics K, Pierri JN, et al: Gene expression profiling reveals alterations of specific metabolic pathways in schizophrenia. J Neurosci 22:2718–2729, 2002

Reynolds JH, Desimone R: Interacting roles of attention and visual salience in V4. Neuron 37:853–863, 2003

Zirlinger M, Kreiman G, Anderson DJ: Amygdala-enriched genes identified by microarray technology are restricted to specific amygdaloid subnuclei. Proc Natl Acad Sci USA 98:5270–5275, 2001

CHAPTER 2

A NEW INTELLECTUAL FRAMEWORK FOR PSYCHIATRY

Eric R. Kandel, M.D.

When historians of science turn their attention to the emergence of molecular medicine in the last half of the twentieth century, they will undoubtedly note the peculiar position occupied throughout this period by psychiatry. In the years following World War II, medicine was transformed from a practicing art into a scientific discipline based on molecular biology (Pauling et al. 1949). During that same period, psychiatry was transformed from a medical discipline into a practicing therapeutic art. In the 1950s and in some academic centers extending into the 1960s, academic psychiatry transiently

This article was originally published in the *American Journal of Psychiatry,* Volume 155, Number 4, 1998, pp. 457–469.

This paper is an extended version of an address given on the hundredth anniversary of the New York State Psychiatric Institute of Columbia University. Received July 21, 1997; revision received November 4, 1997; accepted November 11, 1997. From the Howard Hughes Medical Institute and Center for Neurobiology and Behavior, Departments of Psychiatry and Biochemistry and Molecular Biophysics, Columbia University College of Physicians and Surgeons.

The author thanks James H. Schwartz and Thomas Jessell for discussions of ideas considered in this article in the course of work on our joint textbook, *Principles of Neural Science.*

abandoned its roots in biology and experimental medicine and evolved into a psychoanalytically based and socially oriented discipline that was surprisingly unconcerned with the brain as an organ of mental activity.

This shift in emphasis had several causes. In the period after World War II, academic psychiatry began to assimilate the insights of psychoanalysis. These insights provided a new window on the richness of human mental processes and created an awareness that large parts of mental life, including some sources of psychopathology, are unconscious and not readily accessible to conscious introspection. Initially, these insights were applied primarily to what were then called neurotic illnesses and to some disorders of character. However, following the earlier lead of Eugen Bleuler (1911/1950) and Carl Jung (1906/1936), the reach of psychoanalytic therapy soon extended to encompass almost all of mental illness, including the major psychoses: schizophrenia and the major depressions (Day and Semrad 1978; Fromm-Reichmann 1948, 1959; Rosen 1963; Rosenfeld 1965).

Indeed, the extension of psychoanalytic psychiatry did not stop here; it next expanded to include specific *medical* illnesses (Alexander 1950; Sheehan and Hackett 1978). Influenced in part by their experience in World War II, many psychiatrists came to believe that the therapeutic efficacy of psychoanalytic insights might solve not only the problems of mental illness but also otherwise intractable medical illnesses such as hypertension, asthma, gastric ulcers, and ulcerative colitis—diseases that did not readily respond to the pharmacological treatments available in the late 1940s. These illnesses were thought to be psychosomatic and to be induced by unconscious conflicts.

Thus, by 1960 psychoanalytically oriented psychiatry had become the prevailing model for understanding all mental and some physical illnesses. When in 1964 Harvard Medical School celebrated the twentieth year of the psychoanalytically oriented Department of Psychiatry at Beth Israel Hospital, Ralph Kahana, a member of the faculty of that department, summarized the leadership role of psychoanalytically oriented psychiatry in the following way: "In the past 40 years, largely under the impact of psychoanalysis, dynamic psychotherapy has become the principal and essential curative skill of the American psychiatrist and, increasingly, a focus of his training" (Kahana 1968).

By merging the descriptive psychiatry of the period before World War II with psychoanalysis, psychiatry gained a great deal in explanatory power and clinical insight. Unfortunately, this was achieved at the cost of weakening its ties with experimental medicine and with the rest of biology.

The drift away from biology was not due simply to changes in psychiatry; it was in part due to the slow maturation of the brain sciences. In the late 1940s, the biology of the brain was neither technically nor conceptually ma-

ture enough to deal effectively with the biology of most higher mental processes and their disorders. The thinking about the relationship between
brain and behavior was dominated by a view that different mental functions
could not be localized to specific brain regions. This view was espoused by
Karl Lashley (1929), who argued that the cerebral cortex was equipotential;
all higher mental functions were presumed to be represented diffusely
throughout the cortex. To most psychiatrists and even to many biologists,
the notion of the equipotentiality of the cerebral cortex made behavior seem
intractable to empirical biological analysis.

In fact, the separation of psychiatry from biology had its origins even earlier. When Sigmund Freud (1954) first explored the implications of unconscious mental processes for behavior, he tried to adopt a neural model of
behavior in an attempt to develop a scientific psychology. Because of the im-
maturity of brain science at the time, he abandoned this biological model for
a purely mentalistic one based on verbal reports of subjective experiences.
Similarly, in the 1930s B. F. Skinner rejected neurological theories in his
studies of operant conditioning in favor of objective descriptions of observable acts (Skinner 1938).

Initially, this separation may have been as healthy for psychiatry as it was
for psychology. It permitted the development of systematic definitions of behavior and of disease that were not contingent on still-vague correlations
with neural mechanisms. Moreover, by incorporating the deep concern of
psychoanalysis for the integrity of an individual's personal history, psychoanalytic psychiatry helped develop direct and respectful ways for physicians
to interact with mentally ill patients, and it led to a less stigmatized social
perspective on mental illness.

However, the initial separation of psychoanalysis from neural science advocated by Freud was stimulated by the realization that a merger was premature. As psychoanalysis evolved after Freud—from being an investigative
approach limited to a small number of innovative thinkers to becoming the
dominant theoretical framework in American psychiatry—the attitude toward neural science also changed. Rather than being seen as premature, the
merger of psychoanalysis and biology was seen as unnecessary, because neural science was increasingly considered irrelevant.

Moreover, as the limitations of psychoanalysis as a system of rigorous,
self-critical thought became apparent, rather than confronting these limitations in a systematic, questioning, experimental manner, and perhaps
rejoining biology in searching for newer ways of exploring the brain, psychoanalytic psychiatry spent most of the decades of its dominance—the period from 1950 to 1980—on the defensive. Although there were important
individual exceptions, as a group, psychoanalysts devalued experimental inquiry. Consequently, psychoanalysis slid into an intellectual decline that has

had a deleterious effect on psychiatry, and because it discouraged new ways of thought, it has had a particularly deleterious effect on the training of psychiatrists.

Let me illustrate with a personal example the extent to which this unquestioning attitude came to influence my own psychiatry training. In the summer of 1960, I left my postdoctoral training in neural science at the National Institutes of Health (NIH) to begin residency training at the Massachusetts Mental Health Center, the major psychiatric teaching hospital of Harvard Medical School. I entered training together with 20-odd other young physicians, many of whom went on to become leaders in American psychiatry: Judith Livant Rapoport, Anton Kris, Dan Buie, Ernest Hartmann, Paul Wender, Joseph Schildkraut, Alan Hobson, and George Vaillant. Yet in the several years in which this outstanding group of physicians was in training, at a time when training was leisurely and there was still a large amount of spare time, there were no required or even recommended readings. We were assigned no textbooks; rarely was there a reference to scientific papers in conferences or in case supervision. Even Freud's papers were not recommended reading for residents.

Much of this attitude came from our teachers, from the heads of the residency program. They made a point of encouraging us *not* to read. Reading, they argued, interfered with a resident's ability to listen to patients and therefore biased his or her perception of the patients' life histories. One famous, much quoted remark was that "there are those who care about people and there are those who care about research." Through the efforts of the heads of the residency program, the whole thrust of psychoanalytic psychiatry at the Massachusetts Mental Health Center, and perhaps at Harvard Medical School in general, was not simply to develop better psychiatrists but to develop better therapists—therapists prepared to understand and empathize with the patients' existential problems.

This view was summarized in 1978 by Day and Semrad in the following terms:

> The essence of therapy with the schizophrenic patient is the interaction between the creative resources of both therapist and patient. The therapist must rely on his own life experience and translate his knowledge of therapeutic principles into meaningful interaction with the patient while recognizing, evoking, and expanding the patient's experience and creativity; both then learn and grow from the experience.
>
> In order to engage a schizophrenic patient in therapy, the therapist's basic attitude must be an acceptance of the patient as he is—of his aims in life, his values, and his modes of operating, even when they are different and very often at odds with his own. Loving the patient as he is, in his state of decompensation, is the therapist's primary concern in approaching the patient. As a result the therapist must find his personal satisfactions elsewhere. His job

is extremely taxing in its contradictions, for he must love the patient, expect him to change, and yet derive his additional satisfactions elsewhere and tolerate frustration.

In small measure this advice was sound, even in retrospect. A humane and compassionate perspective taught one to listen carefully and insightfully to one's patients. It helped us to develop the empathy essential for all aspects of a therapeutic relationship. But as a framework for a psychiatric education designed to train leaders in academic psychiatry, it was incomplete. For almost all residents it was intellectually limiting, and for some talented residents it proved stifling.

The almost unrealistic demand for empathy left little room for intellectual content. There were, for example, no grand rounds at the Massachusetts Mental Health Center. No outside speakers were invited to address the house officers on a regular basis to discuss current clinical or scientific issues. The major coordinated activity for the residents was a weekly group therapy session (with a wonderful and experienced group leader) in which the residents constituted the members of the group—the patients, so to speak.

It was only through the insistence of the house staff and their eagerness for knowledge that the first grand rounds were established at the Massachusetts Mental Health Center in 1965. To initiate these rounds, several of us tried to recruit a psychiatrist in the Boston area to speak about the genetic basis of mental illness. We could find *no one;* not a single psychiatrist in all of Boston was concerned with or even had thought seriously about that issue. We finally imposed on Ernst Mayr, the great Harvard biologist and a friend of Franz Kallmann, a founder of psychiatric genetics, to come and talk to us.

I am providing here an oversimplified description of the weakness of an environment that had many excellent qualities and many strengths. The intellectual quality of the house officers was remarkable, and the commitment of the faculty to the training of the house staff and to the treatment of the patients was admirable. Moreover, I am describing the predominant trend at the center; there were countervailing ones. While the heads of the training program actively discouraged both reading and research, the director of the center, Jack Ewalt, strongly encouraged research. Moreover, I have been assured that during this period Harvard psychiatry was remarkably out of step with the rest of the country, and that a lack of scholarly concern was not universal within academic psychiatry nationally. Clearly, scholarly concerns were not lacking at Washington University under Eli Robins, at a number of other centers in the Midwest, or at Johns Hopkins University under Seymour Kety (1959). But a lack of critical questioning seemed to be widespread in Boston and at many other institutions on the east and west coasts of the country.

Our residency years—the decade of the 1960s—marked a turning point in American psychiatry. To begin with, new and effective treatments, in the form of psychopharmacological drugs, began to be available. Initially, a number of supervisors discouraged us from using them, believing that they were designed more to aid our anxiety than that of the patients. By the mid-1970s, the therapeutic scene had changed so dramatically that psychiatry was forced to confront neural science if only to understand how specific pharmacological treatments were working.

With the advent of psychopharmacology, psychiatry was changed, and that change brought it back into the mainstream of academic medicine. There were three components to this progress. First, whereas psychiatry once had the least effective therapeutic armamentarium in medicine, it now had effective treatments for the major mental illnesses and something that began to approach a practical cure for two of the three most devastating diseases: depression and manic-depressive illness. Second, led first by Eli Robins at Washington University and then by Robert Spitzer at Columbia University's New York State Psychiatric Institute, new clinically validated and objective criteria were established for diagnosing mental illness. Third, Seymour Kety used his leadership position at NIH to spark a renewed interest in the biology of mental illness and specifically in the genetics of schizophrenia and depression.

In parallel, the years since 1980 have witnessed major developments in brain sciences, in particular in the analysis of how different aspects of mental functioning are represented by different regions of the brain. Thus, psychiatry is now presented with a new and unique opportunity. When it comes to studying mental function, biologists are badly in need of guidance. It is here that psychiatry, and cognitive psychology, as guide and tutor, can make a particularly valuable contribution to brain science. One of the powers of psychiatry, of cognitive psychology, and of psychoanalysis lies in their perspectives. Psychiatry, cognitive psychology, and psychoanalysis can define for biology the mental functions that need to be studied for a meaningful and sophisticated understanding of the biology of the human mind. In this interaction, psychiatry can play a double role. First, it can seek answers to questions on its own level, questions related to the diagnosis and treatment of mental disorders. Second, it can pose the behavioral questions that biology needs to answer if we are to have a realistically advanced understanding of human higher mental processes.

A Common Framework for Psychiatry and the Neural Sciences

As a result of advances in neural science in the last several years, both psychiatry and neural science are in a new and better position for a rapprochement, a rapprochement that would allow the insights of the psychoanalytic perspective to inform the search for a deeper understanding of the biological basis of behavior. As a first step toward such a rapprochement, I here outline an intellectual framework designed to align current psychiatric thinking and the training of future practitioners with modern biology.

This framework can be summarized in five principles that constitute, in simplified form, the current thinking of biologists about the relationship of mind to brain.

Principle 1.   All mental processes, even the most complex psychological processes, derive from operations of the brain. The central tenet of this view is that what we commonly call mind is a range of functions carried out by the brain. The actions of the brain underlie not only relatively simple motor behaviors, such as walking and eating, but all of the complex cognitive actions, conscious and unconscious, that we associate with specifically human behavior, such as thinking, speaking, and creating works of literature, music, and art. As a corollary, behavioral disorders that characterize psychiatric illness are disturbances of brain function, even in those cases where the causes of the disturbances are clearly environmental in origin.

Principle 2.   Genes and their protein products are important determinants of the pattern of interconnections between neurons in the brain and the details of their functioning. Genes, and specifically combinations of genes, therefore exert a significant control over behavior. As a corollary, one component contributing to the development of major mental illnesses is genetic.

Principle 3.   Altered genes do not, by themselves, explain all of the variance of a given major mental illness. Social or developmental factors also contribute very importantly. Just as combinations of genes contribute to behavior, including social behavior, so can behavior and social factors exert actions on the brain by feeding back upon it to modify the expression of genes and thus the function of nerve cells. Learning, including learning that results in dysfunctional behavior, produces alterations in gene expression. Thus all of "nurture" is ultimately expressed as "nature."

Principle 4.   Alterations in gene expression induced by learning give rise to changes in patterns of neuronal connections. These changes not only contribute to the biological basis of individuality but presumably are responsible for initiating and maintaining abnormalities of behavior that are induced by social contingencies.

Principle 5.   Insofar as psychotherapy or counseling is effective and produces long-term changes in behavior, it presumably does so through learning, by producing changes in gene expression that alter the strength of synaptic connections and structural changes that alter the anatomical pattern of interconnections between nerve cells of the brain. As the resolution of brain imaging increases, it should eventually permit quantitative evaluation of the outcome of psychotherapy.

I now consider each of these principles in turn and illustrate the experimental basis of this new framework and its implications for the theory and practice of psychiatry.

All Functions of Mind Reflect Functions of Brain

This principle is so central in traditional thinking in biology and medicine (and has been so for a century) that it is almost a truism and hardly needs restatement. This principle stands as the basic assumption underlying neural science, an assumption for which there is enormous scientific support. Specific lesions of the brain produce specific alterations in behavior, and specific alterations in behavior are reflected in characteristic functional changes in the brain (Kandel et al. 1991). Nevertheless, two points deserve emphasis.

First, although this principle is now accepted among biologists, the details of the relationship between the brain and mental processes—precisely how the brain gives rise to various mental processes—is understood poorly, and only in outline. The great challenge for biology and psychiatry at this point is to delineate that relationship in terms that are satisfying to both the biologist of the brain and the psychiatrist of the mind.

Second, the relationship of mind to brain becomes less obvious, more nuanced, and perhaps more controversial when we appreciate that biologists apply this principle to all aspects of behavior, from our most private thoughts to our most public expression of emotion. The principle applies to behaviors by single individuals, to behaviors between individuals, and to social behavior in groups of individuals. Viewed in this way, all sociology must to some degree be sociobiology; social processes must, at some level, reflect biological functions. I hasten to add that formulating a relationship between social processes (or even psychological processes) and biological functions might not necessarily prove to be optimally insightful in elucidating social dynamics. For many aspects of group or individual behavior, a biological analysis might not prove to be the optimal level or even an informative level of analysis, much as subatomic resolution is often not the optimal level for the analysis of biological problems. Nevertheless, it is important to appreciate that there are critical biological underpinnings to all social actions.

This aspect of the principle has not been readily accepted by all, especially not by all sociologists, as can be illustrated by one example from the Center for Advanced Studies in the Behavioral Sciences in Palo Alto, California, probably the country's premier think tank in the social sciences. In its annual report of 1996, the center described the planning of a special project entitled Culture, Mind, and Biology. As plans for this project progressed, it became clear that many social scientists had a deep and enduring antipathy toward the biological sciences because they equated biological thinking with a view of human nature that they found simplistic, misguided, and socially and ethically dangerous. Since two earlier and influential biological approaches to the social sciences—scientifically argued racism and social Darwinism—had proven to be intellectually sterile and socially destructive, many social scientists objected to the idea. They objected to the notion

> that a *living organism's properties* (not only its physical form but also its behavioral inclinations, abilities, and life prospects) *are material* and hence *reducible to its genes.* The conception of human nature that many social scientists associate with biological thinking asserts that individual and group differences as well as individual and group similarities in physical form, behavioral inclination, abilities, and life prospects can similarly be understood and explained by genes. . . . As a result of this understanding, many disclaim the relevance of biological thinking for behavior and instead embrace some type of radical *mind-body dualism* in which it is *assumed* that the processes and products of the mind have very little to do with the processes and products of the body. (Annual Report, Center for Advanced Studies in the Behavioral Sciences, 1996; italics added)

What is the basis of this unease among social scientists? Like all knowledge, biological knowledge is a double-edged sword. It can be used for ill as well as for good, for private profit or public benefit. In the hands of the misinformed or the malevolent, natural selection was distorted to social Darwinism, and genetics was corrupted into eugenics. Brain sciences have also been and can again be misused for social control and manipulation. How can we ensure that the advances of the brain sciences will serve to enrich our lives and to elevate our understanding of ourselves and each other? The only way to encourage the responsible use of this knowledge is to base the uses of biology in social policy on an *understanding* of biology.

The unease of social scientists derives in part from two misapprehensions (not unique to social scientists): first, that biologists think that biological processes are strictly determined by genes, and second, that the sole function of genes is the inexorable transmission of hereditary information from one generation to another. These profoundly wrong ideas lead to the notion that invariant, unregulated genes, not modifiable by external events, exert an inevitable influence on the behavior of individuals and their progeny. In this view, social forces as such have little influence on human behavior. They are powerless in the face of the predetermined, relentless actions of the genes.

This fatalistic and fundamentally wrong view was behind the eugenics movements of the 1920s and 1930s. As a basis for social policy, this view justifiably elicits fear and distrust in clear-thinking people. However, this view is based on a fundamental misconception of how genes work, which even some psychiatrists may not fully appreciate. The key concept of importance here is that genes have dual functions.

First, genes serve as stable templates that can replicate reliably. This *template function* is exercised by each gene, in each cell of the body, including the gametes. It is this function that provides succeeding generations with copies of each gene. The fidelity of the template replication is high. Moreover, the template is not regulated by social experience of any sort. It can only be altered by mutations, and these are rare and often random. This function of the gene, its template (transmission) function, is indeed beyond our individual or social control.

Second, genes determine the phenotype; they determine the structure, function, and other biological characteristics of the cell in which they are expressed. This second function of the gene is referred to as its *transcriptional function.* Although almost every cell of the body has all of the genes that are present in every other cell, in any given cell type (be it a liver cell or a brain cell) only a fraction of genes, perhaps 10%–20%, are expressed (transcribed). All of the other genes are effectively repressed. A liver cell is a liver cell and a brain cell is a brain cell because each of these cell types expresses only a particular subset of the total population of genes. When a gene is expressed in a cell, it directs the phenotype of that cell: the manufacture of specific proteins that specify the character of that cell.

Whereas the template function, the sequence of a gene—and the ability of the organism to replicate that sequence—is not affected by environmental experience, the transcriptional function of a gene—the ability of a given gene to direct the manufacture of specific proteins in any given cell—is, in fact, highly regulated, and *this regulation is responsive to environmental factors.*

A gene has two regions (Figure 2–1). A coding region encodes mRNA, which in turn encodes a specific protein. A regulatory region usually lies upstream of the coding region and consists of two DNA elements. The *promoter* element is a site where an enzyme, called RNA polymerase, will begin to read and transcribe the DNA coding region into mRNA. The *enhancer* element recognizes protein signals that determine in which cells, and when, the coding region will be transcribed by the polymerase. Thus, a small number of proteins, or transcriptional regulators, that bind to different segments of the enhancer element determine how often RNA polymerase binds to the promoter element and transcribes the gene. Internal and external stimuli—steps in the development of the brain, hormones, stress, learning, and social interaction—alter the binding of the transcriptional regulators to the enhancer element, and in this way different combinations of transcriptional regulators are recruited. This aspect of gene regulation is sometimes referred to as *epigenetic* regulation.

Stated simply, the regulation of gene expression by social factors makes all bodily functions, including all functions of the brain, susceptible to social influences. These social influences will be biologically incorporated in the

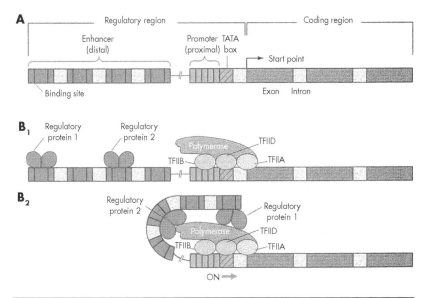

FIGURE 2-1.   Genetic transcriptional control.

A: The typical eukaryotic gene has two regions. The coding region is transcribed by RNA polymerase II into an mRNA and is then translated into a specific protein. The regulatory region, consisting of enhancer elements and a promoter element, which contains the TATA box (T=thymidine, A=adenine), regulates the initiation of transcription of the structural gene.

Transcriptional regulatory proteins bind both the promoter and the enhancer regions. B_1: A set of proteins (such as TATA box factors IIA, IIB, IID, and others) binds to the TATA box, to the promoter, and to the distal enhancer regions. B_2: Proteins that bind to the enhancer region cause looping of the DNA, thereby allowing the regulatory proteins that bind to distal enhancers to contact the polymerase.

Source.   Adapted from Schwartz and Kandel 1995.

altered expressions of specific genes in specific nerve cells of specific regions of the brain. These socially influenced alterations are transmitted culturally. They are not incorporated in the sperm and egg and therefore are not transmitted genetically. In humans, the modifiability of gene expression through learning (in a nontransmissible way) is particularly effective and has led to a new kind of evolution: cultural evolution. The capability of learning is so highly developed in humans that humankind changes much more by cultural evolution than by biological evolution. Measurements of skulls found in the fossil record suggest that the size of the human brain has not changed since *Homo sapiens* first appeared approximately 50,000 years ago; yet clearly, human culture has evolved dramatically in that same time.

Genes Contribute Importantly to Mental Function and Can Contribute to Mental Illness

Let us consider the contribution of the template functions of DNA—the heritable aspects of gene action. Here we first need to ask, How do genes contribute to behavior? Clearly, genes do not code for behavior in a direct way. A single gene encodes a single protein; it cannot by itself encode for a single behavior. Behavior is generated by neural circuits that involve many cells, each of which expresses specific genes that direct the production of specific proteins. The genes expressed in the brain encode proteins that are important in one or another step of the development, maintenance, and regulation of the neural circuits that underlie behavior. A wide variety of proteins— structural, regulatory, and catalytic—are required for the differentiation of a single nerve cell, and many cells and many more genes are required for the development and function of a neural circuit.

To account for what we now appreciate as variations in the template functions of a gene, Darwin and his followers first postulated that variations in human behavior may, in part, be due to natural selection. If this is so, some element of the behavioral variation in any population will necessarily have a genetic basis. Some portion of this variation in turn should show up as clearly heritable differences. Control studies of heritable factors in human behavior have proven difficult to devise, because it is not possible or desirable to control an individual's environment for experimental purposes except in some very limited situations. Thus, behavioral studies of identical twins provide important information not otherwise available.

Identical twins share an identical genome and are therefore as alike genetically as is possible for two individuals. Similarities between identical twins who have been separated early in life and raised in different households, as occasionally happens, will therefore be more attributable to genes than to environment. Identical twins, compared with a group of individuals matched in age, sex, and socioeconomic status, share a remarkable number of behavioral traits. These include tastes, religious preferences, and vocational interests that are commonly considered to be socially determined and distinctive features of an individual. These findings argue that human behavior has a significant hereditary component. But the similarity is far from perfect. Twins can and do vary a great deal. Thus, twin studies also emphasize the importance of environmental influences; they indicate quite clearly that environmental factors are very important (Kandel et al. 1991).

A similar situation applies to disturbances of behavior and to mental illness. The first direct evidence that genes are important in the development of schizophrenia was provided in the 1930s by Franz Kallmann (1938). Kallmann was impressed with the fact that the incidence of schizophrenia throughout the world is uniformly about 1%, even though the social and environmental factors vary dramatically. Nevertheless, he found that the incidence of schizophrenia among parents, children, and siblings of patients with the disease is 15%, strong evidence that the disease runs in families. However, a genetic basis for schizophrenia cannot simply be inferred from the increased incidence in families. Not all conditions that run in families are necessarily genetic: wealth and poverty, habits, and values also run in families, and in earlier times even nutritional deficiencies such as pellagra ran in families.

To distinguish genetic from environmental factors, Kallmann turned to twin studies and compared the rates of illness in identical (monozygotic) and fraternal (dizygotic) twins. As we have seen, monozygotic twins share almost all of each other's genes. By contrast, dizygotic twins share only 50% of their genes and are genetically equivalent to siblings. Therefore, if schizophrenia is caused entirely by genetic factors, monozygotic twins should be identical in their tendency to develop the disease. Even if genetic factors were necessary but not sufficient for the development of schizophrenia, because environmental factors were involved, a monozygotic twin of a patient with schizophrenia should be at substantially higher risk than a dizygotic twin. The tendency for twins to have the same illness is called concordance. Studies on twins have established that the concordance for schizophrenia in monozygotic twins is about 45%, compared to only about 15% in dizygotic twins, which is about the same as for other siblings.

To disentangle further the effects of nature and nurture, Heston (1970) studied patients in the United States and Rosenthal and colleagues (1971) studied patients in Denmark. In both sets of studies, the rate of schizophrenia was higher among the biological relatives of adopted children who had schizophrenia than among those of adopted children who were normal. The difference in rate, about 10%–15%, was the same as that observed earlier by Kallmann.

This familial pattern of schizophrenia is most dramatically evident in an analysis of the data from Denmark by Gottesman (1991). Gottesman examined the data from 40 Danish patients with schizophrenia, identifying all relatives with schizophrenia for whom good family pedigrees were available. He then ranked the relatives in terms of the percentage of genes shared with the schizophrenic patient. He found a higher incidence of schizophrenia among first-order relatives—those who share 50% of the patient's genes, including siblings, parents, and children—than among second-order relatives—those who share 25% of the patient's genes, including aunts, uncles, nieces, nephews, and grandchildren. Even the third-degree relatives, who share only 12.5% of the patient's genes, had a higher incidence of schizophrenia than the 1% found in the population at large. These data strongly suggest a genetic contribution to schizophrenia.

If schizophrenia were caused entirely by genetic abnormalities, the concordance rate for monozygotic twins, who share almost all of each other's genes, would be nearly 100%. The fact that the rate is 45% clearly indicates that genetic factors are not the only cause. Multiple causality is also evident from studies of the genetic transmission of the disease. Relatively routine studies of pedigrees are sufficient to pinpoint whether a disease is transmitted by dominant or recessive Mendelian inheritance, but this has not proven to be the mode of transmission of schizophrenia. The most likely explanation for the unusual genetic transmission of schizophrenia is that it is a multigenic disease involving allelic variations in perhaps as many as 10–15 loci in the population worldwide, and that perhaps combinations of three to five loci are needed to cause the disease in an individual. Moreover, these several genes can vary in the degree of *penetrance*.

In a natural population, any gene at any locus will exist in a number of different, clearly related forms called *alleles*. The penetrance of an allele depends on the interaction between that allele and the remainder of the genome, as well as with environmental factors. One twin can inherit a set of genes that program tall growth, but without good nutrition that twin may never grow tall. Similarly, not all people with the same dominant and abnormal Huntington's disease gene will have the full-blown movement disorders and accompanying cognitive disturbances; a few may have a more moderate form of the disease.

As in other polygenic diseases, such as diabetes and hypertension, most forms of schizophrenia are thought to require not only the accumulation of several genetic defects but also the actions of developmental and environmental factors. To understand schizophrenia, it will be essential to learn how several genes combine to predispose an individual to a disease and to determine how the environment influences the expression of these genes.

The fact that many genes are involved does not mean, however, that in some cases single genes are not essential for the expression of a behavior. The importance of specific genes to behavior can best be demonstrated in simple animals, such as fruit flies or mice, in which mutations in a single gene can be more easily studied. Mutations of single genes in *Drosophila* or in mice can produce abnormalities in a variety of behaviors, including learned behavior as well as innate behavior such as courtship and locomotion.

Behavior Itself Can Also Modify Gene Expression

I have considered the template function of the gene, which is transmissible but not regulated. I now turn to that aspect of genetic function that is regulated but not transmitted. Studies of learning in simple animals provided the first evidence that experience produces sustained changes in the effectiveness of neural connections by altering gene expression. This finding has profound ramifications that should revise our view of the relationship between social and biological processes in the shaping of behavior.

To appreciate the importance of this relationship, consider for a moment the situation in American psychiatry as recently as 1968, when DSM-II appeared. A common view in psychiatry at that time was that biological and social determinants of behavior act on separate levels of the mind: one level had a clear empirical basis, and the other was unspecified. As a result, until the 1970s psychiatric illnesses were traditionally classified into two major categories: organic and functional. Thus, Seltzer and Frazier wrote in 1978, "organic brain syndrome is a general term used to describe those conditions of impaired function of the nervous system that are manifest by psychiatric symptoms. This contrasts with the majority of psychiatric syndromes called 'functional'."

These organic mental illnesses included the dementias, such as Alzheimer's disease, and the toxic psychoses, such as those that follow the chronic use of cocaine, heroin, and alcohol. Functional mental illnesses included not only the neurotic illnesses but also the depressive illnesses and the schizophrenias.

This distinction originally derived from the observations of nineteenth-century neuropathologists, who examined the brains of patients at autopsy and found gross and readily demonstrable distortions in the architecture of the brain in some psychiatric diseases but not in others. Diseases that produced anatomical evidence of brain lesions were called organic; those lacking these features were called functional.

This distinction, now clearly outdated, is no longer tenable. There can be no changes in behavior that are not reflected in the nervous system and no persistent changes in the nervous system that are not reflected in structural changes on some level of resolution. Everyday sensory experience, sensory deprivation, and learning can probably lead to a weakening of synaptic connections in some circumstances and a strengthening of connections in others. We no longer think that only certain diseases, the organic diseases, affect mentation through biological changes in the brain and that others, the functional diseases, do not. The basis of the new intellectual framework for psychiatry is that all mental processes are biological, and therefore any alteration in those processes is necessarily organic.

As is now evident in DSM-IV, the classification of mental disorders must be based on criteria other than the presence or absence of *gross* anatomical abnormalities. The absence of detectable structural changes does not rule out the possibility that more subtle but nonetheless important biological changes are occurring. These changes may simply be below the level of detection with the still-limited techniques available today. Demonstrating the biological nature of mental functioning requires more sophisticated anatomical methodologies than the light-microscopic histology of nineteenth-century pathologists. To clarify these issues it will be necessary to develop a neuropathology of mental illness that is based on anatomical function as well as anatomical structure. Imaging techniques such as positron emission tomography and functional magnetic resonance imaging have opened the door to the noninvasive exploration of the human brain at a level of resolution that begins to approach that which is required to understand the physical mechanisms of mentation and therefore of mental disorders. This approach is now being pursued in the study of schizophrenia, depression, obsessive-compulsive disorders, and anxiety disorders (Jones 1995).

We now need to ask, How do the biological processes of the brain give rise to mental events, and how in turn do social factors modulate the biological structure of the brain? In the attempt to understand a particular mental illness, it is more appropriate to ask, To what degree is this biological process determined by genetic and developmental factors? To what degree is it environmentally or socially determined? To what degree is it determined by a toxic or infectious agent? Even the mental disturbances that are considered to be most heavily determined by social factors must have a biological component, since it is the activity of the brain that is being modified.

A New View of the Relationship Between Inherited and Acquired Mental Illnesses

In the few instances where it has been possible to examine rigorously the persistent changes in mental functions, these functions have been shown to involve alterations in gene expression. Thus, in studying the specific changes that underlie persistent mental states, normal as well as disturbed, we should also look for altered gene expression. As we have seen, there is now substantial evidence that the susceptibility to major psychotic illnesses (schizophrenia and manic-depressive disorders) is heritable. These illnesses in part reflect alterations in the template function of the gene—in the nucleotide sequence of a number of different genes—leading to abnormal mRNAs and abnormal proteins. It is therefore tempting to think that insofar as psychiatric illnesses such as posttraumatic stress syndrome are acquired by experience, they are likely to involve alterations in the transcriptional function of the gene—in the regulation of gene expression. Nonetheless, some individuals may be much more susceptible to this syndrome because of the *combination* of genes they have inherited.

Development, stress, and social experience are all factors that can alter gene expression by modifying the binding of transcriptional regulators to each other and to the regulatory regions of genes. It is likely that at least some neurotic illnesses (or components of them) result from reversible defects in gene regulation, which may be due to altered binding of specific proteins to certain upstream regions that control the expression of certain genes (Figure 2–2).

Maintenance of Learned Alterations in Gene Expression by Structural Alterations in Neural Circuits of the Brain

How does altered gene expression lead to the stable alterations of a mental process? Animal studies of alterations in gene expression induced by learning indicate that one major consequence of such alterations in gene activation is the growth of synaptic connections. This growth was first delineated by studies in simple invertebrate animals such as the snail *Aplysia* (Bailey and Kandel 1993). Animals subjected to controlled learning that gave rise to long-term memory had twice as many presynaptic terminals as untrained animals. Some forms of learning, such as long-term habituation, produce the opposite changes; they lead to a regression and pruning of synaptic connections. These morphological changes seem to be a signature of the long-term memory process. They do not occur with short-term memory.

In mammals, and especially in humans, each functional component of the nervous system is represented by hundreds of thousands of nerve cells. In such complex systems a specific instance of learning is likely to lead to alterations in a large number of nerve cells insofar as the interconnections of the various sensory and motor systems involved in the learning are changed. Indeed, studies have shown that such vast changes do occur. The most detailed evidence has come from studies of the somatic sensory system.

The primary somatic sensory cortex contains four separate maps of the surface of the body in four areas in the postcentral gyrus (Brodmann's areas 1, 2, 3a, and 3b). These cortical maps differ among individuals in a manner that reflects their use. Moreover, the cortical maps for somatic sensations are dynamic, not static, even in mature animals (Merzenich et al. 1988). The distribution of these functional connections can expand and retract, depending on the particular uses or activities of the peripheral sensory pathways. Since each of us is brought up in a somewhat different environment, exposed to different combinations of stimuli, and we develop motor skills in different ways, each brain is modified in unique ways. This distinctive modification of brain architecture, along with a unique genetic makeup, constitutes the biological basis for individuality.

Two studies provide evidence for this view (Merzenich et al. 1988). One study found that the somatosensory maps vary considerably among normal animals. However, this study did not separate the effects of different experiences from the consequences of different genetic endowment. Another study

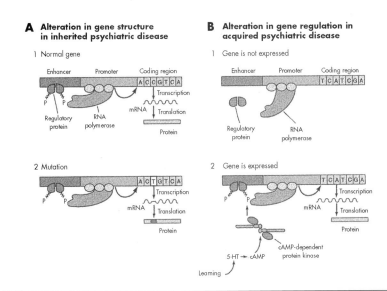

FIGURE 2-2. There is a genetic component to both inherited and acquired psychiatric illness.

Genetic and acquired illnesses both have a genetic component. Genetic illnesses (e.g., schizophrenia) are expressions of altered genes, whereas illnesses acquired as learned behavior (neuroses) involve the modulation of gene expression by environmental stimuli, leading to the transcription of a previously inactive gene. The gene is illustrated as having two segments. A coding region is transcribed into an mRNA by an RNA polymerase. The mRNA in turn is translated into a specific protein. A regulatory segment consists of an enhancer region and a promoter region. In this example the RNA polymerase can transcribe the gene when the regulatory protein binds to the enhancer region. For gene activation to occur, the regulatory protein must first be phosphorylated.

A(1): Under normal conditions the phosphorylated regulatory protein binds to the enhancer region, thereby activating the transcription of the gene, leading to the production of the protein (P=phosphorus, A=adenine, C=cytosine, G=guanine, T=thymidine). A(2): A mutant form of the coding region of the structural gene, in which a T has been substituted for a C, leads to transcription of an altered mRNA. This in turn produces an abnormal protein, giving rise to the disease state. This alteration in gene structure becomes established in the germ line and is heritable.

B(1): If the regulatory protein for a normal gene is not phosphorylated, it cannot bind to the enhancer site, and thus gene transcription cannot be initiated. B(2): In this case a specific experience leads to the activation of serotonin (5-HT) and cAMP, which activate the cAMP-dependent protein kinase. The catalytic unit phosphorylates the regulatory protein, which then can bind to the enhancer segment and thus initiate gene transcription. By this means, an abnormal learning experience could lead to the expression of a protein that gives rise to symptoms of a neurotic disorder. *Source.* Adapted from Kandel 1995.

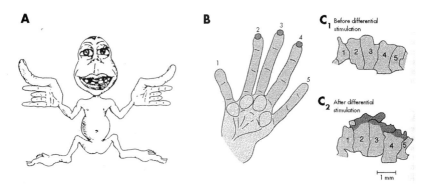

FIGURE 2–3. The representation of the body on the surface of the cerebral cortex is modified by experience.

A: Penfield's somatic-sensory homunculus redrawn as a complete body, showing the overrepresentation of certain parts of the skin surface.

B: Training expands existing afferent inputs in the cortex. A monkey was trained for 1 hour per day to perform a task that required repeated use of the tips of fingers 2, 3, and occasionally 4 (dark shading).

C_1: Representation of the tips of the digits of an adult monkey in Brodmann's cortical area 3b three months before training. C_2: After a period of repeated stimulation, the portion of area 3b representing the tips of the stimulated fingers is substantially enlarged (dark shading).

Source. (A) Adapted from Blakemore 1977. (B) Adapted from Jenkins et al. 1990.

was conducted to see whether activity is important in determining the topographic organization of the somatosensory cortex. Adult monkeys were encouraged to use three middle fingers at the expense of two other fingers of the hand to obtain food. After several thousand trials, the area of cortex devoted to the three fingers was greatly expanded at the expense of the area normally devoted to the other fingers (Figure 2–3). Practice alone, therefore, may not only strengthen the effectiveness of existing patterns of connections but also change cortical connections to accommodate new patterns of actions.

Psychotherapy and Pharmacotherapy May Induce Similar Alterations in Gene Expression and Structural Changes in the Brain

As these arguments make clear, it is intriguing to suggest that insofar as psychotherapy is successful in bringing about substantive changes in behavior, it does so by producing alterations in gene expression that produce new structural changes in the brain. This obviously should also be true of psychopharmacological treatment. Treatment of neurosis or character disorders by psychotherapeutic intervention should, if successful, also produce functional and structural changes. We face the interesting possibility that as brain imaging techniques improve, these techniques might be useful not only for diagnosing various neurotic illnesses but also for monitoring the progress of psychotherapy. The joint use of pharmacological and psychotherapeutic interventions might be especially successful because of a potentially interactive and synergistic—not only additive—effect of the two interventions. Psychopharmacological treatment may help consolidate the biological changes caused by psychotherapy.

One example of this congruence is now evident in obsessive-compulsive disorder (OCD). This common debilitating psychiatric illness is characterized by recurrent unwanted thoughts, obsessions, and conscious ritualized acts and compulsions that are usually attributed to attempts to deal with the anxiety generated by the obsessions. Medications that are selective serotonin reuptake inhibitors (SSRIs) and specific behavioral therapies that use the principles of deconditioning, involving exposure and response prevention, are effective in reducing the symptoms of many patients with OCD.

Many investigators have postulated a role for the cortical-striatal-thalamic brain system in the mediation of OCD symptoms. OCD is associated with functional hyperactivity of the head of the right caudate nucleus. After effective treatment of OCD with either an SSRI (such as fluoxetine) alone or with behavioral modification alone (with exposure and response prevention techniques), there is a substantial decrease in activity (measured as glucose metabolic rate) in the head of the right caudate nucleus. In one study (J.M. Schwartz et al. 1996), patients who responded to behavior therapy had a significant decrease in glucose metabolic rate in the caudate nucleus bilaterally compared to those who did not respond to treatment.

These arguments suggest that when a therapist speaks to a patient and the patient listens, the therapist is not only making eye contact and voice contact, but the action of neuronal machinery in the therapist's brain is having an indirect and, one hopes, long-lasting effect on the neuronal machinery in the patient's brain; and quite likely, vice versa. Insofar as our words produce changes in our patient's mind, it is likely that these psychotherapeutic interventions produce changes in the patient's brain. From this perspective, the biological and sociopsychological approaches are joined.

Implications of a New Framework for the Practice of Psychiatry

The biological framework that I have outlined here is not only important conceptually; it is also important practically. To function effectively in the future, the psychiatrists we are training today will need more than just a nodding familiarity with the biology of the brain. They will need the knowledge of an expert, a knowledge perhaps different from but fully comparable to that of a well-trained neurologist. In fact, it is likely that in the decades ahead we will see a new level of cooperation between neurology and psychiatry. This cooperation is likely to have its greatest impact on patients for whom the two approaches—neurological and psychiatric—overlap, such as those in treatment for autism, mental retardation, and the cognitive disorders due to Alzheimer's and Parkinson's diseases.

It can be argued that an intellectual framework so fully embedded in biology and aligned with neurology is premature for psychiatry. In fact, we are only beginning to understand the simplest mental functions in biological terms; we are far from having a realist neurobiology of clinical syndromes and even farther from having a neurobiology of psychotherapy. These arguments have some validity. Thus, the decision for psychiatry revolves around the question, When will the time be optimal for a more complete rapprochement between psychiatry and biology? Is it when the problem is still premature—when the biology of mental illness still confronts us as deep mysteries—or is it when the problem is already postmature—when mental illness is on the way to being understood? If psychiatry will join the intellectual fray in full force only when the problems are largely solved, then psychiatry will deprive itself of one of its main functions, which is to provide leadership in the attempts to understand the basic mechanisms of mental processes and their disorders. Since the presumed function of academic psychiatry is to train people who advance knowledge—people who can not only benefit from the insights of the current biological revolution but also contribute to it—psychiatry must take its commitment to the training of biological scientists more seriously. It must put its own oars into the water and pull its own weight. If the biology of mental processes continues to be solved by others without the active participation of psychiatrists, we may well ask, What is the purpose of a psychiatric education?

While psychiatrists debate the degree to which they should immerse themselves in modern molecular biology, most of the remaining scientific community has resolved that issue for itself. Most biologists sense that we are in the midst of a remarkable scientific revolution, a revolution that is transforming our understanding of life's processes—the nature of disease and of medical therapeutics. Most biologists believe that this revolution will have a profound impact on our understanding of mind. This view is shared by students just beginning their scientific training. Many of the very best graduate students in biology and the best M.D.-Ph.D. students are drawn to neural science and particularly to the biology of mental processes for this very reason. If the progress of the past few years and the continued influx of talented people are any guide, we can expect a major growth in our understanding of mental processes.

We thus are confronting an interesting paradox. While the scientific community at large has become interested in the biology of mental processes, the interest of medical students in a psychiatric career is declining. Thus, from an educational point of view, psychiatry is in a trough. One reason for the loss of interest, beyond the economic issue of managed care, is the current intellectual scene in psychiatry. Medical students realize that insofar as the teaching of psychiatry is often based primarily on doing psychotherapy, a major component of psychiatry as it is now taught does not require a medical education. As Freud so clearly emphasized, psychotherapy can be carried out effectively by nonmedical specialists. Why, then, go to medical school?

As a greater emphasis on biology begins to change the nature of psychiatry, it also is likely to draw an increasing number of talented medical students into psychiatry. In addition, it will make psychiatry a more technologically sophisticated and more scientifically rigorous medical discipline. A biological orientation can help revitalize the teaching and practice of psychiatry by bringing to bear on the problems of mental illness a critical understanding of brain processes, a familiarity with therapeutics, and an understanding of both neurological and psychiatric diseases—in short, an ability to encompass mental and emotional life within a framework that includes biological as well as social determinants. A renewed involvement of psychiatry with biology and with neurology, therefore, not only is scientifically important but also emphasizes the scientific competence that, I would argue, should be the basis for the clinical specialty of psychiatry in the twenty-first century.

Biology and the Possibility of a Renaissance of Psychoanalytic Thought

It would be unfortunate, even tragic, if the rich insights that have come from psychoanalysis were to be lost in the rapprochement between psychiatry and the biological sciences. With the perspective of time, we can readily see what has hindered the full intellectual development of psychoanalysis during the last century. To begin with, psychoanalysis has lacked any semblance of a scientific foundation. Even more, it has lacked a scientific tradition, a questioning tradition based not only on imaginative insights but on creative and critical experiments designed to explore, support, or, as is often the case, falsify those insights. Many of the insights from psychoanalysis are derived from clinical studies of individual cases. Insights from individual cases can be powerful, as we have learned from Paul Broca's study of the patient Leborgne (Schiller 1992). The analysis of this patient is a historical landmark; it marks the origin of neuropsychology. Study of this one patient led to the discovery that the expression of language resides in the left hemisphere and specifically in the frontal cortex of that hemisphere. But as Broca's cases illustrate, clinical insights, especially those based on individual cases, need to be supported by independent and objective methods. Broca achieved this by studying Leborgne's brain at autopsy and by subsequently discovering eight other patients with similar lesions and similar symptoms. It is, I believe, the lack of a scientific culture more than anything else that led to the insularity and anti-intellectualism which characterized psychoanalysis in the last 50 years and which in turn influenced the training of psychiatrists in the period of World War II, the period in which psychoanalysis was the dominant mode of thought in American psychiatry.

But the sins of the fathers (and mothers) need not be passed on to succeeding generations. Other disciplines have recovered from similar periods of decline. American psychology, for example, went through a period of insularity and myopia in the 1950s and 1960s despite its being a rigorous and experimental discipline. Under the leadership of Hull, Spence, and Skinner, the behaviorist tradition they espoused focused only on the reflexive and observable aspects of behavior and dealt with these as if they represented all there is to mental life.

With the emergence of computers to model and test ideas about mind, and with the development of more controlled ways of examining human mental processes, psychology reemerged in the 1970s in its modern form as a cognitive psychology that has explored language, perception, memory, motivation, and skilled movements in ways that have proven stimulating, insightful, and rigorous. Modern psychology is still evolving. The recent merger of cognitive psychology with neural science—the discipline we now call cognitive neural science—is proving to be one of the most exciting areas in all of biology. What is the aspiration of psychoanalysis if not to be the most cognitive of neural sciences? The future of psychoanalysis, if it is to have a future, is in the context of an empirical psychology, abetted by imaging techniques, neuroanatomical methods, and human genetics. Embedded in the sciences of human cognition, the ideas of psychoanalysis can be tested, and it is here that these ideas can have their greatest impact.

The following is but one example from my own field, the cognitive neural science of memory. One of the great insights of modern cognitive neural science in the study of memory is the realization that memory is not a unitary function of mind but has at least two forms, called explicit and implicit: a memory for *what* things are as compared to a memory for *how* to do something. *Explicit memory* encodes *conscious* information about autobiographical events and factual knowledge. It is a memory about people, places, facts, and objects, and it requires for its expression the hippocampus and the medial temporal lobe. *Implicit memory* involves for its recall an *unconscious* memory for motor and perceptual strategies. It depends on the specific sensory and motor systems as well as on the cerebellum and the basal ganglia.

Patients with lesions of the medial temporal lobe—or the hippocampus, which lies deep in it—cannot acquire new explicit memories for people, places, and objects. But they are fully able to learn motor skills and are also able to improve their performance on perceptual tasks. Implicit memory is not limited to simple tasks. It also includes a sophisticated form of memory called *priming*, in which recognition of words or objects is facilitated by prior exposure to the words or visual clues. Thus, a subject can recall the cued item better than other items for which no cues have been provided. Similarly, when shown the first few letters of previously studied words, a subject with temporal lobe lesions often responds by selecting correctly the previously presented word, even though he cannot remember ever seeing the word before!

The tasks that patients who lack explicit memory are capable of learning have in common that they do not require conscious awareness. The patient need not deliberately remember anything. Thus, when given a highly complex mechanical puzzle to solve, the patient may learn it as quickly as a normal person, but on questioning will not remember seeing the puzzle or having worked on it previously. When asked why his performance on a task is much better after several days of practice than on the first day, the patient may respond, "What are you talking about? I've never done this task before."

What a momentous discovery! Here we have, for the first time, the neural basis for a set of unconscious mental processes. Yet this unconscious bears no resemblance to Freud's unconscious. It is not related to instinctual strivings or to sexual conflicts, and the information never enters consciousness. These sets of findings provide the first challenge to a psychoanalytically oriented neural science. Where, if it exists at all, is the other unconscious? What are its neurobiological properties? How do unconscious strivings become transformed to enter awareness as a result of analytic therapy?

There are other challenges, of course. But at the very least, a biologically based psychoanalysis would redefine the usefulness of psychoanalysis as an effective perspective on certain specific disorders. At its best, psychoanalysis could live up to its initial promise and help revolutionize our understanding of mind and brain.

References

Alexander F: Psychosomatic Medicine: Its Principles and Applications. New York, WW Norton, 1950

Bailey CH, Kandel ER: Structural changes accompanying memory storage. Annu Rev Physiol 55:397–426, 1993

Blakemore C: Mechanisms of the Mind. Cambridge, England, Cambridge University Press, 1977

Bleuler E: Dementia Praecox or the Group of Schizophrenias (1911). Translated by Zinkin J. New York, International Universities Press, 1950

Day M, Semrad EV: Schizophrenic reactions, in The Harvard Guide to Modern Psychiatry. Edited by Nicholi AM Jr. Cambridge, MA, Harvard University Press, 1978, pp 199–241

Freud S: Project for a scientific psychology, in The Origins of Psycho-Analysis (1895). Edited by Bonaparte M, Freud A, Kris E. Translated by Mosbacher E, Strachey J. London, Imago, 1954

Fromm-Reichmann F: Notes on the development of treatment of schizophrenia by psychoanalytic therapy. Psychiatry 11:263–273, 1948

Fromm-Reichmann F: Psychoanalysis and Psychotherapy: Selected Papers of Frieda Fromm-Reichmann. Edited by Bullard DM, Weigert EV. Chicago, IL, University of Chicago Press, 1959

Gottesman II: Schizophrenia Genesis: The Origins of Madness. New York, WH Freeman, 1991

Heston LL: The genetics of schizophrenic and schizoid disease. Science 167:249–256, 1970

Jenkins WM, Merzenich MM, Ochs MT, et al: Functional reorganization of primary somatosensory cortex in adult owl monkeys after behaviorally controlled tactile stimulation. J Neurophysiol 63:82–104, 1990

Jones EG: Cortical development and neuropathology in schizophrenia, in Development of the Cerebral Cortex: Ciba Foundation Symposium 193. Chichester, England, Wiley, 1995, pp 277–295

Jung C: The Psychology of Dementia Praecox (1906) (Nervous and Mental Disease Monograph Series, No 3). Translated by Brill AA. New York, Nervous and Mental Disease Publishing, 1936

Kahana RJ: Psychotherapy: models of the essential skill, in The Teaching of Dynamic Psychiatry: A Reappraisal of the Goals and Techniques in the Teaching of Psychoanalytic Psychiatry. Edited by Bibring GL. Madison, CT, International Universities Press, 1968, pp 87–103

Kallmann FJ: The Genetics of Schizophrenia. New York, Augustin, 1938

Kandel ER: Cellular mechanisms of learning and memory, in Essentials of Neural Science and Behavior. Edited by Kandel ER, Schwartz JH, Jessell TM. East Norwalk, CT, Appleton & Lange, 1995, pp 667–694

Kandel ER, Schwartz JH, Jessell TM (eds): Principles of Neural Science, 3rd Edition. East Norwalk, CT, Appleton & Lange, 1991

Kety SS: Biochemical theories of schizophrenia, I and II. Science 129:1528–1532, 1590–1596, 1959

Lashley KS: Brain Mechanisms and Intelligence: A Quantitative Study of Injuries to the Brain. Chicago, IL, University of Chicago Press, 1929

Merzenich MM, Recanzone EG, Jenkins WM, et al: Cortical representational plasticity, in Neurobiology of Neocortex. Edited by Rakic P, Singer W. New York, Wiley, 1988, pp 41–67

Pauling L, Itano HA, Singer SJ, et al: Sickle cell anemia: a molecular disease. Science 110:543–548, 1949

Rosen JN: The Concept of Early Maternal Environment in Direct Psychoanalysis. Doylestown, PA, Doylestown Foundation, 1963

Rosenfeld HA: Psychotic States: A Psychoanalytic Approach. New York, International Universities Press, 1965

Rosenthal D, Wender PH, Kety SS, et al: The adopted-away offspring of schizophrenics. Am J Psychiatry 128:307–311, 1971

Schiller F: Paul Broca. Oxford, England, Oxford University Press, 1992

Schwartz JH, Kandel ER: Modulation of synaptic transmission: second-messenger systems, in Essentials of Neural Science and Behavior. Edited by Kandel ER, Schwartz JH, Jessell TM. East Norwalk, CT, Appleton & Lange, 1995, pp 243–267

Schwartz JM, Stoessel PW, Baxter LR, et al: Systematic changes in cerebral glucose metabolic rate after successful behavior modification treatment of obsessive-compulsive disorders. Arch Gen Psychiatry 53:109–113, 1996

Seltzer B, Frazier SH: Organic mental disorders, in The Harvard Guide to Modern Psychiatry. Edited by Nicholi AM Jr. Cambridge, MA, Harvard University Press, 1978, pp 297–318

Sheehan DV, Hackett TP: Psychosomatic disorders, in The Harvard Guide to Modern Psychiatry. Edited by Nicholi AM Jr. Cambridge, MA, Harvard University Press, 1978, pp 319–353

Skinner BF: The Behavior of Organisms: An Experimental Analysis. New York, Appleton-Century-Crofts, 1938

"BIOLOGY AND THE FUTURE OF PSYCHOANALYSIS"

Arnold M. Cooper, M.D.

Today more than ever before in its 100-year history, psychoanalysis is in a state of theoretical and clinical excitement, uncertainty, and open debate. Under the term *theoretical pluralism* psychoanalysts have acknowledged that there are multiple competing views concerning the nature of mental life, the origins of psychopathology, the centrality of intrapsychic conflict, the sources of the resistance to change, and the relationship of present to past. There is also a sharp continuing debate over whether psychoanalysis should aspire to accommodate any variety of scientific methodology or whether it should confine itself to being a hermeneutic discipline. In this setting, the attempts to adjudicate the superiority of one or another viewpoint by detailed reporting of analyst-patient interactions have failed. Each of the various schools of analytic thought—for example, Kleinian, ego-psychological, relational, and self-psychological—finds little difficulty in claiming its point of view as the superior explanatory base for clinical observation. From the hermeneutic viewpoint, psychoanalysis cannot and will not advance greatly beyond the initial discoveries of Freud and their later elaboration and enrichment by others. In contrast, Eric Kandel calls upon psychoanalysis to find ways to invigorate itself, to become a source of new ideas, and to enrich the neurosciences by becoming part of the neuroscientific community, while preserving and expanding its knowledge and skill in mapping the conscious and unconscious mental life of human beings.

Psychoanalysis may be following a destiny typical of many sciences: an early period of discovery and innovation followed by a decline to a baseline level of scientific activity. If that is the case, psychoanalysis can surely benefit by being a partner of neuroscience, which has only begun its upward arc. Kandel has provided a brief course in neurobiology for psychoanalysts in this remarkable paper filled with suggestions for future combined research. He vividly describes the advances of neuroscience that converge with psychoanalytic interests. The psychoanalytic map of the mind—the most complete and interesting map available—helps to set an agenda for neurobiology. After all, what is most interesting about the brain is how it generates mental life. What is the biology of subjectivity, consciousness, selfhood, and conflict? Kandel points to promising starts in these directions. Simultaneously, the advances in neurobiology have begun to demonstrate aspects of what psychoanalytic treatment must achieve biologically if its effects are to be significant. As has been demonstrated, talking cures are reflected in brain changes, as are pharmacologic cures. Neuroscience has embarked on an amazing voyage of discovery and psychoanalysis has an opportunity and obligation to participate and contribute.

There can be no question that another century of neuroscience will produce advances that seem unimaginable today, including a richer, more nuanced understanding of such human qualities as emotional responsiveness, unconscious mental processing, chronic resentment, self-damaging behaviors, self-pity, persistent avoidance of loving and gratifying relationships, and resistance to change. Some fear that this is a road to an Orwellian world where a pill will make us all feel good and that it will reduce individuality and the quest for knowledge. Psychoanalysis can play an important role in ensuring that neuroscience does not go down that path but joins the psychoanalytic effort for expanded self-awareness and choice. Kandel offers multiple examples of why analysts should interest themselves in and avail themselves of newer neuroscience findings. While at the moment the two enterprises are in many respects parallel, we can be certain that the continuing advances of neuroscience are likely to demonstrate that certain psychoanalytic ideas are wrong (as is already the case for infantile amnesia, once attributed to repression and now known to be the result of the lack of necessary memory pathways in a young child), while other ideas are better understood; for example, the understanding of implicit memory and its extraordinary fixity and ubiquity place the concepts of repetition-compulsion and resistance in a new light. Shifts in philosophy of mind may be required on both sides of the current divide: an acceptance of mind-brain as a single entity that can be investigated from different perspectives, rejection of a computational model in favor of nonlinear complexity, and full recognition of the power of both environment and molecular alteration to influence complex social behaviors.

The psychoanalytic enterprise has suffered hugely from its isolation from the academic setting. To my knowledge, no analytic institute gives a priority to empirical scientific research in the way that is routine in medical schools. Few analytic institutes are even in a position to begin to muster the resources that are required to provide their students with research experience. The International Psychoanalytical Association has been making a major and successful effort to recruit and train researchers, and there is now a thriving psychoanalytic empirical research effort in such areas as the development of mentalization, the effects of early mother-infant attunement, the consequences of different attachment patterns, outcome studies of short-term dynamic psychotherapies, among many others. However, this is a small beginning and does not touch upon the problem of institutes that are often satisfied with their clinical expertise, untested though it is. An appropriate task for psychoanalytic educators is to accomplish what medical education is finding so difficult: inculcating the necessary reductionism of scientific thought while retaining the humanism required for empathic understanding of another's experience.

For all of these reasons, Kandel's views have aroused both fervent support and opposition among psychoanalysts. I look forward to an era of enhanced cooperation of psychoanalysis and neuroscience and can imagine the possibility that interpretation may be accompanied by imaging techniques that will demonstrate differences between one interpretive pathway and another and that analytic progress may be checked against brain changes, and we will know better what must change if pathology is being successfully dealt with. It would be similar to the internist's use of laboratory findings to check the progress of a treatment. The advent of pharmacologic agents for depression, anxiety, tics, and obsessional symptoms, to mention just a few syndromes, was initially regarded by many psychoanalysts as an interference and obstruction to the conduct of analytic work. Time has shown that these agents often make analysis possible for patients whose symptom predominance overwhelms their capacity for free associative exploration of their emotional lives. Relief from overwhelming anxiety, depression, or obsession opens the possibilities for a search for a better understanding or a reconstruction of one's selfhood, one's history, and one's decision making. The problem for psychoanalysts is how to absorb the findings of neuroscience, whose precision concerns very limited parts of human mental function, into the theories of psychoanalysis, which are concerned with much larger, less precisely defined areas of mind.

Psychoanalysts are, or should be, grateful to Kandel for his invitation to us to bring our knowledge of the mind to the neuroscientists in their astonishing journey through the brain. Eric Kandel has offered a challenge to psychoanalysis that we must meet.

CHAPTER 3

BIOLOGY AND THE FUTURE OF PSYCHOANALYSIS

A New Intellectual Framework for Psychiatry Revisited

Eric R. Kandel, M.D.

> We must recollect that all of our provisional ideas in psychology will presumably one day be based on an organic substructure.
>
> —*Sigmund Freud,*
> *"On Narcissism" (S. Freud 1914/1957)*

> The deficiencies in our description would probably vanish if we were already in a position to replace the psychological terms with physiological or chemical ones....We may expect [physiology and chemistry] to give the most surprising information and we cannot guess what answers it will return in a few dozen years of questions we have put to it. They may be of a kind that will blow away the whole of our artificial structure of hypothesis.
>
> —*Sigmund Freud,*
> *"Beyond the Pleasure Principle" (S. Freud 1920/1955)*

This article was originally published in the *American Journal of Psychiatry*, Volume 156, Number 4, 1999, pp. 505–524.

The *American Journal of Psychiatry* has received a number of letters in response to my earlier "Framework" article (Kandel 1998). Some of these are reprinted elsewhere in this issue, and I have answered them briefly there. However, one issue raised by some letters deserves a more detailed answer, and that relates to whether biology is at all *relevant* to psychoanalysis. To my mind, this issue is so central to the future of psychoanalysis that it cannot be addressed with a brief comment. I therefore have written this article in an attempt to outline the importance of biology for the future of psychoanalysis.

During the first half of the twentieth century, psychoanalysis revolutionized our understanding of mental life. It provided a remarkable set of new insights about unconscious mental processes, psychic determinism, infantile sexuality, and, perhaps most important of all, about the irrationality of human motivation. In contrast to these advances, the achievements of psychoanalysis during the second half of this century have been less impressive. Although psychoanalytic thinking has continued to progress, there have been relatively few brilliant new insights, with the possible exception of certain advances in child development (for a review of recent progress, see Isenstadt 1998; Levin 1998; Shapiro and Emde 1995; Shevrin 1998). Most important, and most disappointing, psychoanalysis has not evolved scientifically. Specifically, it has not developed objective methods for testing the exciting ideas it had formulated earlier. As a result, psychoanalysis enters the twenty-first century with its influence in decline.

This decline is regrettable, since psychoanalysis still represents the most coherent and intellectually satisfying view of the mind. If psychoanalysis is to regain its intellectual power and influence, it will need more than the stimulus that comes from responding to its hostile critics. It will need to be engaged constructively by those who care for it and who care for a sophisticated and realistic theory of human motivation. My purpose in this article is to suggest one way that psychoanalysis might reenergize itself, and that is by developing a closer relationship with biology in general and with cognitive neuroscience in particular.

A closer relationship between psychoanalysis and cognitive neuroscience would accomplish two goals for psychoanalysis, one conceptual and the other experimental. From a conceptual point of view, cognitive neuroscience could provide a new foundation for the future growth of psychoanalysis, a foundation that is perhaps more satisfactory than metapsychology. David Olds has referred to this potential contribution of biology as "rewriting metapsychology on a scientific foundation." From an experimental point of view, biological insights could serve as a stimulus for research, for testing specific ideas about how the mind works.

Others have argued that psychoanalysis should be satisfied with more modest goals; it should be satisfied to strive for a closer interaction with cognitive psychology, a discipline that is more immediately related to psycho-analysis and more directly relevant to clinical practice. I have no quarrel with this argument. It seems to me, however, that what is most exciting in cognitive psychology today and what will be even more exciting tomorrow is the merger of cognitive psychology and neuroscience into one unified dis-cipline, which we now call cognitive neuroscience (for one example of this merger, see Milner et al. 1998). It is my hope that by joining with cognitive neuroscience in developing a new and compelling perspective on the mind and its disorders, psychoanalysis will regain its intellectual energy.

Meaningful scientific interaction between psychoanalysis and cognitive neuroscience of the sort that I outline here will require new directions for psychoanalysis and new institutional structures for carrying them out. My purpose in this article, therefore, is to describe points of intersection be-tween psychoanalysis and biology and to outline how those intersections might be investigated fruitfully.

The Psychoanalytic Method and the Psychoanalytic View of the Mind

Before I outline the points of congruence between psychoanalysis and biol-ogy, it is useful to review some of the factors that have led to the current cri-sis in psychoanalysis, a crisis that has resulted in good part from a restricted methodology. Three points are relevant here.

First, at the beginning of the twentieth century, psychoanalysis intro-duced a new method of psychological investigation, a method based on free association and interpretation. Freud taught us to listen carefully to patients and in new ways, ways that no one had used before. Freud also outlined a provisional schema for interpretation, for making sense out of what other-wise seemed to be unrelated and incoherent associations of patients. This approach was so novel and powerful that for many years, not only Freud but also other intelligent and creative psychoanalysts could argue that psycho-therapeutic encounters between patient and analyst provided the best con-text for scientific inquiry. In fact, in the early years, psychoanalysts could and did make many useful and original contributions to our understanding of the mind simply by listening to patients, or by testing ideas from the an-alytic situation in observational studies, a method that has proved particu-larly useful for studying child development. This approach may still be useful clinically because, as Anton Kris has emphasized, one listens differ-ently now. Nevertheless, it is clear that as a research tool this particular method has exhausted much of its novel investigative power. One hundred years after its introduction, there is little new in the way of theory that can be learned by merely listening carefully to individual patients. We must, at last, acknowledge that at this point in the modern study of mind, clinical observation of individual patients, in a context like the psychoanalytic situation that is so susceptible to observer bias, is not a sufficient basis for a science of mind.

This view is shared even by senior people within the psychoanalytic community. Thus, Kurt Eissler (1969) wrote, "the decrease in momentum of psychoanalytic research is due not to subjective factors among the analysts, but rather to historical facts of wider significance: the psychoanalytic situation has already given forth everything it contains. It is depleted with regard to research possibilities, at least as far as the possibility of new paradigms is concerned."

Second, as these arguments make clear, although psychoanalysis has historically been scientific in its aim, it has rarely been scientific in its methods; it has failed over the years to submit its assumptions to testable experimentation. Indeed, psychoanalysis has traditionally been far better at generating ideas than at testing them. As a result of this failure, it has not been able to progress as have other areas of psychology and medicine.

The concerns of modern behavioral science for controlling experimenter bias by means of blind experiments have largely escaped the concern of psychoanalysts (for important exceptions, see Dahl 1974; Luborsky and Luborsky 1995; Teller and Dahl 1995). With rare exception, the data gathered in psychoanalytic sessions are private: the patient's comments, associations, silences, postures, movements, and other behaviors are privileged. In fact, the privacy of communication is central to the basic trust engendered by the psychoanalytic situation. Here is the rub. In almost all cases, we have only the analysts' subjective accounts of what they believe has happened. As the research psychoanalyst Hartvig Dahl (1974) has long argued, hearsay evidence of this sort is not accepted as data in most scientific contexts. Psychoanalysts, however, are rarely concerned that their account of what happened in a therapy session is bound to be subjective and biased.

As a result, what Boring (1950) wrote, nearly 50 years ago, still stands: "We can say, without any lack of appreciation for what has been accomplished, that psychoanalysis has been prescientific. It has lacked experiments, having developed no techniques for control. In the refinement of description without control it is impossible to distinguish semantic specification from fact."

Thus, in the future, psychoanalytic institutes should strive to have at least a fraction of all supervised analyses be accessible to this sort of scrutiny. This is important not only for the psychoanalytic situation but also for other areas of investigation. Insights gained in therapy sessions have importantly inspired other modes of investigation outside the psychoanalytic situation. A successful example is the direct observation of children and the experimental analysis of attachment and parent-child interaction. Basing future experimental analyses on insights gained from the psychoanalytic situation makes it all the more important that the scientific reliability of these situations be optimized.

Third, unlike other areas of academic medicine, psychoanalysis has a serious institutional problem. The autonomous psychoanalytic institutes that have persisted and proliferated over the last century have developed their own unique approaches to research and training, approaches that have become insulated from other forms of research. With some notable exceptions, the psychoanalytic institutes have not provided their students or faculty with appropriately academic settings for questioning scholarship and empirical research.

To survive as an intellectual force in medicine and in cognitive neuroscience, and indeed in society as a whole, psychoanalysis will need to adopt new intellectual resources, new methodologies, and new institutional arrangements for carrying out its research. Several medical disciplines have grown by incorporating the methodologies and concepts of other disciplines. By and large, psychoanalysis has failed to do so. Because psychoanalysis has not yet recognized itself as a branch of biology, it has not incorporated into the psychoanalytic view of the mind the rich harvest of knowledge about the biology of the brain and its control of behavior that has emerged in the last 50 years. This, of course, raises the question, Why has psychoanalysis not been more welcoming of biology?

The Current Generation of Psychoanalysts Have Raised Arguments for and Against a Biology of Mind

In 1894, Freud argued that biology had not advanced enough to be helpful to psychoanalysis. It was premature, he thought, to bring the two together. One century later, a number of psychoanalysts have a far more radical view. Biology, they argue, is irrelevant to psychoanalysis. To give an example, Marshall Edelson in his book *Hypothesis and Evidence in Psychoanalysis* wrote:

> Efforts to tie psychoanalytic theory to a neurobiological foundation, or to mix hypotheses about mind and hypotheses about brain in one theory, should be resisted as expressions of logical confusion.
>
> I see no reason to abandon the position Reiser takes despite his avowed belief in the "functional unity" of mind and body, when he considers the mind-body relation:
>
> "The science of the mind and the science of the body utilize different languages, different concepts (with differing levels of abstraction and complexity), and different sets of tools and techniques. Simultaneous and parallel

psychological and physiological study of a patient in an intense anxiety state produces of necessity two separate and distinct sets of descriptive data, measurements, and formulations. There is no way to unify the two by translation into a common language, or by reference to a shared conceptual framework, nor are there as yet bridging concepts that could serve...as intermediate templates, isomorphic with both realms. For all practical purposes, then, we deal with mind and body as separate realms; virtually, all of our psychophysiological and psychosomatic data consist in essence of covariance data, demonstrating coincidence of events occurring in the two realms within specified time intervals at a frequency beyond chance." (Reiser 1975, p. 479)

I think it is at least possible that scientists may eventually conclude that what Reiser describes does not simply reflect the current state of the art, methodologically, or the inadequacy of our thought but represents, rather, something that is logically or conceptually necessary, something that no practical or conceptual developments will ever be able to mitigate. (Edelson 1984)

In my own numerous interactions with Reiser, I have never sensed him to have difficulty relating brain to mind. Nevertheless, I have quoted Edelson at length because his view is representative of that shared by a surprisingly large number of psychoanalysts, and even by Freud in some of his later writings. This view, often referred to as the hermeneutic as opposed to the scientific view of psychoanalysis, reflects a position that has hindered psychoanalysis from continuing to grow intellectually (M.S. Roth 1998; Shapiro 1996).

Now, psychoanalysis could, if it wanted to do so, easily rest on its hermeneutic laurels. It could continue to expound on the remarkable contributions of Freud and his students, on the insights into the unconscious mental processes and motivations that make us the complex, psychologically nuanced individuals we are (Bowlby 1960; Erikson 1963; A. Freud 1936; S. Freud 1933[1932]/1964; Hartmann 1939/1958; Klein 1957; Kohut 1971; Spitz 1945; Winnicott 1954/1958). Indeed, in the context of these contributions, few would challenge Freud's position as the great modern thinker on human motivation or would deny that our century has been permanently marked by Freud's deep understanding of the psychological issues that historically have occupied the Western mind from Sophocles to Schnitzler.

But if psychoanalysis is to rest on its past accomplishments, it must remain, as Jonathan Lear (1998) and others have argued, a philosophy of mind, and the psychoanalytic literature—from Freud to Hartmann to Erikson to Winnicott—must be read as a modern philosophical or poetic text alongside Plato, Shakespeare, Kant, Schopenhauer, Nietzsche, and Proust. On the other hand, if the field aspires, as I believe most psychoanalysts do aspire, to be an evolving, active contributor to an emerging science of the mind, then psychoanalysis is falling behind.

I therefore agree with the sentiment expressed by Lear (1998): "Freud is dead. He died in 1939, after an extraordinary productive and creative life....It is important not to get stuck on him, like some rigid symptom, either to idolize him or to denigrate him."

Biology in the Service of Psychoanalysis

My focus in this article is on ways that biology might reinvigorate the psychoanalytic exploration of mind. I should say at the outset that although we have the outlines of what could evolve into a meaningful biological foundation for psychoanalysis, we are very much at the beginning. We do not yet have an intellectually satisfactory biological understanding of any complex mental processes. Nevertheless, biology has made remarkable progress in the last 50 years, and the pace is not slacking. As biologists come to focus more of their efforts on the brain-mind, most of them have become convinced that the mind will be to the biology of the twenty-first century what the gene has been to the biology of the twentieth century. Thus, Francois Jacob (1998) writes, "the century that is ending has been preoccupied with nucleic acids and proteins. The next one will concentrate on memory and desire. Will it be able to answer the questions they pose?"

My key argument is that the biology of the next century is, in fact, in a good position to answer some of the questions about memory and desire, that these answers will be all the richer and more meaningful if they are forged by a synergistic effort of biology and psychoanalysis. In turn, answers to these questions, and the very effort of providing them in conjunction with biology, will provide a more scientific foundation for psychoanalysis.

In the next century, biology is likely to make deep contributions to the understanding of mental processes by delineating the biological basis for the various unconscious mental processes, for psychic determinism, for the role of unconscious mental processes in psychopathology, and for the therapeutic effect of psychoanalysis. Now, biology will not immediately enlighten these deep mysteries at their core. These issues represent, together with the nature of consciousness, the most difficult problems confronting all of biology—in fact, all of science. Nevertheless, one can begin to outline how biology might at least clarify some central psychoanalytic issues, at least at their margins. Here I outline eight areas in which biology could join with psychoanalysis to make important contributions: 1) the nature of unconscious mental processes, 2) the nature of psychological causality, 3) psychological causality and psychopathology, 4) early experience and the predisposition to mental illness, 5) the preconscious, the unconscious, and the prefrontal cortex, 6) sexual orientation, 7) psychotherapy and structural changes in the brain, and 8) psychopharmacology as an adjunct to psychoanalysis.

1. Unconscious Mental Processes

Central to psychoanalysis is the idea that we are unaware of much of our mental life. A great deal of what we experience—what we perceive, think, dream, fantasize—cannot be directly accessed by conscious thought. Nor can we explain what often motivates our actions. The idea of unconscious mental processes not only is important in its own right, but it is critical for understanding the nature of psychic determinism. Given the centrality of unconscious psychic processes, what can biology teach us about them?

In 1954 Brenda Milner made the remarkable discovery, based on studies of the amnestic patient H.M., that the medial temporal lobe and the hippocampus mediate what we now call declarative (explicit) memory storage, a conscious memory for people, objects, and places (Scoville and Milner 1957). In 1962 she made the further discovery that even though H.M. had no conscious recall of new memories about people, places, and objects, he was nonetheless fully capable of learning new perceptual and motor skills (for a recent review, see Milner et al. 1998). These memories—what we now call procedural or implicit memory—are completely unconscious and are evident only in performance rather than in conscious recall.

Using the two memory systems together is the rule rather than the exception. These two memory systems overlap and are commonly used together so that many learning experiences recruit both of them. Indeed, constant repetition can transform declarative memory into a procedural type. For example, learning to drive an automobile at first involves conscious recollection, but eventually driving becomes an automatic and nonconscious motor activity. Procedural memory is itself a collection of processes involving several different brain systems: priming, or recognition of recently encountered stimuli, is a function of sensory cortices; the acquisition of various cued feeling states involves the amygdala; formation of new motor (and perhaps cognitive) habits requires the neostriatum; learning new motor behavior or coordinated activities depends on the cerebellum. Different situations and learning experiences recruit different subsets of these and other procedural memory systems, in variable combination with the explicit memory system of the hippocampus and related structures (Squire and Kandel 1999; Squire and Zola-Morgan 1991) (Figure 3–1).

In procedural memory, then, we have a biological example of one component of unconscious mental life. How does this biologically delineated unconscious relate to Freud's unconscious? In his later writings, Freud used the concept of the unconscious in three different ways (for a review of Freud's ideas on consciousness, see Solms 1997). First, he used the term in a strict or structural way to refer to the *repressed* or *dynamic unconscious*. This unconscious is what the classical psychoanalytic literature refers to as

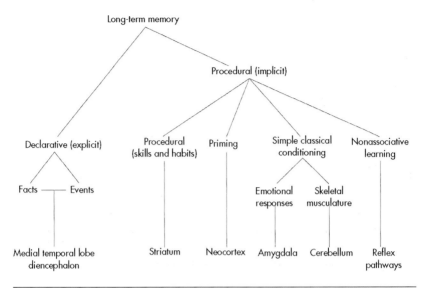

FIGURE 3-1. A taxonomy of the declarative and procedural memory systems.

This taxonomy lists the brain structures and connections thought to be especially important for each kind of declarative and nondeclarative memory.

Source. Reprinted from Milner B, Squire LR, Kandel ER: "Cognitive Neuroscience and the Study of Memory." *Neuron* 20:445–468, 1998. Used with permission of Cell Press.

the unconscious. It includes not only the id but also that part of the ego which contains unconscious impulses, defenses, and conflicts and therefore is similar to the dynamic unconscious of the id. In this dynamic unconscious, information about conflict and drive is prevented from reaching consciousness by powerful defensive mechanisms such as repression.

Second, in addition to the repressed parts of the ego, Freud proposed that still another part of the ego is unconscious. Unlike the unconscious parts of the ego that are repressed and therefore resemble the dynamic unconscious, the unconscious part of the ego that is not repressed is not concerned with unconscious drives or conflicts. Moreover, unlike the preconscious unconscious, this unconscious part of the ego is never accessible to consciousness even though it is not repressed. Since this unconscious is concerned with habits and perceptual and motor skills, it maps onto procedural memory. I shall therefore refer to it as the *procedural unconscious*.

Finally, Freud used the term descriptively, in a broader sense—the *preconscious unconscious*—to refer to almost all mental activities, to most thoughts and all memories that enter consciousness. According to Freud, an individual is not aware of almost all of the mental processing events themselves yet can have ready conscious access to many of them by an effort of attention. From this perspective, most of mental life is unconscious much of the time and becomes conscious only as sensory percepts: as words and images.

Of these three unconscious mental processes, only the procedural unconscious, the unconscious part of the ego that is not conflicted or repressed, appears to map onto what neuroscientists call *procedural memory* (for a similar argument, see also Lyons-Ruth 1998). This important correspondence between cognitive neuroscience and psychoanalysis was first recognized in a thoughtful article by Robert Clyman (1991), who considered procedural memory in the context of emotion and its relevance for transference and for treatment. This idea has been developed further by Louis Sander, Daniel Stern, and their colleagues in the Boston Process of Change Study Group (1998), who have emphasized that many of the changes that advance the therapeutic process during an analysis are not in the domain of conscious insight but rather in the domain of unconscious procedural (non-verbal) knowledge and behavior. To encompass this idea, Sander (1998), Stern (1998), and their colleagues have developed the idea that there are *moments of meaning*—moments in the interaction between patient and therapist—that represent the achievement of a new set of implicit memories that permits the therapeutic relationship to progress to a new level. This progression does not depend on conscious insights; it does not require, so to speak, the unconscious becoming conscious. Rather, moments of meaning are thought to lead to changes in behavior that increase the patient's range of procedural strategies for doing and being. Growth in these categories of knowledge leads to strategies for action that are reflected in the ways in which one person interacts with another, including ways that contribute to transference.

Marianne Goldberger (1996) has extended this line of thought by emphasizing that moral development also is advanced by procedural means. She points out that people do not generally remember, in any conscious way, the circumstances under which they assimilated the moral rules that govern their behavior; these rules are acquired almost automatically, like the rules of grammar that govern our native language.

I illustrate this distinction between procedural and declarative memory that comes from cognitive neuroscience to emphasize the utility for psychoanalytic thought of a fundamentally neurobiological insight. But in addition, I would suggest that as applied to psychoanalysis, these biological ideas are still only ideas. What biology offers is the opportunity to carry these ideas one important step further. We now know a fair bit about the biology of this procedural knowledge, including some of its molecular underpinnings (Milner et al. 1998).

The interesting convergence of psychoanalysis and biology on the problem of procedural memory confronts us with the task of testing these ideas in a systematic way. We will need to examine, from both a psychoanalytic and a biological perspective, the range of phenomena we have subsumed under the term *procedural memory* and see how they map onto different neural systems. In so doing we will want to examine, in behavioral, observational, and imaging studies, to what degree different components of a given moment of meaning or different moments of this sort recruit one or another anatomical subsystem of procedural memory.

As these arguments make clear, one of the earlier limitations to the study of unconscious psychic processes was that no method existed for directly observing them. All methods for studying unconscious processes were indirect. Thus, a key contribution that biology can now make—with its ability to image mental processes and its ability to study patients with lesions in different components of procedural memory—is to change the basis of the study of unconscious mental processes from indirect inference to direct observation. By these means we might be able to determine which aspects of psychoanalytically relevant procedural memory are mediated by which of the subcortical systems concerned. In addition, imaging methods may also allow us to discern which brain systems mediate the two other forms of unconscious memory, the dynamic unconscious and the preconscious unconscious.

Before I turn to the preconscious unconscious and its possible relation to the prefrontal cortex, I first want to consider three other features related to the procedural unconscious: its relation to psychic determinism, to conscious mental processes, and to early experience.

2. The Nature of Psychological Determinacy: How Do Two Events Become Associated in the Mind?

In Freud's mind, unconscious mental processes provided an explanatory mechanism for psychic determinism. The fundamental idea of psychic determinism is that little, if anything, in one's psychic life occurs by chance. Every psychic event, whether procedural or declarative, is determined by an event that precedes it. Slips of the tongue, apparently unrelated thoughts, jokes, dreams, and all images within each dream are related to preceding psychological events and have a coherent and meaningful relationship to the rest of one's psychic life. Psychological determinacy is similarly important in psychopathology. Every neurotic symptom, no matter how strange it may seem to the patient, is not strange in the unconscious mind but is related to preceding mental processes. The connections between symptoms and causative mental processes or between the images of a dream and their preceding psychically related events are obscured by the operation of ubiquitous and dynamic unconscious processes.

The development of many ideas within psychoanalytic thought and its core methodology, free association, derives from the concept of psychic determinism (Kris 1982). The purpose of free association is to have the patient report to the psychoanalyst all thoughts that come to mind and to refrain from exercising over them any degree of censorship or direction (Brenner 1978; Kris 1982). The key idea of psychic determinism is that any mental event is causally related to its preceding mental event. Thus, Brenner (1978) wrote, "In the mind, as in physical nature about us, nothing happens by chance, or in a random way. Each psychic event is determined by the ones which precede it."

Although we do not have a rich biological model of psychic declarative explicit knowledge, we have in biology a good beginning of an understanding of how associations develop in procedural memory (for a review, see Squire and Kandel 1999). Insofar as aspects of procedural knowledge are relevant to moments of meaning, these biological insights should prove useful for understanding the procedural unconscious.

In the last decade of the nineteenth century, at the time that Freud was working on his theory of psychological determinacy, Ivan Pavlov was developing an empirical approach to a particular instance of psychic determinism at the level of what we now call procedural knowledge: learning by association. Pavlov sought to elucidate an essential feature of learning that had been known since antiquity. Western thinkers since Aristotle had appreciated that memory storage requires the temporal association of contiguous thoughts, a concept later developed systematically by John Locke and the British empiricist philosophers. Pavlov's brilliant achievement was to develop an animal model of learning by association that could be studied rigorously in the laboratory. By changing the timing of two sensory stimuli and observing changes in simple reflex behavior, Pavlov (1927) established a procedure from which reasonable inferences could be made about how changes in the association between two stimuli could lead to changes in behavior—to learning (for more recent reviews, see Dickinson 1980; Domjan and Burkhard 1986; Rescorla 1988; Squire and Kandel 1999). Pavlov thus developed powerful paradigms for associative learning that led to a permanent shift in the study of behavior, moving it from an emphasis on introspection to an objective analysis of stimuli and responses. This is exactly the sort of shift we are looking for in psychoanalytic investigations of psychic determinism.

I have described this familiar paradigm because I want to emphasize three points relevant to psychoanalytic thought. First, in learning to associate two stimuli, a subject does not simply learn that one stimulus *precedes* the other. Instead, in learning to associate two stimuli, a subject learns that one stimulus comes to *predict* the other (for a discussion of this point, see Fanselow 1998; Rescorla 1988). Second, as we shall see below, classical conditioning is a superb paradigm for analyzing how knowledge can move from being unconscious to entering consciousness (Shevrin et al. 1996). Finally, classical conditioning can be used to acquire not only appetitive responses but also aversive ones and thus can give us insight into the emergence of psychopathology. I now turn to each of these points.

The psychic determinism of classical conditioning is probabilistic

For many years psychologists thought that classical conditioning followed rules of psychic determinism similar to those outlined by Freud. They thought that classical conditioning depended only on contiguity, on a critical minimum interval between the conditioned and the unconditioned stimulus, so that the two were experienced as connected. According to this view, each time a conditioned stimulus is followed by a reinforcing or unconditioned stimulus, a neural connection is strengthened between the stimulus and the response or between one stimulus and another, until eventually the bond becomes strong enough to change behavior. The only relevant variable determining the strength of conditioning was thought to be the number of pairings of the conditioned stimulus and unconditioned stimulus. In 1969, Leon Kamin made what now is generally considered the most significant empirical discovery in conditioning since Pavlov's initial findings at the turn of the century. Kamin found that animals learn more than contiguity; they learn contingencies. They do not simply learn that the conditioned stimulus precedes the unconditioned stimulus but rather that the conditioned stimulus predicts the unconditioned stimulus (Kamin 1969). Thus, associative learning does not depend on a critical number of pairings of conditioned stimulus and unconditioned stimulus but on the power of the conditioned stimulus to predict a biologically significant unconditioned stimulus (Rescorla 1988).

These considerations suggest why animals and people acquire classical conditioning so readily. Classical conditioning, and perhaps all forms of associative learning, likely evolved to enable animals to learn to distinguish events that regularly occur together from those that are only randomly associated. In other words, the brain seems to have evolved a simple mechanism that "makes sense" out of events in the environment by assigning a predictive function to some events. What environmental conditions might have shaped or maintained a common learning mechanism in a wide variety of species? All animals must be able to recognize and avoid danger; they must search out rewards such as food that is nutritious and avoid food that is spoiled or poisoned. An effective way to achieve this knowledge is to be able to detect regular relationships between stimuli or between behavior and stimuli. It is possible that by examining this relationship in cell biological terms, we may well be looking at the elementary mechanism of psychic determinism.

Classical conditioning and the relationship of conscious procedural to unconscious declarative mental processes

Conventional classical conditioning is usually carried out in a form called *delay conditioning*, in which the onset of the conditioned stimulus typically precedes the onset of the unconditioned stimulus by about 500 msec, and both the conditioned stimulus and the unconditioned stimulus terminate together (Figure 3–2). This form of conditioning is prototypically procedural (Clark and Squire 1998; Squire and Kandel 1999). When a normal human subject learns an eyeblink response to a weak tactile stimulus on his brow, that subject is unaware that he or she is being conditioned. Patients with damage to the hippocampus and the medial temporal neocortex, who therefore lack explicit (declarative) memory altogether, can be conditioned like normal subjects in a delay-conditioning paradigm.

A slight variation, *trace conditioning*, converts implicit conditioning into explicit memory. With trace conditioning the conditioned stimulus terminates before the unconditioned stimulus occurs, so that the conditioned stimulus is brief, and there is a 500-msec gap between the termination of the conditioned stimulus and the onset of the unconditioned stimulus (Figure 3–2). Richard Thompson and his colleagues found that trace conditioning depends on the hippocampus and is eliminated in experimental animals with lesions of the hippocampus (Kim et al. 1995; Solomon et al. 1986). Clark and Squire (1998) extended these experiments to humans and found that trace conditioning requires conscious recall. In the course of trace conditioning, normal subjects usually become consciously aware of the temporal gap in the relationship between the conditioned stimulus and unconditioned stimulus. Those subjects who do not become aware of this gap do not acquire trace conditioning. Moreover, this task cannot be mastered by people who suffer from amnesia—from a defect in declarative memory—as a result of lesions to the medial temporal lobe.

Thus, a small shift in temporal sequence changes an instance of psychic determinism from being unconscious to being conscious! This is consistent with the idea that the two memory systems, procedural and declarative, are often jointly recruited by a common task and encode different aspects of the sensory pattern of stimuli (or of the external world) present to the subject. Where in the medial temporal lobe is this shift from one type of memory

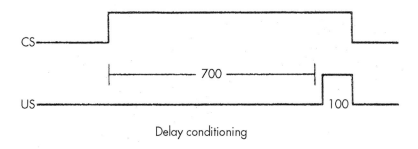

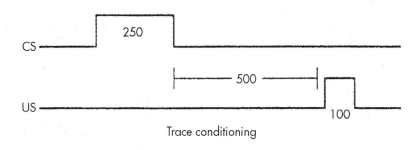

FIGURE 3–2. The different temporal relationships between the conditioned stimulus (CS) and the unconditioned stimulus (US) for delay conditioning and trace conditioning.

During delay conditioning, a tone-conditioned stimulus is presented and remains on until a 100-msec air puff to the eye (the unconditioned stimulus) is presented, and both stimuli terminate together. The word *delay* refers to the interval between the onset of the conditioned stimulus and the onset of the unconditioned stimulus (in this example, about 700 msec). During trace conditioning, the presentation of the conditioned stimulus and the presentation of the unconditioned stimulus are separated by an interval (in this example, 500 msec) during which no stimulus is present.

Source. Reprinted from Clark RE, Squire LR: "Classical Conditioning and Brain Systems: The Role of Awareness." *Science* 280:77–81, 1998. Used with permission of the American Association for the Advancement of Science.

storage to the other occurring? Eichenbaum (1998) has argued that the hippocampus functions to associate noncontiguous events over space and time. We in fact now know that trace conditioning recruits the hippocampus and the circuitry of the medial temporal lobe. Which parts of the hippocampal circuitry are key for trace conditioning? Do other regions become involved? Does the prefrontal cortex (which we shall consider below)—an area concerned with working memory that is thought to represent an aspect of the preconscious unconscious—mediate associations between unconscious and conscious memories that are the subject of analysis?

3. Psychological Causality and Psychopathology

We have seen that one point of convergence between biology and psychoanalysis is the relevance of procedural memory for early moral development, for aspects of transference, and for moments of meaning in psychoanalytic therapy. We have considered a second point of convergence in examining the relationship between the associative characteristic of classical conditioning and psychological determinacy. Here, I want to illustrate a third point of convergence: that between Pavlovian *fear conditioning,* a form of procedural memory mediated by the amygdala, signal anxiety, and posttraumatic stress syndromes in humans.

Early in his work on classical conditioning, Pavlov appreciated that conditioning is appetitive when the unconditioned stimulus is rewarding, but the same procedure will produce defensive conditioning when the unconditioned stimulus is aversive. Pavlov next found that defensive conditioning provides a particularly good experimental model of signal anxiety, a form of learned fear that can be advantageous.

> It is pretty evident that under natural conditions the normal animal must respond not only to stimuli which themselves bring immediate benefit or harm, but also to other physical or chemical agencies…which in themselves only *signal* the approach of these stimuli; though it is not the sight or the sound of the beast of prey which is itself harmful to smaller animals, but its teeth and claws. (Pavlov 1927, p. 14)

A similar proposal was made independently by Freud. Because painful stimuli are often associated with neutral stimuli, symbolic or real, Freud postulated that repeated pairing of neutral and noxious stimuli can cause the neutral stimulus to be perceived as dangerous and to elicit anxiety. Placing this argument in a biological context, Freud wrote:

> The individual will have made an important advance in his capacity for self-preservation if he can foresee and expect a traumatic situation of this kind which entails helplessness, instead of simply waiting for it to happen. Let us call a situation which contains the determinant for such expectation a *danger situation.* It is in this situation that the *signal of anxiety is given.* (S. Freud 1926/1959, p. 166; italics added)

Thus, both Pavlov and Freud appreciated that it is biologically adaptive to have the ability to respond defensively to danger signals before the real danger is present. *Signal* or *anticipatory anxiety* prepares the individual for fight or flight if the signal is from the environment. Freud suggested that mental defenses substitute for actual flight or withdrawal in response to internal danger. Signal anxiety therefore provides an opportunity for studying how mental defenses are recruited: how psychic determinism gives rise to psychopathology.

We know that the amygdala is important for emotionally charged memory, as in classical conditioning of fear by pairing a neutral tone with a shock (LeDoux 1996). The amygdala coordinates the flow of information between the areas of the thalamus and the cerebral cortex that process the sensory cues and areas that process the expression of fear: the hypothalamus, which regulates the autonomic response to fear, and the limbic neocortical association areas, the cingulate cortex and prefrontal cortex, which are thought to be involved in evaluating the conscious evaluation of emotion. LeDoux has argued that in anxiety, the patient experiences the autonomic arousal as something threatening happening, an arousal mediated by the amygdala. LeDoux attributes the absence of awareness to a shutting down of the hippocampus by stress, a mechanism considered below. We now have excellent methods for imaging these structures in both experimental animals and humans in order to address the question of how these linkages are established and, once established, how they are maintained (Breiter et al. 1996; LeDoux 1996; Whalen et al. 1996).

4. Early Experience and Predisposition to Psychopathology

Signal anxiety represents a simple example of an acquired psychopathology. But, as is the case with all things acquired, some people have a greater constitutional disposition than others to acquire neurotic anxiety. What factors predispose an individual to associate a variety of neutral stimuli with threatening ones?

In "Mourning and Melancholia" and in his other writings, Freud emphasized two components in the etiology of acquired psychopathology: constitutional (including genetic) predispositions and early experiential factors, especially loss. Indeed, there is evidence in the development of many forms of mental illness for both genetic components and experiential factors (both early developmental factors and later acute precipitating factors). As one example, while there is a clear genetic contribution to susceptibility to depression, many patients with major depression have experienced stressful life events during childhood, including abuse or neglect, and these stressors are important predictors of depression (Agid et al. 1999; Bremner et al. 1995; Brown et al. 1997; Heim et al. 1997a, 1997b; Kendler et al. 1992). The case is most clear for posttraumatic stress disorder (PTSD), which requires for its diagnosis the presence of stressful experience so severe as to be outside the range of usual human experience. About 30% of individuals traumatized in this way subsequently develop the full syndrome of PTSD (Heim et al. 1997a, 1997b). This incomplete penetrance raises the question, What (besides genes) predisposes people to developing PTSD and other stress-related disorders?

The component of the early environment thought to be most important for humans, and in fact for all mammals, is the infant's major caretaker, usually the mother. Psychoanalysis has long argued that the manner in which a mother and her infant interact creates within the child's mind the first internal representation not only of another person but of an interaction, of a relationship. This initial representation of people and of relationships is thought to be critical for the subsequent psychological development of the child. The interaction goes both ways. The way the infant behaves toward the mother exerts a considerable influence on the mother's behavior. Secure attachment of mother and infant is thought to foster in the infant comfort with itself and basic trust in others, whereas insecure attachment is thought to foster anxiety.

One of the key initial ideas to emerge from both cognitive and neurobiological study of development is that the development of these internal representations can only be induced during certain early and critical periods in the infant's life. During these critical periods, and only during these periods, the infant (and its developing brain) *must* interact with a responsive environment (an "average expectable environment," to use Heinz Hartmann's term) if the development of the brain and of the personality is to proceed satisfactorily.

The first compelling evidence for the importance of early relationships between parents and offspring came from Anna Freud's studies on the traumatic effects of family disruption during World War II (A. Freud and Burlingham 1973). The importance of family disruption was further developed by René Spitz (1945), who compared two groups of infants separated from their mothers. One group was raised in a foundling home where the infants were cared for by nurses, each of whom was responsible for seven infants; the other group was in a nursing home attached to a women's prison, where the infants were cared for daily by their mothers. By the end of the first year, the motor and intellectual performance of the children in the orphanage had fallen far below that of the children in the nursing home; those children were withdrawn and showed little curiosity or gaiety.

Harry Harlow extended this work one important step further by developing an animal model of infant development (Harlow 1958; Harlow et al. 1965). He found that when newborn monkeys were isolated for 6 months to 1 year and then returned to the company of other monkeys, they were physically healthy but behaviorally devastated. These monkeys crouched in a corner of their cages and rocked back and forth like severely disturbed or autistic children. They did not interact with other monkeys, nor did they fight, play, or show any sexual interest. Isolation of an older animal for a comparable period was innocuous. Thus, in monkeys, as in humans, there is a critical period for social development. Harlow next found that the syndrome could be partially reversed by giving the isolated monkey a surrogate mother, a cloth-covered wooden dummy. This surrogate elicited clinging behavior in the isolated monkey but was insufficient for the development of fully normal social behavior. Normal social development could only be rescued if, in addition to a surrogate mother, the isolated animal had contact for a few hours each day with a normal infant monkey who spent the rest of the day in the monkey colony.

The work of Anna Freud, Spitz, and Harlow was importantly extended by John Bowlby, who began to think about the interaction of the infant and its caregiver in biological terms. Bowlby (1960, 1969) formulated the idea that the defenseless infant maintains a closeness to its caretaker by means of a system of emotive and behavioral response patterns that he called the *attachment system*. Bowlby conceived of the attachment system as an inborn instinctual or motivational system, much like hunger or thirst, that organizes the memory processes of the infant and directs it to seek proximity to and communication with the mother. From an evolutionary point of view, the attachment system clearly enhances the infant's chances for survival by allowing the immature brain to use the parents' mature functions to organize its own life processes. The infant's attachment mechanism is mirrored in the parents' emotionally sensitive responses to the infant's signals. Parental responses serve both to amplify and reinforce the infant's positive emotional state and attenuate the infant's negative emotional states by giving the infant secure protection when upset. These repeated experiences become encoded in procedural memory as expectations that help the infant feel secure.

It should be noted that during the first 2–3 years of life, when an infant's interaction with its mother is particularly important, the infant relies primarily on its procedural memory systems. Both in humans and in experimental animals, declarative memory develops later. Thus, infantile amnesia, which results in the fact that very few memories from early childhood are accessible to later recall, is evident not only in humans but also in other mammals, including rodents. This amnesia presumably occurs not because of the powerful repression of memories during resolution of the oedipal complex, but because of slow development of the declarative memory system (Clyman 1991).

Bowlby described the response to separation as occurring in two phases: protest and despair. Events that disturb the proximity of the infant to the attachment object elicit protest: clinging, following, searching, crying, and acute physiological arousal lasting minutes to hours. These behaviors serve to restore proximity. When contact is regained, these clinging behaviors are shut off, according to Bowlby, by a feedback mechanism, and alternative behavioral systems, most notably exploratory behavior, become activated. If separation is prolonged, despair gradually replaces the early responses as the infant recognizes that separation may be prolonged or permanent and shifts from anxiety and anger to sadness and despair. Whereas protest is thought to be adaptive by increasing the likelihood that the parent and infant find each other again, despair is thought to prepare the infant for prolonged passive survival, achieved by conserving energy and withdrawing from danger.

We owe to Levine and colleagues (1957, 1962; Levine et al. 1967), Ader and Grota (1969), and Hofer (1981, 1994) the discovery that a similar attachment system exists in rodents. The extension of this research to a rodent model system, which is much simpler, but still mammalian, holds great power. For example, in mice, individual genes can be expressed or ablated, which allows a powerful approach for relating individual genes to behavior. Levine found that rat pups show an immediate protest to separation, consisting of repeated high-intensity vocalization, agitated searching, and high levels of self-grooming. If the mother fails to return and the separation continues, the protest behaviors wane over a period of hours and are replaced by a number of slower-developing behaviors—akin to despair—as the pups become progressively less alert and responsive, and their body temperature and heart rate drop. Much as Harlow was able to dissect the components of the caregiver that were essential for normal character development, so Hofer was able to show that three different aspects of pups' protest-despair responses were triggered by three different hidden regulators within the mother-infant interaction: loss of warmth, loss of food, and loss of tactile stimulation.

Levine and his colleagues (1967) were the first to carry the analysis to a molecular level by studying how varying degrees of infant attachment affected the animals' subsequent ability to respond to stress. Hans Selye had pointed out as early as 1936 that humans and experimental animals respond to stressful experiences by activating their hypothalamic-pituitary-adrenal (HPA) axis. The end product of the HPA system is the release of glucocorticoid hormones by the adrenal gland. These hormones serve as major regulators of homeostasis—of intermediary metabolism, muscle tone, and cardiovascular function. Together with catecholamines released by the autonomic nervous system and by the adrenal medulla, the secretion of glucocorticoids is essential for survival in the face of stress.

Levine therefore asked the question, Can the long-term response of the HPA system to stress be modulated by experience? If so, is it particularly sensitive to early experience? Levine discovered that when, during the first

2 weeks of life, pups were removed from their mothers for only a few minutes, the pups showed increased vocalization, which elicited increased maternal care. The mothers responded by licking, grooming, and carrying these pups around more often than if they had not been removed. This increase in the mother's attachment behavior reduced, *for the rest of the animal's life,* the pup's HPA response—its plasma levels of glucocorticoid—to a variety of stressors! Concomitantly, it reduced the pup's fearfulness and vulnerability to stress-related disease (Liu et al. 1997; Plotsky and Meaney 1993). By contrast, when, during the same 2-week period of life, pups were separated from their mothers for prolonged periods of time (3–6 hours per day for 2 weeks), the opposite reaction ensued. Now the mothers ignored the pups, and the pups showed an increase in plasma ACTH and glucocorticoid responses to stress as adults. Thus, differences in an infant's interactions with its mother—differences that fall in the range of naturally occurring individual differences in maternal care—are crucial risk factors for an individual's future response to stress. Here we have a remarkable example of how early experience alters the set point for a biological response to stress.

Studies by Charles Nemeroff and Paul Plotsky have found that these early adverse life experiences result in increased gene expression for corticotropin-releasing factor (CRF), the hormone released from the hypothalamus to initiate the HPA response. Daily maternal separation during the first 2 weeks is associated in the rat with profound and persistent increases in the expression of the mRNA for CRF, not only in the hypothalamus but also in limbic areas, including the amygdala and the bed nucleus of the stria terminalis (Meaney et al. 1991; Nemeroff 1996; Plotsky and Meaney 1993).

However, the biological insights into attachment theory do not stop here. Bruce McEwen, Robert Sapolsky, and their colleagues have discovered that the increases in glucocorticoids which follow prolonged separation have adverse effects on the hippocampus (McEwen and Sapolsky 1995; Sapolsky 1996). There are two types of receptors for glucocorticoids: type 1 (the mineralocorticoid receptors) and type 2 (the glucocorticoid receptors). The hippocampus is one of the few sites in the body that has both! Thus, repeated stress (or exposure to elevated glucocorticoids over a number of weeks) causes atrophy of neurons of the hippocampus, which is reversible when the stress or glucocorticoid exposure is discontinued. However, when stress or elevated glucocorticoid exposure is prolonged over many months or even years, permanent damage occurs, and there is a loss of hippocampal neurons. As we might predict from the key role of the hippocampus in declarative memory, both reversible atrophy and permanent damage result in significant impairment of memory. This deficit in memory is detectable at the cellular level; it is evident in a weakening of a process called long-term potentiation, an intrinsic mechanism that is thought to be critical for learning-related strengthening of synaptic connections (McEwen and Sapolsky 1995; Squire and Kandel 1999) (Figure 3–3). Thus, what may initially appear as repression may actually prove to be a true amnesia: damage to the medial temporal lobe system of the brain.

This set of experiments has deep significance for the relationship of early unconscious mental processes to later conscious mental processes. Stress early in life produced by separation of the infant from its mother produces a reaction in the infant that is stored primarily by the procedural memory system, the only well-differentiated memory system that the infant has early in its life, but this action of the procedural memory system leads to a cycle of changes that ultimately damages the hippocampus and thereby results in a persistent change in declarative memory.

This rodent model has direct clinical relevance. Patients with Cushing's syndrome overproduce glucocorticoids as a result of having a tumor in the adrenal gland, the pituitary gland, or the part of the hypothalamus that controls the pituitary. Starkman and her colleagues (1992) have studied these patients and found that those who have had the disease for over 1 year have selective atrophy of the hippocampus and concomitant memory loss. Similar atrophy and memory loss are thought to occur with posttraumatic stress. Bremner and his colleagues (1995, 1997) have found that patients with combat-related PTSD have deficits in declarative memory as well as an 8% reduction in the volume of the right hippocampus (Figure 3–3). Here, however, the atrophy and memory loss are not secondary to increased glucocorticoids but are due to some other mechanisms, since in these patients the glucocorticoid levels are lower than normal.

In the 1970s, Sachar first showed that similar events occur in the hypothalamic-pituitary axis of patients with depression (Sachar 1976). Over 50% of depressed patients have sustained levels of glucocorticoids. Subsequent studies showed that elevated glucocorticoids are associated with a decrease in the number of glucocorticoid receptors and with resistance to cortisol suppression by dexamethasone. Consistent with the data from rodents, patients with depression have a significant reduction in the volume of the hippocampus and an elevated loss of declarative memory.

Nemeroff and his colleagues (reviewed in Nemeroff 1998) have found that in depressed patients, the secretion of CRF is markedly increased. This has suggested the interesting idea that in depressed patients, the neurons in the brain that secrete CRF are hyperactive. Consistent with this idea, when CRF is injected directly into the central nervous system of mammals, it produces many of the signs and symptoms of depression, including decreased appetite, altered autonomic nervous system activity, decreased libido, and disrupted sleep. In view of the evidence that early untoward life experience increases the likelihood in adulthood of suffering from depression or certain

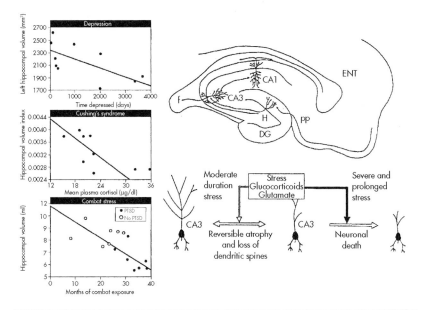

FIGURE 3–3.   Schematic summary of actions of adrenal steroids that affect hippocampal function and alter cognitive performance.

Left: Do stress-induced glucocorticoids cause brain atrophy? Relation between hippocampal volume and (top) duration of depression among individuals with a history of major depression, (middle) extent of cortisol hypersecretion among patients with Cushing's syndrome, and (bottom) duration of combat exposure among veterans with or without a history of posttraumatic stress disorder. *Cortisol* is another term for the human glucocorticoid hydrocortisone.

Right: (top): Hippocampal circuitry is diagrammed showing some of the main connections between entorhinal cortex (ENT), Ammon's horn (H), and dentate gyrus (DG). f=fornix; pp=perforant pathway; CA1 and CA3 are subregions of the hippocampus. (bottom): Moderate-duration stress, acting through both glucocorticoids and excitatory amino acids (especially glutamate), causes reversible atrophy of apical dendrites of CA3 pyramidal neurons; severe and prolonged stress causes pyramidal cell loss that is especially apparent in CA3, but spreads to CA1 as well. The mechanistic relationship between reversible atrophy and permanent neuron loss is not presently known, although both glucocorticoids and excitatory amino acids are involved. *Source.*   (Left): Reprinted from Sapolsky RM: "Why Stress is Bad for Your Brain." *Science* 273:749–750, 1996. Used with permission of the American Association for the Advancement of Science. (Right): Reprinted from McEwen BS, Sapolsky RM: "Stress and Cognitive Function." *Curr Opin Neurobiol* 5:205–216, 1995. Used with permission of Elsevier.

anxiety disorders, Nemeroff has suggested that this vulnerability is probably mediated by the hypersecretion of CRF.

These insights are likely to have several applications. First is the development of progressively more refined animal models for the factors that predispose to stress and depression, models that may allow one to identify—in experimental animals and perhaps later in humans—the genes that are activated by CRF and that predispose to anxiety. Second, drugs that block the actions of CRF on its receptors in target tissue may prove useful for certain types of depression. Finally, with increased resolution, one might conceivably be able to follow the therapeutic responses of patients by imaging the hippocampus and seeing to what degree anatomical changes are halted, or even reversed, and by seeing how responses to psychotherapy correlate with levels of CRF and glucocorticoids.

5. The Preconscious Unconscious and the Prefrontal Cortex

We have so far only considered the implicit unconscious. What about the preconscious unconscious concerned with all memories and thought capable of reading consciousness and the repressed or unconscious? We have reasons to believe that aspects of the preconscious unconscious may be mediated by the prefrontal cortex. Perhaps the strongest argument is that the prefrontal cortex is involved in bringing a variety of explicit knowledge to conscious awareness. The prefrontal association cortex has two major functions: it integrates sensory information, and it links it to planned movement. Because the prefrontal cortex mediates these two functions, it is thought to be one of the anatomical substrates of goal-directed action in long-term planning and judgment. Patients with damaged prefrontal association areas have difficulty in achieving realistic goals. As a result, they often achieve little in life, and their behavior suggests that their ability to plan and organize everyday activities is diminished (Damasio 1994, 1996).

Over the last two decades, it has become clear that the prefrontal cortex subserves as one component of a system that serves as a critical short-term holding function for information, including information that is stored in or recalled from declarative memory stores. This idea emerged from the discovery that lesions in the prefrontal cortex produce a specific deficit in a short-term component of explicit memory called working memory. The cognitive psychologist Alan Baddeley (1986), who developed the idea of *working memory*, suggested that this type of memory integrates moment-to-moment perceptions across time, rehearses them, and combines them with stored information about past experience, actions, or knowledge. This memory mechanism is crucial for many apparently simple aspects of everyday life: carrying on a conversation, adding a list of numbers, driving a car. Baddeley's idea was further developed in neurobiological experiments by Joaquin Fuster (1997) and Patricia Goldman-Rakic (1996), who first suggested that some aspects of working memory are represented in the prefrontal association cortex and that the recall of any explicit information from memory—the recall from preconscious to conscious—requires working memory. A prediction of this finding is that in trace conditioning, the unconditioned stimulus might activate the working memory system of the dorsolateral prefrontal cortex, and thereby it acts, often together with the hippocampus, to render into consciousness the otherwise procedural associative process. Clinical studies of patients with lesions suggest that the prefrontal cortex also seems to represent some aspects of moral judgments; it governs our ability to plan intelligently and responsibly (Damasio 1996). This raises the interesting possibility that the recall of explicit knowledge may depend on an adaptive and realistic evaluation of the information to be recalled. In this sense the prefrontal cortex may, as suggested by Solms (1998), be involved in coordinating functions psychoanalysts attribute to the executive functions of the ego on the one hand and the superego on the other.

6. Sexual Orientation and the Biology of Drives

Freud conceived of drives as the energetic components of mind. A drive, he argued, leads to a state of tension or excitation, a state that cognitive psychologists now call the motivational state. Motivational states impel actions with the goal of reducing tension.

Early in his career, perhaps influenced by Havelock Ellis (1901), Magnus Hirschfeld (1899), and Richard Krafft-Ebing (1901), Freud believed that a person's sexual orientation was significantly influenced by innate developmental processes and that all humans were constitutionally bisexual. This constitutional bisexuality was a key factor in both male and female homosexuality. Later, however, he came to think of sexual orientation as an acquired characteristic. Freud (1905/1953) specifically thought of male homosexuality as representing a failure of normal sexual development, a failure of the developing male child to separate himself adequately from an intense sexual bond with his mother. As a result, the grown boy identifies with his mother and seeks to play her role in an attempt to reenact the relationship that existed between them. Freud proposed that the boy's failure to separate from his mother might be the result of several factors, including a close, binding relationship to a possessive mother and a weak, hostile, or absent father. In terms of his three phases of psychosexual development, Freud saw male homosexuality, with its emphasis on anal intercourse, as a failure to progress normally from the anal to the genital phase. Female homosexuality was defined less clearly in Freud's mind, but he thought of it as the mirror image of the process he outlined for men. Freud also saw a latent homosexual component in the development of paranoia, alcoholism, and drug addiction.

Freud's views on sexuality are now at least 50 years old, and in some cases 90 years old. Some have understandably been abandoned by modern psychoanalytic thought, and all have been modified. But I recount them not to hold Freud or the psychoanalytic community responsible for outdated ideas but to illustrate that any psychological or clinical insight into sexuality, no matter how modern, will almost certainly be clarified by a better biological understanding of gender identification and sexual orientation, even though at the moment we know little. As homosexuality has become more openly accepted by society at large, there has been active discussion within the homosexual community, the psychoanalytic community, and society about the degree to which sexual orientation is inborn or acquired. The observation by Freud and other analysts that some gay men tend to recollect their fathers as hostile or distant and their mothers as unusually close has more recent corroboration (LeVay 1997). However, other studies suggest a genetic contribution to sexual orientation.

This is a complex area, because genotypic gender, phenotypic gender, gender identification, and sexual orientation are distinct from one another but interrelated. Indeed, the recognition of this complexity can render standard terms such as *male, female, masculine,* and *feminine* imprecise and in need of qualification (Bell et al. 1981).

Genotypic gender is determined by the genes, whereas phenotypic gender is defined by the development of the internal and external genitalia (Bell et al. 1981; Gorski 2000; Green 1985). Gender identification is more subtle and complex and refers to the subjective perception of one's sex. Finally, sexual orientation refers to the preference for sexual partners. The factors that contribute to the various aspects of gender are not fully understood, but I discuss them because historically this is an area that is central to psychoanalysis; and since the nurture–nature dichotomy is one that biology has repeatedly confronted and sometimes enlightened, this is an area in which biology could make a distinctive contribution. Although gender identification and sexual orientation are complex and have features that are distinctively human and may well not be amenable to study in experimental animals, many other aspects of sexual behavior are much like feeding and drinking behavior—so essential to survival that they are extremely conserved among mammals, involving common brain and hormonal systems and even aspects of stereotypic behavior. As a result, we have learned a good deal about the neural control of sex hormones and behavior from experimental animals such as rats and mice.

Early embryonic development of the gonad is identical in males and females. Genotypic gender is determined by an individual's complement of sex chromosomes: females have two X chromosomes, whereas males have one X and one Y. Male phenotypic gender is determined by a single gene, called testis determining factor, on the Y chromosome. This gene initiates the development of the bisexual early gonad into a testis, which produces testosterone; in the absence of testis determining factor, the gonad develops into an ovary and produces estrogen. All of the other phenotypic sexual characteristics result from the effects of gonadal hormones on other tissues. Of particular interest both to biologists and to psychoanalysts is that sexual dimorphism extends to the brain and thereby to behavior.

The behavior of males and females differs, even before puberty. Since many aspects of sexuality are conserved among all mammals, sexual behavior relevant to human sexuality can be studied in primates and even in rodents. Young male monkeys participate in more rough-and-tumble play than do female monkeys, a difference related to testosterone levels. Human girls who have been exposed prenatally to unusually high levels of androgens as a result of congenital adrenal hyperplasia prefer the same play as boys (Gorski 1996, 2000; Schiavi et al. 1988). It seems likely that sex differences in the play behavior of children are influenced at least in part by the organizational effects of the level of prenatal androgens.

The level of testosterone has other dramatic effects on behavior (Gladue and Clemens 1978; Gorski 1996; Imperato-McGinley et al. 1991; Knobil and Neil 1994). Male rats castrated at or prior to birth fail as adults to show the mounting behavior typical of males in the presence of receptive females, even if they are given testosterone. Furthermore, if these rats are given estrogen and progesterone in adulthood, mimicking the hormonal milieu of the adult female rats, they display the same sexually receptive posture typical of females in heat. If castration is performed a few days after birth, neither of these effects occurs. Thus, like perceptual skills and motor coordination, sex-typical behavior is organized during a critical period, around the time of birth, even though the behavior itself is not seen until much, much later.

Sex differences in behavior, to the extent that they manifest differences in brain function, must at least partly result from sex differences in the structure of the central nervous system. One possible anatomical site for these differences is the hypothalamus, which is concerned with sexual behavior as well as a variety of other homeostatic drives (for a review, see Knobil and Neil 1994). Electrical stimulation of the hypothalamus in intact, awake rhesus monkeys and rats generates sex-typical sexual behavior (Perachio et al. 1979). Biologists have found a striking sexually dimorphic difference in the medial preoptic area of the hypothalamus in rodents (Allen and Gorski 1992; Allen et al. 1989). Here there are four functional groups of neurons—of unknown function so far—called the interstitial nuclei of the anterior hypothalamus (INAH-1 to INAH-4). One of these nuclei, INAH-3, is five times larger in the male rat than in the female. Many cells in this nucleus die during female development; these cells are rescued in male pups by circulating testosterone and can be rescued in females by testosterone injections during a critical developmental window (Davis et al. 1996; Dodson and Gorski 1993).

There are also sexual dimorphisms in the thickness of various regions of the cerebral cortex in the rat. For example, there is greater asymmetry in the male: the thickness of the left side of a male rat cortex is greater than the right. Perhaps as a consequence, the splenium of the corpus callosum contains more neurons in the female. Other brain regions also show sexual dimorphisms, and doubtless there are more to be found.

The finding of a biological basis for gender genotype and phenotype raises the question, What is the biological basis for sexual orientation? To begin with, it is obvious that as the development of gender is multifactorial, so the etiology of sexual orientation must also be multifactorial; presumably, it is determined by hormones, genes, and environmental factors. A behavioral trait such as sexual orientation almost certainly is not caused by a single gene, a single alteration in a hormone or in brain structure, or a single life experience. The continuing progress in studies of sexually dimorphic characteristics will no doubt help psychoanalysts better understand gender identity and sexual orientation.

Anatomical studies on sexual orientation are just beginning, and we will need much more information before we can have confidence in the published findings on anatomical differences. At the moment they should rather be considered as interesting possibilities. Simon LeVay (1991, 1997) obtained brains of gay men and presumed heterosexual men, all of whom died of AIDS, and the brains of women. INAH-3, the most prominent of the sexually dimorphic nuclei in the rat hypothalamus, was on average two to three times bigger in the presumed heterosexual men than in the women. However, in the gay men INAH-3 was on average the same size as in the women. None of the other three INAH nuclei showed any difference between the groups. In addition to potential problems with the sample under study, it is not possible on the basis of LeVay's observations to say whether the structural differences are present at birth, whether they influence men to become gay or straight, or whether the dimorphism is a result of differences in sexual behavior. But with better sampling and improvements in brain imaging techniques, it may be possible to answer these questions.

Allen and Gorski (1992) described still another difference between gay and straight men in the anterior commissure, a pathway between the left and right sides of the brain that is generally larger in women than in men. Allen and Gorski found that the anterior commissure is on average larger in gay men than in straight men. In fact, it is larger in gay men than in women (see also Zhou et al. 1995).

Another question that is now being addressed is whether sexual orientation is inherited or acquired (Bailey and Pillard 1991; Bailey et al. 1993; Dörner et al. 1991; Eckert et al. 1986; Hamer et al. 1993; Pillard and Weinrich 1986; Whitman et al. 1993). Sexual orientation seems to be influenced by genes, and this influence is, as one would expect, complex. Sexual orientation runs in families. If a male is gay, the chances of a twin brother being gay increase substantially. In the case of monozygotic twins, individuals who share the same genes, the concordance rate is 50%. For dizygotic twins, the concordance rate is about 25%. By contrast, in the general population, the incidence of male homosexuality is less than 10%. For female homosexuality, the genetic relationship is weaker—about 30% of monozygotic twins and about 15% of dizygotic twins. These numbers seem roughly similar to those for other complex traits, indicating that both genetic and important nongenetic factors operate.

These are all early findings, and their consistency over groups of people, both heterosexual and homosexual, is still being questioned. But the methods are at hand for establishing whether there are reliable anatomical differences between people with different sexual orientations. As I suggested above, either outcome should greatly influence psychoanalytic thinking about the dynamics of sexual orientation.

7. Outcome of Therapy and Structural Changes in the Brain

Recent work in experimental animals indicates that long-term memory leads to alterations in gene expression and to subsequent anatomical changes in the brain. Anatomical changes in the brain occur throughout life and are likely to shape the skills and character of an individual. The representation of body parts in the sensory and motor areas of the cerebral cortex depends on their use and, thus, on the particular experience of the individual. Edward Taub and his colleagues scanned the brains of string instrument players. During performance, string players are continuously engaged in skillful hand movement. The second to fifth fingers of the left hand, which contact the strings, are manipulated individually, while the fingers of the right hand, which move the bow, do not express as much patterned, differentiated movement. Brain images of these musicians revealed that their brains were different from the brains of nonmusicians. Specifically, the cortical representation of the fingers of the left hand, but not of the right, was larger in the musicians (for review, see Ebert et al. 1995; Squire and Kandel 1999) (Figure 3–4).

Such structural changes are more readily achieved in the early years of life. Thus, Johann Sebastian Bach was Bach not simply because he had the

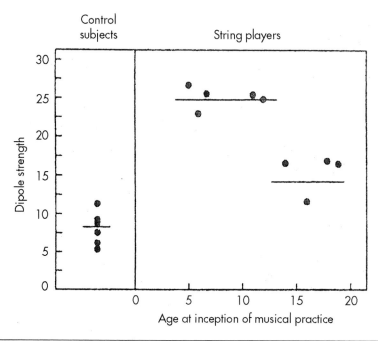

FIGURE 3–4. Larger size of the cortical representation of the fifth finger of the left hand in string players than in nonmusicians.

The figure shows the size of cortical representations measured by magnetoencephalography as the dipole strength, which is thought to be an index of total neuronal activity. Among string players, those who begin musical practice before age 13 have a larger representation than do those who begin later. Horizontal lines indicate means.
Source. Based on Ebert et al. 1995 as modified by Squire LR, Kandel ER: *Memory: From Mind to Molecules.* New York: Scientific American Library, 2000. Figure reprinted by permission of Scientific American Inc.

right genes but probably also because he began practicing musical skills at a time when his brain was most sensitive to being modified by experience. Taub and his colleagues found that musicians who learned to play their instruments by the age of 12 years had a larger representation of the fingers of the left hand, their important playing hand, than did those who started later in life (Figure 3–4) (Ebert et al. 1995).

These considerations raise a question central to psychoanalysis: Does therapy work in this way? If so, where do these psychotherapeutically induced changes occur? Do the therapeutically induced structural changes occur at the same sites altered by the mental disorder itself, or are the therapeutically induced changes independent compensatory changes that occur at other related sites?

Long-lasting changes in mental functions involve alteration in gene expression (Ebert et al. 1995; Squire and Kandel 1999). Thus, in studying the specific changes that underlie persistent mental states, normal as well as disturbed, we should also look for altered gene expression. How does altered gene expression lead to long-lasting alteration of a mental process? Animal studies of alterations in gene expression associated with learning indicate that such alterations are followed by changes in the pattern of connections between nerve cells, in some cases the growth and retraction of synaptic connections.

It is intriguing to think that insofar as psychoanalysis is successful in bringing about persistent changes in attitudes, habits, and conscious and unconscious behavior, it does so by producing alterations in gene expression that produce structural changes in the brain. We face the interesting possibility that as brain imaging techniques improve, these techniques might be useful not only for diagnosing various neurotic illnesses but also for monitoring the progress of psychotherapy.

8. Psychopharmacology and Psychoanalysis

As early as 1962, Mortimer Ostow, a psychoanalyst trained in neurology who had a long interest in the relationship of neurobiology to psychoanalysis (Ostow 1954a, 1954b), pointed to the utility of using drugs in the course of psychoanalysis (Ostow 1962). He argued even then that in addition to its therapeutic value, pharmacological intervention can serve as a biological tool for investigating aspects of affective function. Ostow observed that one of the principal effects of psychopharmacological agents is on affect, which led him to argue that affect often is a more important determinant of behavior and of illness than ideation or conscious interpretation. This idea reinforces that of Sander, Stern, and the Boston Process of Change Study Group on the relative importance of unconscious affect over conscious insight, and stresses once again the importance of changes in unconscious procedural knowledge (such as those that occur during the moments of meaning considered above) as indices of therapeutic progress, indices that the Boston group considers as important as conscious insight. Both the arguments of Ostow and those of the Boston group make clear that changes in the patient's unconscious internal representations can be beneficial for progress even without reaching consciousness. Perhaps, in these cases, the unconscious is more important than even Freud appreciated! Thus, the theme that emerges from Ostow's study on the actions of psychopharmacological agents on the psychoanalytic process echoes the ideas of Sanders and Stern, which stress that progress in psychotherapy has an important procedural component and that much of what happens in therapy need not be directly related to insight.

A Genuine Dialogue Between Biology and Psychoanalysis Is Necessary If We Are to Achieve a Coherent Understanding of Mind

As I have suggested earlier, most biologists believe that the mind will be to the twenty-first century what the gene was to the twentieth century. I have briefly discussed how the biological sciences in general and cognitive neuroscience in particular are likely to contribute to a deeper understanding of a number of key issues in psychoanalysis. An issue that is often raised is that a neurobiological approach to psychoanalytic issues would reduce psychoanalytic concepts to neurobiological ones. If that were so, it would deprive psychoanalysis of its essential texture and richness and change the character of therapy. Such a reduction is not simply undesirable but impossible. The agendas for psychoanalysis, cognitive psychology, and neural science overlap, but they are by no means identical. The three disciplines have different perspectives and aims and would converge only on certain critical issues.

The role of biology in this endeavor is to illuminate those directions that are most likely to provide deeper insights into specific paradigmatic processes. Biology's strength is its rigorous way of thinking and its depth of analysis. Our understanding of heredity, gene regulation, the cell, antibody diversity, the development of the body plan and of the brain, and the generation of behavior has been profoundly expanded as biology has probed progressively deeper into the molecular dynamics of life processes. The strengths of psychoanalysis are its scope and the complexity of the issues it addresses, strengths that cannot be diminished by biology. Just as medicine has time and again provided direction to biology, and psychiatry to neuroscience, so can psychoanalysis serve as a skillful and reality-oriented tutor for a sophisticated understanding of the mind-brain.

During the past half century, we have repeatedly seen successful unifications within the biological sciences without the disappearance of the core disciplines. For example, classical genetics and molecular biology have merged into a common discipline, molecular genetics. We now know that the traits that Gregor Mendel described and the genes on specific locations on chromosomes that Thomas Hunt described are stretches of double-stranded DNA. This insight has allowed us to understand how genes replicate and how they control cellular function. These insights have revolutionized biology, but this has hardly abolished the discipline of genetics. To the contrary, with the human genome sequence expected to be completed in the year 2003, genetics is flourishing. It has used the powerful insight of molecular biology, applied it effectively to its own agenda, and moved on. So be it with psychoanalysis.

Are We Seeing the Beginnings of a Dialogue?

As we have seen, biology could help psychoanalysis in two ways: conceptually and experimentally. We are in fact already beginning to see signs of conceptual progress. A number of psychoanalytic institutes, or at least a number of people within psychoanalysis, have struggled to make psychoanalysis more rigorous and to align it more closely with biology. Freud argued for this position at the beginning of his career. More recently, Mortimer Ostow of the Neuroscience Project of the New York Psychoanalytic Institute and David Olds and Arnold Cooper at the Columbia Institute (Olds and Cooper 1997), as well as others across the country, have earlier expressed ideas similar to those I outline here.

For many years both the Association for Psychoanalytic Medicine at Columbia and the New York Psychoanalytic Institute, to use but two examples, have instituted (with the help of my colleague James H. Schwartz) neuropsychoanalytic centers that address interests common to psychoanalysis and neuroscience, including consciousness, unconscious processing, autobiographical memory, dreaming, affect, motivation, infantile mental development, psychopharmacology, and the etiology and treatment of mental illness. The prospectus of the New York Psychoanalytic Institute now reads as follows:

> The explosion of new insights into numerous problems of vital interest to psychoanalysis needs to be integrated in meaningful ways with the older concepts and methods as do the burgeoning research technologies and pharmacological treatments. Similarly neuroscientists exploring the complex problems of human subjectivity for the first time have much to learn from a century of analytic inquiry. (New York Psychoanalytic Institute 1999)

Thus, psychoanalysts are beginning to learn about neural science and psychopharmacology, an exciting step forward, a step that should lead in the long run to a new curriculum for the analytic clinician.

As a result of these efforts, there has been a bit of progress in the second function of biology, the experimental function. Several investigators have seen the exciting possibility of merging psychoanalysis and biology experimentally. Most commendable are the important attempts by Karen Kaplan-Solms and Mark Solms to delineate anatomical systems in the brain that are relevant to psychoanalysis by studying alterations in the mental functioning of patients with brain lesions (Kaplan-Solms and Solms 2000). Kaplan-Solms and Solms believe that the power of psychoanalysis derives from its ability to investigate mental processes from a subjective perspective. However, as they point out, this very strength is also its greatest weakness. Subjective phenomena do not readily lend themselves to objective empirical analysis. We need to develop creative ways of studying subjective phenomena. As a result, these investigators argue that only by connecting psychoanalytic thought to objective neurobiological phenomena, as in personality changes following focal lesions of the brain, can one derive empirical correlates of the subjectively derived constructs of psychoanalysis. Similarly, there is also the important and long-standing tradition of work by Howard Shevrin, correlating the perception of subliminal and supraliminal stimuli with event-related potentials in the brain in an attempt to analyze aspects of unconscious mental processes (Shevrin 1998; Shevrin et al. 1996).

These beginnings are extremely encouraging. But for psychoanalysis to be reinvigorated, it will need to match its intellectual restructuring with institutional changes. For biology to help, two aspects of psychoanalysis require particular attention: therapeutic outcome and the role of psychoanalytic institutes.

The Evaluation of Psychoanalytic Outcome

As a mode of therapy, psychoanalysis is no longer as widely practiced as it was 50 years ago. Jeffrey (1998) claims that the number of patients seeking psychoanalysis steadily decreased by 10% a year over the last 20 years, as has the number of gifted psychiatrists seeking training in psychoanalytic institutes. This decline is disappointing, because psychoanalytic therapy seems to have become more realistically focused and therefore is more likely to be efficacious. During the last several decades, psychoanalysis has largely abandoned the unrealistic goals of the 1950s, when it attempted to treat by itself autism, schizophrenia, and severe bipolar illness, illnesses for which it had little, if anything, to offer. Nowadays, psychoanalysis is thought to be most successful for people with the nonpsychotic character disorders, people who have major deficits in working effectively or maintaining satisfactory relationships and who want to acquire better ways of managing their lives. A substantial number of these patients suffer from borderline personality disorder with concomitant disturbances of affect. In these cases, psychoanalysis and psychoanalytically oriented psychotherapy are thought to be an important adjunct to pharmacotherapy (see Friedman et al. 1998 for the distribution of patients seen in psychoanalysis). As a result of this narrower focus on patients who are not psychotic, psychoanalysis and psychoanalytically oriented psychotherapy may in the best of hands be more effective today than ever before.

I am here reminded of Kay Jamison's (1996) haunting discussion of her own manic-depressive illness and her effective response to combined lithium medication and psychotherapy:

At this point in my existence, I cannot imagine leading a normal life without both taking lithium and having had the benefits of psychotherapy. Lithium prevents my seductive but disastrous highs, diminishes my depressions, clears out the wool and webbing from my disordered thinking, slows me down, gentles me out, keeps me from ruining my career and relationships, keeps me out of a hospital, alive, and makes psychotherapy possible. But, ineffably, psychotherapy heals. It makes some sense of the confusion, reins in the terrifying thoughts and feelings, returns some control and hope and possibility of learning from it all. Pills cannot, do not, ease one back into reality; they only bring one back headlong, careening, and faster than can be endured at times. Psychotherapy is a sanctuary; it is a battleground; it is a place I have been psychotic, neurotic, elated, confused, and despairing beyond belief. But, always, it is where I have believed or have learned to believe—that I might someday be able to contend with all of this.

No pill can help me deal with the problem of not wanting to take pills; likewise, no amount of psychotherapy alone can prevent my manias and depressions. I need both. It is an odd thing, owing life to pills, one's own quirks and tenacities, and this unique, strange, and ultimately profound relationship called psychotherapy.

Given these advances, why is the practice of psychoanalysis no longer thriving? This decline in the use of psychoanalytic therapy is mostly attributable to causes outside psychoanalysis: the proliferation of different forms of short-term psychotherapy (almost all of which are, to varying degrees, derived from psychoanalysis), the emergence of pharmacotherapy, and the economic impact of managed care. But one important cause derives from psychoanalysis itself. One full century after its founding, psychoanalysis still has not made the required effort to obtain objective evidence to convince an increasingly skeptical medical profession that it is a more effective mode of therapy than placebo. Thus, unlike various forms of cognitive therapy and other psychotherapies, for which compelling objective evidence now exists—both as therapies in their own right and as key adjuncts to pharmacotherapy—there is as yet no compelling evidence, outside subjective impressions, that psychoanalysis works better than nonanalytically oriented therapy or placebo (Bachrach et al. 1991; Cooper 1995; Doidge 1997; Fonagy 1999; Kantrowitz 1993; Roth and Fonagy 1996; Seligman 1995; Weissman and Markowitz 1994; Weissman et al. 1979).

The failure of psychoanalysis to provide objective evidence that it is effective as a therapy can no longer be accepted. Psychoanalysts must be persuaded by Arnold Cooper's (1995) realistic and critical view:

To the extent that psychoanalysis lays claim to being a method of treatment, we are, for better or worse, drawn into the orbit of science, and we cannot then escape the obligations of empirical research. As long as we develop practitioners who are members of a profession and charge for their services, it is incumbent upon us to study what we do and how we affect our patients.

As Cooper points out, a number of the major studies initially designed to evaluate the outcome of therapy—Wallerstein's (1995) study and the studies reviewed by Kantrowitz (1993) and by Bachrach (1995)—have abandoned their long-term goal for a more accessible short-term aim unrelated to outcome. Despite their cost and complexity, rigorous outcome studies, with comparison to short-term nonanalytically oriented psychotherapy and placebo, need to be at the top of any list of priorities if psychoanalysis is to continue to be a well-recognized therapeutic option.

A Flexner Report for the Psychoanalytic Institutes?

But the much more difficult step is to go beyond an appreciation of biology and of having a tiny cadre of full-time researchers to the development within psychoanalysis of an intellectual climate that will make a significant fraction of psychoanalysts technically competent in cognitive neuroscience and eager to test their own ideas with new methods. The challenge for psychoanalysts is to become active participants in the difficult joint attempt of biology and psychology, including psychoanalysis, to understand the mind. If this transformation in the intellectual climate of psychoanalysis is to occur, as I believe it must, the psychoanalytic institutes themselves must change from being vocational schools—guilds, as it were—to being centers of research and scholarship.

At the cusp of the twenty-first century, the psychoanalytic institutes in the United States resemble the proprietary medical schools that populated this country in the early 1900s. At the turn of the last century, the United States experienced a great proliferation of medical schools—155 all told—most of which had no laboratories for teaching the basic sciences. At these schools, medical students were taught by private practitioners who often were busy with their own practices.

To examine this problem, the Carnegie Foundation commissioned Abraham Flexner to study medical education in the United States. The Flexner Report, which was completed in 1910, emphasized that medicine is a science-based profession and requires a structured education in both basic science and its application to clinical medicine (Flexner 1910). To promote a quality education, the Flexner Report recommended limiting the medical schools in this country to those that were integral to a university. As a consequence of this report, many inadequate schools were closed, and credentialed standards for the training and practice of medicine were established. To return to its former vigor and contribute importantly to our future understanding of mind, psychoanalysis needs to examine and restructure the intellectual context in which its scholarly work is done and to develop a more critical way of training the psychoanalysts of the future. Thus, what psychoanalysis may need, if it is to survive as an intellectual force into the twenty-first century, is something akin to a Flexner Report for the psychoanalytic institutes.

What drew so many of us to psychoanalysis in the late 1950s and early 1960s was its bold curiosity—its investigative zeal. I myself was drawn to the neurobiological study of memory because I saw memory as central to a deeper understanding of the mind, an interest first sparked by psychoanalysis. One would hope that the excitement and success of current biology would rekindle the investigative curiosities of the psychoanalytic community and that a unified discipline of neurobiology, cognitive psychology, and psychoanalysis would forge a new and deeper understanding of mind.

Acknowledgments

In the course of working on this article, I have benefited greatly from insightful discussions with Marianne Goldberger, who also gave critical comments on earlier drafts of this manuscript. In addition, I have received helpful suggestions from Nancy Andreasen, Mark Barad, Robert Glick, Jack Gorman, Myron Hofer, Anton O. Kris, Charles Nemeroff, Russell Nichols, David Olds, Mortimer Ostow, Chris Pittenger, Stephen Rayport, Michael Rogan, James Schwartz, Theodore Shapiro, Mark Solms, Anna Wolff, and Marc Yudkoff.

References

Ader R, Grota LJ: Effects of early experience on adrenocortical reactivity. Physiol Behav 4:303–305, 1969

Agid O, Shapira B, Zislin J, et al: Environment and vulnerability to major psychiatric illness: a case control study of early parental loss in major depression, bipolar disorder and schizophrenia. Mol Psychiatry 4:163–172, 1999

Allen LS, Gorski RA: Sexual orientation and size of the anterior commissure in the human brain. Proc Natl Acad Sci USA 89:7199–7202, 1992

Allen LS, Hines M, Shryne JE, et al: Two sexually dimorphic cell groups in the human brain. J Neurosci 9:497–506, 1989

Bachrach HM: The Columbia Records Project and the evolution of psychoanalytic outcome research, in Research in Psychoanalysis: Process, Development, Outcome. Edited by Shapiro T, Emde RN. Madison, CT, International Universities Press, 1995, pp 279–297

Bachrach HM, Galatzer-Levy R, Skolnikoff A, et al: On the efficacy of psychoanalysis. J Am Psychoanal Assoc 39:871–916, 1991

Baddeley A: Working Memory. New York, Oxford University Press, 1986

Bailey JM, Pillard RC: A genetic study of male sexual orientation. Arch Gen Psychiatry 48:1089–1096, 1991

Bailey JM, Pillard RC, Neale MC, et al: Heritable factors influence sexual orientation in women. Arch Gen Psychiatry 50:217–223, 1993

Bell AP, Weinberg MS, Hammersmith SK: Sexual Preference: Its Development in Men and Women. New York, Simon & Schuster, 1981

Boring EG: A History of Experimental Psychology. New York, Appleton-Century-Crofts, 1950, p 713

Boston Process of Change Study Group: Interventions that effect change in psychotherapy: a model based on infant research. Infant Ment Health J 19:277–353, 1998

Bowlby J: Grief and mourning in infancy and early childhood. Psychoanal Study Child 15:9–52, 1960

Bowlby J: Attachment and Loss, Vols 1, 2. New York, Basic Books, 1969, 1973

Breiter HC, Etcoff NL, Whalen PJ, et al: Response and habituation of the human amygdala during visual processing of facial expression. Neuron 17:875–887, 1996

Bremner JD, Randall P, Scott TM, et al: MRI-based measurement of hippocampal volume in patients with combat-related posttraumatic stress disorder. Am J Psychiatry 152:973–981, 1995

Bremner JD, Randall P, Vermetten E, et al: Magnetic resonance imaging-based measurement of hippocampal volume in posttraumatic stress disorder related to childhood physical and sexual abuse—a preliminary report. Biol Psychiatry 41:23–32, 1997

Brenner C: An Elementary Textbook of Psychoanalysis, 2nd Edition. New York, International Universities Press, 1978

Brown GW, Harris T, Copeland JR, et al: Depression and loss. Br J Psychiatry 130:1–18, 1997

Clark RE, Squire LR: Classical conditioning and brain systems: the role of awareness. Science 280:77–81, 1998

Clyman R: The procedural organization of emotion: a contribution from cognitive science to the psychoanalytic therapy of therapeutic action. J Am Psychoanal Assoc 39:349–381, 1991

Cooper A: Discussion: on empirical research, in Research in Psychoanalysis: Process, Development, Outcome. Edited by Shapiro T, Emde RN. Madison, CT, International Universities Press, 1995, pp 381–391

Dahl H: The measurement of meaning in psychoanalysis by computer analysis of verbal contexts. J Am Psychoanal Assoc 22:37–57, 1974

Damasio AR: Descartes' Error: Emotion, Reason and the Human Brain. New York, Putnam, 1994

Damasio AR: The somatic marker hypothesis and the possible functions of the prefrontal cortex: review. Philos Trans R Soc Lond B Biol Sci 351:1413–1420, 1996

Davis EC, Popper P, Gorski RA: The role of apoptosis in sexual differentiation of the rat sexually dimorphic nucleus of the preoptic area. Brain Res 734:10–18, 1996

Dickinson A: Contemporary Animal Learning Theory. Cambridge, England, Cambridge University Press, 1980

Dodson RE, Gorski RA: Testosterone propionate administration prevents the loss of neurons within the central part of the medial preoptic nucleus. J Neurobiol 24:80–88, 1993

Doidge N: Empirical evidence for the efficacy of psychoanalytic psychotherapies and psychoanalysis: an overview. Psychoanal Inquiry Suppl 184:102–150, 1997

Domjan M, Burkhard B: The Principles of Learning and Behavior, 2nd Edition. Monterey, CA, Brooks/Cole, 1986

Dörner G, Poppe I, Stahl F, et al: Gene- and environment-dependent neuroendocrine etiogenesis of homosexuality and transsexualism. Exp Clin Endocrinol 98:141–150, 1991

Ebert T, Panter C, Wienbruch C, et al: Increased use of the left hand in string players associated with increased cortical representation of the fingers. Science 220:21–23, 1995

Eckert ED, Bouchard TJ, Bohlen J, et al: Homosexuality in monozygotic twins reared apart. Br J Psychiatry 148:421–425, 1986

Edelson M: Hypothesis and Evidence in Psychoanalysis. Chicago, IL, University of Chicago Press, 1984

Eichenbaum H: Amnesia, the hippocampus, and episodic memory (editorial). Hippocampus 8:197, 1998

Eissler KR: Irreverent remarks about the present and future of psychoanalysis. Int J Psychoanal 50:461–471, 1969

Ellis H: The development of the sexual instinct. The Alienist and Neurologist 22:500–521, 615–623, 1901

Erikson E: Childhood and Society. New York, WW Norton, 1963

Fanselow MS: Pavlovian conditioning, negative feedback, and blocking: mechanisms that regulate association formation. Neuron Minireview 20:625–627, 1998

Flexner A: Medical Education in the United States and Canada. A Report to the Carnegie Foundation for the Advancement of Teaching (Bulletin No 4). Boston, MA, Updyke, 1910

Fonagy P (ed): An Open Door Review of Outcome Studies in Psychoanalysis. London, International Psychoanalytical Association, Research Committee, 1999

Freud A: The Ego and the Mechanisms of Defense. London, Hogarth Press, 1936

Freud A, Burlingham D: Infants Without Families: The Writings of Anna Freud, Vol 3. New York, International Universities Press, 1973

Freud S: Three essays on the theory of sexuality (1905), in Standard Edition of the Complete Psychological Works of Sigmund Freud, Vol 7. Edited and translated by Strachey J. London, Hogarth Press, 1953, pp 125–243

Freud S: On narcissism: an introduction (1914), in Standard Edition of the Complete Psychological Works of Sigmund Freud, Vol 14. Edited and translated by Strachey J. London, Hogarth Press, 1957, pp 67–102

Freud S: Beyond the pleasure principle (1920), in Standard Edition of the Complete Psychological Works of Sigmund Freud, Vol 18. Edited and translated by Strachey J. London, Hogarth Press, 1955, pp 7–64

Freud S: Inhibitions, symptoms and anxiety (1926 [1925]), in Standard Edition of the Complete Psychological Works of Sigmund Freud, Vol 20. Edited and translated by Strachey J. London, Hogarth Press, 1959, pp 75–175

Freud S: New introductory lectures on psycho-analysis (1933 [1932]), in Standard Edition of the Complete Psychological Works of Sigmund Freud, Vol 22. Edited and translated by Strachey J. London, Hogarth Press, 1964, pp 1–182

Friedman RC, Bucci W, Christian C, et al: Private psychotherapy patients of psychiatrist psychoanalysts. Am J Psychiatry 155:1772–1774, 1998

Fuster JM: The Prefrontal Cortex: Anatomy, Physiology, and Neurophysiology of the Frontal Lobe, 3rd Edition. Philadelphia, PA, Lippincott-Raven, 1997

Gladue BA, Clemens LG: Androgenic influences on feminine sexual behavior in male and female rats: defeminization blocked by prenatal androgen. Endocrinology 103:1702–1709, 1978

Goldberger M: Daydreams: even more secret than dreams, in Symposium: The Secret of Dreams, Western New England Psychoanalytic Society. New Haven, CT, Yale University Press, 1996

Goldman-Rakic PS: Regional and cellular fractionation of working memory. Proc Natl Acad Sci USA 93:13473–13480, 1996

Gorski RA: Gonadal hormones and the organization of brain structure and function, in The Lifespan Development of Individuals: Behavioral, Neurobiological, and Psychosocial Perspectives. Edited by Magsnusson D. New York, Cambridge University Press, 1996, pp 315–340

Gorski RA: Sexual differentiation of the nervous system, in Principles of Neural Science, 4th Edition. Edited by Kandel ER, Schwartz JH, Jessell T. New York, McGraw-Hill, 2000

Green R: Gender identity in childhood and later sexual orientation: follow-up of 78 males. Am J Psychiatry 142:339–341, 1985

Hamer DH, Hu S, Magnuson VL, et al: A linkage between DNA markers on the X chromosome and male sexual orientation. Science 261:321–327, 1993

Harlow HF: The nature of love. Am J Psychol 13:673–686, 1958

Harlow HF, Dodsworth RO, Harlow MK: Total social isolation in monkeys. Proc Natl Acad Sci USA 54:90–97, 1965

Hartmann H: Ego Psychology and the Problem of Adaptation (1939). Translated by Rapaport D. New York, International Universities Press, 1958

Heim C, Owens MJ, Plotsky PM, et al: Persistent changes in corticotropin-releasing factor systems due to early life stress: relationship to the pathophysiology of major depression and post-traumatic stress disorder, I: endocrine factors in the pathophysiology of mental disorders. Psychopharmacol Bull 33:185–192, 1997a

Heim C, Owens MJ, Plotsky PM, et al: The role of early adverse life events in the etiology of depression and posttraumatic stress disorder: focus on corticotropin-releasing factor. Ann NY Acad Sci 821:194–207, 1997b

Hirschfeld M: Die Objective Diagnose der Homosexualität. Jahrbuch für sexuelle Zwischenstufen 1:4–35, 1899

Hofer MA: The Roots of Human Behavior. New York, WH Freeman, 1981

Hofer MA: Hidden regulators in attachment, separation, and loss. Monogr Soc Res Child Dev 59:192–207, 1994

Imperato-McGinley J, Pichardo M, Gautier T, et al: Cognitive abilities in androgen-insensitive subjects: comparison with control males and females from the same kindred. Clin Endocrinol 34:341–347, 1991

Isenstadt L: The neurobiology of childhood emotion: anxiety. American Psychoanalyst 32:29–37, 1998

Jacob F: Of Flies, Mice and Men. Cambridge, MA, Harvard University Press, 1998

Jamison K: An Unquiet Mind. New York, Vintage Books, 1996

Jeffrey DW: Lead article. Am Psychoanalyst 32(1), 1998

Kamin L: Predictability, surprise, attention, and conditioning, in Punishment and Aversive Behavior. Edited by Campbell BA, Church RM. New York, Appleton-Century-Crofts, 1969, pp 279–296

Kandel ER: A new intellectual framework for psychiatry. Am J Psychiatry 155:457–469, 1998

Kantrowitz JL: The uniqueness of the patient-analyst pair: approaches for elucidating the analyst's role. Int J Psychoanal 74:893–904, 1993

Kaplan-Solms K, Solms M: Clinical Studies in Neuro-Psychoanalysis. Madison, CT, International Universities Press, 2000

Kendler KS, Neale MC, Kessler RC, et al: Childhood parental loss and adult psychopathology in women: a twin study perspective. Arch Gen Psychiatry 49:109–116, 1992

Kim JJ, Clark RE, Thompson RF: Hippocampectomy impairs the membrane of recently but not remotely acquired trace eyeblink conditioned responses. Behav Neurosci 109:195–203, 1995

Klein J: Envy and Gratitude. London, Tavistock, 1957

Knobil E, Neil J (eds): Physiology of Reproduction. Philadelphia, PA, Lippincott-Raven, 1994

Kohut H: The Analysis of the Self: A Systematic Approach to the Psychoanalytic Treatment of Narcissistic Personality Disorders. New York, International Universities Press, 1971

Krafft-Ebing R: Neue Studien auf dem Gebiete der Homosexualität. Jahrbuch für sexuelle Zwischenstufen 3:1–36, 1901

Kris AO: Free Association, Method and Practice. New Haven, CT, Yale University Press, 1982

Lear J: Open Minded: Working Out the Logic of the Soul. Cambridge, MA, Harvard University Press, 1998

LeDoux J: The Emotional Brain. New York, Simon & Schuster, 1996

LeVay S: A difference in hypothalamic structure between heterosexual and homosexual men. Science 253:1034–1037, 1991

LeVay S: The Sexual Brain. Cambridge, MA, MIT Press, 1997

Levin FM: A brief history of analysis and cognitive neuroscience. American Psychoanalyst 32(3):26–27, 35, 1998

Levine S: Infantile experience and resistance to physiological stress. Science 126:405–406, 1957

Levine S: Plasma-free corticosteroid response to electric shock in rats stimulated in infancy. Science 135:795–796, 1962

Levine S, Haltmeyer GC, Kaas GG, et al: Physiological and behavioral effects of infantile stimulation. Physiol Behav 2:55–63, 1967

Liu D, Diorio J, Tannenbaum B, et al: Maternal care, hippocampal glucocorticoid receptors, and hypothalamic-pituitary-adrenal responses to stress. Science 277:1659–1662, 1997

Luborsky L, Luborsky E: The era of measures of transference: the CCRT and other measures, in Research in Psychoanalysis: Process, Development, Outcome. Edited by Shapiro T, Emde RN. Madison, CT, International Universities Press, 1995, pp 329–351

Lyons-Ruth K: Implicit relational knowing: its role in development and psychoanalytic treatment. Infant Ment Health J 19:282–289, 1998

McEwen BS, Sapolsky RM: Stress and cognitive function. Curr Opin Neurobiol 5:205–216, 1995

Meaney MJ, Aitken DH, Sapolsky RM: Environmental regulation of the adrenocortical stress response in female rats and its implications for individual differences in aging. Neurobiol Aging 12:31–38, 1991

Milner B, Squire LR, Kandel ER: Cognitive neuroscience and the study of memory. Neuron 20:445–468, 1998

Nemeroff CB: The corticotropin-releasing factor (CRF) hypothesis of depression: new findings and new directions. Mol Psychiatry 1:326–342, 1996

Nemeroff CB: The neurobiology of depression. Sci Am 278:28–35, 1998

Olds D, Cooper AM: Dialogues with other sciences: opportunities for mutual gain. Int J Psychoanal 78:219–225, 1997

Ostow M: The psychoanalytic contribution to the study of brain function, I: frontal lobes. Psychoanal Q 23:317–338, 1954a

Ostow M: The psychoanalytic contribution to the study of brain function, II: the temporal lobes; III: synthesis. Psychoanal Q 24:383–423, 1954b

Ostow M: Drugs in Psychoanalysis and Psychotherapy. New York, Basic Books, 1962

Pavlov I: Conditioned Reflexes: An Investigation of the Physiological Activity of the Cerebral Cortex. Translated by Anrep GV. London, Oxford University Press, 1927

Perachio AA, Mar LD, Alexander M: Sexual behavior in male rhesus monkeys elicited by electrical stimulation of preoptic and hypothalamic areas. Brain Res 177:127–144, 1979

Pillard RC, Weinrich JD: Evidence of familial nature of male homosexuality. Arch Gen Psychiatry 43:808–812, 1986

Plotsky PM, Meaney MJ: Early, postnatal experience alters hypothalamic corticotropin-releasing factor (CRF) mRNA, median eminence CRF content and stress-induced release in adult rats. Brain Res Mol Brain Res 18:195–200, 1993

Reiser M: Changing theoretical concepts in psychosomatic medicine, in American Handbook of Psychiatry, 2nd Edition, Vol IV. Edited by Reiser M. New York, Basic Books, 1975, pp 477–500

Rescorla RA: Behavioral studies of Pavlovian conditioning. Annu Rev Neurosci 11:329–352, 1988

Roth A, Fonagy P: What Works for Whom? A Critical Review of Psychotherapy Research. New York, Guilford, 1996

Roth MS (ed): Freud: Conflict and Culture: Essays on His Life, Work, and Legacy. New York, Knopf, 1998

Sachar EJ: Neuroendocrine dysfunction in depressive illness. Annu Rev Med 27:389–396, 1976

Sander L: Introductory comment. Infant Ment Health J 19:280–281, 1998

Sapolsky RM: Why stress is bad for your brain. Science 273:749–750, 1996

Schiavi RC, Theilgaard A, Owen DR, et al: Sex chromosome anomalies, hormones, and sexuality. Arch Gen Psychiatry 45:19–24, 1988

Scoville WB, Milner B: Loss of recent memory after bilateral hippocampal lesions. J Neurol Neurosurg Psychiatry 20:11–21, 1957

Seligman MEP: The effectiveness of psychotherapy: the Consumer Reports study. Am Psychol 50:965–974, 1995

Selye H: A syndrome produced by diverse nocuous agents. Nature 138:22–36, 1936

Shapiro T: Discussion of the structural model in relation to Solm's neuroscience-psychoanalysis integration: the ego. Journal of Clinical Psychoanalysis 5:369–379, 1996

Shapiro T, Emde RN (eds): Research in Psychoanalysis: Process, Development, Outcome. Madison, CT, International Universities Press, 1995

Shevrin H: Psychoanalytic and neuroscience research. American Psychoanalyst 32(3), 1998

Shevrin H, Bond J, Brakel LAW, et al: Conscious and Unconscious Processes: Psychodynamic, Cognitive and Neurophysiological Convergences. New York, Guilford, 1996

Solms M: What is consciousness? Charles Fischer Memorial Lecture to the New York Psychoanalytic Society. J Am Psychoanal Assoc 45:681–703, 1997

Solms M: Preliminaries for an integration of psychoanalysis and neuroscience. Br Psychoanal Soc Bull 34:23–37, 1998

Solomon PR, Vander Schaaf ER, Thompson RF, et al: Hippocampal and trace conditioning of the rabbit's classically conditioned nictitating membrane response. Behav Neurosci 100:729–744, 1986

Spitz RA: Hospitalism: an inquiry into the genesis of psychiatric conditions in early childhood. Psychoanal Study Child 1:53–74, 1945

Squire LR, Zola-Morgan S: The medial temporal lobe memory system. Science 253:1380–1386, 1991

Squire LR, Kandel ER: Memory: From Mind to Molecules. New York: Scientific American Library, 1999

Starkman MN, Gebarski SS, Berent S, et al: Hippocampal formation volume, memory dysfunction, and cortisol levels in patients with Cushing's syndrome. Biol Psychiatry 32:756–765, 1992

Stern D: The process of therapeutic change involving implicit knowledge: some implications of developmental observations for adult psychotherapy. Infant Ment Health J 19:300–308, 1998

Teller V, Dahl H: What psychoanalysis needs is more empirical research, in Research in Psychoanalysis: Process, Development, Outcome. Edited by Shapiro T, Emde RN. Madison, CT, International Universities Press, 1995, pp 31–49

Wallerstein RS: The effectiveness of psychotherapy and psychoanalysis: conceptual issues and empirical work, in Research in Psychoanalysis: Process, Development, Outcome. Edited by Shapiro T, Emde RN. Madison, CT, International Universities Press, 1995, pp 299–311

Weissman MM, Markowitz JC: Interpersonal psychotherapy. Arch Gen Psychiatry 51:599–606, 1994

Weissman MM, Prusoff BA, DiMascio A, et al: The efficacy of drugs and psychotherapy in the treatment of acute depressive episodes. Am J Psychiatry 136:555–558, 1979

Whalen PJ, Rauch SL, Etcoff NL, et al: Masked presentations of emotional facial expressions modulate amygdala activity without explicit knowledge. J Neurosci 18:411–418, 1996

Whitman FL, Diamond M, Martin J: Homosexual orientation in twins: a report on 61 pairs and three triplet sets. Arch Sex Behav 22:187–206, 1993

Winnicott DW: The depressive position in normal emotional development (1954), in Through Paediatrics to Psycho-Analysis: Collected Papers. New York, Basic Books, 1958, pp 262–277

Zhou JN, Hofman MA, Gooren LJ, et al: A sex difference in the human brain and its relation to transsexuality. Nature 378:68–70, 1995

"FROM METAPSYCHOLOGY TO MOLECULAR BIOLOGY"

Donald F. Klein, M.D.

Like Eric Kandel, I went to medical school hoping to become a psychoanalyst. I have had substantial exposure to the psychoanalytic process, since I was a candidate at the New York Psychoanalytic Institute for 4 years (1957–1961) and spent 17 years evaluating the effects of pharmacotherapy and psychotherapy in the context of the long-term, psychoanalytic Hillside Hospital. These years were spent under the indulgent guidance of the director, Lew Robbins, an open-minded training analyst from the Topeka School.

Kandel describes how his brilliant cohort of classmates split: some pursued becoming better therapists, while others, hoping to develop a basic science for psychiatry, became immersed in biology. In contrast, my career was molded by the advent of clinical psychopharmacology. In the 1950s, long before we had a clue to the biology of antipsychotics, tricyclic antidepressants, monoamine oxidase inhibitors, and benzodiazepines, the problem of an objective estimate of specific therapeutic benefit had to be addressed. This was largely because the idea that pills could do anything but stun patients into unthinking compliance was the conventional wisdom of the dominant psychoanalytically minded. This need for objective evaluation of specific effects led to the development of comparative, placebo-controlled, double-blind–

rated, randomized clinical trials. During placebo treatment, remarkable gains often occurred. These gains were not directly due to the placebo but rather to a factor we had radically underestimated; that is, many illnesses, after having hit bottom (which led to entering treatment), tend to get better under therapeutic, optimistic, and caring circumstances.

The fact of clinical reconstitution during placebo treatment highlighted the ubiquitous *post hoc, ergo propter hoc* fallacy. Psychotherapy was respected as a powerful tool because it was reported that patients often showed marked improvements during therapy. The placebo findings showed that this causal attribution was, at least, premature.

When the Kefauver-Harrison Drug Amendments of 1962 demanded that medications be proven effective as well as safe, there was a sudden recognition of clinical psychopharmacology as a respectable field. The remarkable discoveries of Julius Axelrod (1972) and Arvid Carlsson (2003) regarding the action of psychotropic drugs on chemical synaptic transmission caused a wave of optimism. Not only were patients having remarkably better treatments but we were coming close to understanding what made them better and, by inference, what had gone wrong. Unfortunately, this optimism was premature. We learned that all novel advances in psychopharmacology have been due to chance and the prepared mind rather than deduced from a basic grasp of pathophysiology and pharmacological structure, function, or relationships. It became clear that neuroscience was the sensible route to an eventual understanding of the mechanisms of psychiatric illnesses and deducing novel therapeutic benefits, but the road is longer than we had hoped.

The important 1983 article "From Metapsychology to Molecular Biology: Explorations Into the Nature of Anxiety" by Eric Kandel served as an icebreaker for more dedicated neuroscientific psychiatric study by addressing how animal models can yield insights, extending from behavioral to molecular levels, relevant to both experimental animals and humans. Experimental neurosis has a distinguished ancestry, starting with Pavlov, but such studies were almost exclusively behavioral. Kandel's extraordinary, fruitful ambition was to lay bare the functioning neurocircuits.

However, Kandel's article must be put into historical context. As he discusses, American psychiatry from the 1950s through the 1980s was largely dominated by psychoanalytic theory, which emphasized dynamic understanding while denigrating descriptive psychiatry as superficial. The discovery that the marked unreliability of descriptive diagnosis was due to criteria variability led directly to the work of the DSM-III Task Force. Its guiding principles were syndromal definition by specific inclusion and exclusion criteria. These consensual definitions were almost entirely derived from clinical experience, amplified somewhat by clinical psychopharmacological trials. Brain physiology and pathophysiology were too undeveloped to play any role in these discussions, which abjured questionable etiologic theory as a basis for diagnosis.

The most pointed conflict with the psychoanalytic establishment came with the realization that the "neuroses" had a common exclusion criterion (loss of contact with reality), but there was no common inclusion criterion, except for dubious psychogenic theory. This led to formulating several syndromally distinct "anxiety disorders."

That imipramine blocked apparently spontaneous clinical panics while concurrently having little effect upon either anticipatory anxiety or phobic avoidance indicated that the single label "anxiety" was simplistic. Furthermore, imipramine did not benefit simple phobia. Having a treatment for panic disorder and a delineation of this syndrome, the question arose regarding what to do about patients who did not have panic attacks but were chronically anxious. Chronic anxiety occurred in several different contexts—for example, social phobia, obsessive-compulsive disorder, generalized anxiety disorder, the inter-panic periods of panic disorder (which often initiated agoraphobia), and posttraumatic stress disorder. Clinical nosology with regard to anxiety is still a work very much in progress.

There seem to be, clinically at least, several types of chronic anxieties. Patients with recurrent, apparently spontaneous panic attacks often develop chronic anxiety between panics, claiming their chronic anxiety is due to fear of the recurrence of a panic attack. Their chronic symptomatology is occasionally associated with tachycardia, a mildly elevated blood cortisol level, sighing, and compensated respiratory alkalosis. In contrast, generalized anxiety disorder, another syndrome with "chronic anxiety," usually has neither cortisol elevation nor autonomic symptomatology nor sighing respiration.

In his 1895 paper "On the Grounds for Detaching a Particular Syndrome From Neurasthenia, Under the Description Anxiety Neurosis," Freud lucidly described the spontaneous panic attack, the existence of limited symptom attacks, and the salience of dyspnea during the attack. Freud initially viewed the attack as a nonpsychological process, whereby ungratified libido was transformed into anxiety. However, the psychoanalytic understanding of anxiety went through many modifications that progressively focused upon the possible eruption of chronically repressed drives that would elicit anxiety, which would then reinforce repression. The chronic conflict brought forth chronic symptomatology, so that the panic attack came to be viewed as the occasional exacerbation of chronic anxiety. The specific phenomenology of the attack, and even references to anxiety attacks, progressively dropped from view. This may be an unfortunate case of how theory can narrow one's field of vision.

The major thrust of Kandel's seminal 1983 paper was to demonstrate animal models of human anxiety that allow explorations at cellular and molecular levels. Chronic and anticipatory anxiety in humans were held to respectively parallel sensitization and aversive classical conditioning, as demonstrated in the sea snail *Aplysia californica.* In *Aplysia,* both of these forms of learned fear rely on presynaptic facilitation; augmentation of presynaptic facilitation accounts for associative conditioning. These findings suggest that a surprisingly simple set of mechanisms, arranged in various combinations, may underlie a wide range of both adaptive and maladaptive behavioral modifications. This simplified paraphrase demonstrates the innovative, rigorous approach Kandel took to illuminate neural functioning, with implications for learning and affect. Using advanced technology, adroit experimental design, and a wonderfully appropriate animal for study, Kandel penetrated Skinner's "black box" by causal experimentation. Perhaps Kandel, like Pavlov, has a sign on his laboratory wall that reads, "From the simple to the complex."

However, Kandel also stated that anxiety "can be adaptive: it prepares us for potential danger....Anxiety can become dysfunctional, by either being inappropriately intense or being displaced by association with neutral events that neither are dangerous themselves nor indicate danger. Thus, anxiety is pathological when it becomes inappropriately severe and persistent or no longer serves only to signal danger" (p. 127). For Kandel to consider anticipatory anxiety as a "learned abnormality" was inconsistent, reflecting the definitional problems of that era. Anticipatory anxiety is brief, triggered by an identifiable signal, and distinguished from fear in that fear requires a present danger. However, is it necessarily learned?

Kandel analogized anticipatory (signaled) anxiety to aversive conditioning, where the lack of a conditioned stimulus predicts the non-occurrence of the unconditioned stimulus. Therefore, the lack of a conditioned stimulus serves as a safety signal, which relieves anxiety. In contrast, if shock is repeated without warning, no safety signal is possible. A state of chronic anxiety or sensitization ensues.

The fact that chronic anxiety is severe and does not serve to signal danger supports a case for abnormality, but the case for necessary learning also requires further evidence. Moreover, recurrent unsignaled shock is often used to model depression, associated with learned helplessness, rather than anxiety. Epidemiological studies indicate that recent-onset generalized anxiety disorder (GAD) is usually transient, whereas GAD lasting for six months often transforms into a depression, indicating clinical complexities.

In *Aplysia,* trauma incites serotonin release, which acts on the presynaptic receptors of the sensory neuron so that, after a complex cascade, the amount of neurotransmitter released by the afferent neuron is chronically increased. This finding was a potent springboard for further work leading to refined synaptic analyses. A further groundbreaking idea, initiated by Kandel and James H. Schwartz, was the suggestion that structural change may depend upon patterns of gene activation and inactivation (Kandel et al. 2000): that experience, and therefore learning, may also cause such patterns.

Kandel emphasized the utility of Freudian insights concerning internal representations for neuroscience. What I would emphasize is the admirable acuteness of the psychopathological observations made by Freud—for example, the importance of anxiety for psychopathology; the continuity between childhood pathological states and adult illnesses; the astute delineation of panic attacks and their antecedent relationship to agoraphobia; the emphasis on dyspnea as a cardinal feature of anxiety attacks; and the importance of separation anxiety both as a normal stage of development and as a manifestation of psychopathology.

However, Freud's explanatory insights about anxiety were frequently misleading. For instance: that behind every phobic fear lay a repressed wish; that attachment to the mother was "anaclitic," derived from need satisfaction; that the infant's stranger anxiety and separation anxiety were both due to anticipation of a rising flood of ungratified libidinal drives; that school phobia was due to the child's unconscious hostility to the mother (A. Freud 1972); that dyspnea is due to the lasting effects of birth trauma; that the panic attack is due to undischarged libido caused by coitus interruptus; that castration anxiety is central to masculine development and superego formation; that social anxiety is due to repressed exhibitionistic desires, and so on. Such claims were based on the flawed but fundamental analytic procedure of "interpretation."

Specific psychoanalytic explanatory insights, as those above, are less and less referred to and have undergone disuse atrophy (rather than invalidation and retraction). As Kandel well knows, contemporary psychoanalytic thinking about anxiety derives more from John Bowlby, whose views were considered heretical (Bowlby 1960), and Harold F. Harlow, who experimentally showed the independence of attachment from oral gratification (Harlow 1964). Current psychological treatment of separation anxiety was initiated by Leon Eisenberg (1958). Contemporary neuroscientific investigations into unconscious processes have little to do with a Freudian, defense-generated, dynamic unconscious responsible for psychopathology. Neuroscientific illumination of the complexities of the descriptive unconscious should not be considered as a psychoanalytic validation.

One of the views expressed in this 1983 paper calls for modification. Kandel emphasized the heritability of major psychotic illnesses, which presumably represent mutations, as opposed to certain neurotic illnesses, "such as chronic anxiety (that are acquired by learning and can respond to psychotherapy)" (p. 145). Recent anxiety disorder studies have shown genetic loadings comparable to the major psychoses. Therefore, this hypothesis reflected the psychoanalytic ideology of the day and appears invalidated. That is irrelevant to the bold, fruitful hypothesis that experiential learning depends on alterations of gene expression. Perhaps most surprising is that panic attacks are not necessarily associated with terror, which leads to the apparent oxymoron of nonfearful panic attacks. Indeed, Freud, in 1895, reported "larval" attacks associated only with physical distress. Patients in general medical practices, rather than psychiatric offices, often have acute episodic distress, chest discomfort and dyspnea, but not fear. It is estimated that half of those whose cardiac catheterization shows no evidence of coronary disease actually have panic disorder.

Furthermore, the spontaneous panic attack is not associated with a sudden sympathetic autonomic crisis, or evidence of hypothalamic-pituitary-adrenal stimulation, as one might expect in terror. Recently, the emphasis on sympathetic discharge has been modified by recent findings on vagal withdrawal, as indicated by heart rate variability studies. Freud's emphasis on dyspnea, which led him to entertain birth trauma as the prototype of anxiety, resurfaced in the recognition that the attack in panic disorder is frequently associated with acute air hunger, which is not a salient feature of other anxiety disorders or realistic fear.

What can be learned (if anything) from this historical survey? My not-to-original take is that 1) brilliant, persuasive insights should not be the end of inquiry but rather provide the promising initiative that fosters the hard labor of meticulous, experimental testing; 2) scientific advances often depend on technical advances, chance observations, and the recognition of fruitful investigative areas that can be put to good use by gifted people; and 3) we should "seek simplicity and distrust it" (Whitehead 1926).

References

Axelrod J: Noradrenaline: fate and control of its biosynthesis, in Nobel Lectures, Physiology or Medicine, 1963–1970. Amsterdam, Elsevier, 1972

Bowlby J: Separation anxiety. Int J Psychoanal 41:89–113, 1960

Carlsson A: A half century of neurotransmitter research: impact on neurology and psychiatry, in Nobel Lectures, Physiology or Medicine, 1996–2000. Edited by Jornvall H. Singapore, World Scientific Publishing Company, 2003

Eisenberg L: School phobia: a study in the communication of anxiety. Am J Psychiatry 114:712–718, 1958

Freud S: On the grounds for detaching a particular syndrome from neurasthenia under the description "anxiety neurosis," in The Standard Edition of the Complete Psychological Works of Sigmund Freud, Vol 3. Edited and translated by Strachey J. London, The Hogarth Press, 1962, pp 90–115

Harlow HF, Rowland GL, Griffin GA: The effect of total social deprivation on the development of monkey behavior. Psychiatr Res Rep Am Psychiatr Assoc 19:116–135, 1964

Kandel ER, Schwartz JH, Jessel TM (eds): Principles of Neural Science, 4th Edition. New York, McGraw-Hill, 2000

Whitehead AN: The Concept of Nature. Cambridge, England, Cambridge University Press, 1926, p 163

FROM ANXIETY TO *APLYSIA* AND BACK AGAIN

Joseph LeDoux, Ph.D.

Anxiety disorders take a huge toll on society. More people visit mental health professionals each year for problems related to anxiety than for any other reason. Although we still know little about the neurobiological basis of anxiety, a new wave of research has emerged attempting to understand anxiety from the point of view of brain mechanisms. This research starts from the assumption that anxiety and the related state of fear are normal functions of the brain in response to threatening stimuli. Fear and anxiety disorders are said to exist when the brain systems that normally process threats are activated inappropriately (when no threat exists) or respond more intensely than the situation warrants. While some environmental stimuli and situations are preordained with threat value by evolution, fear and anxiety in humans are largely products of learning, and they leave their marks on the brain in the form of memory. If we are to understand how fear and anxiety disorders come about and how they might be most effectively treated, we need to know how the brain processes emotions, like fear and anxiety, and especially how the brain learns and stores information about threats. This is an active area of research today, but it was not always so.

In the early 1980s, the study of fear and other emotions was a research backwater; neuroscience was much more enthralled with the idea of studying higher cognition than emotion. Nevertheless, this topic did not escape Eric Kandel, who, as his introduction to this volume tell us, began his medical career as a psychiatrist with special interest in Freudian psychoanalysis. Kandel's research soon turned toward a rigorous scientific analysis of the neurobiology of learning and memory—and not in humans but in the lowly sea snail *Aplysia californica*. Nevertheless, through it all, he remained fascinated with the deep questions of psychiatry, such as the how and why of anxiety. In fact, in 1983 he wrote the following paper in the *American Journal of Psychiatry* that sought a rapprochement between psychoanalysis and modern neuroscience in the understanding of anxiety. I'm honored to make a few comments about this paper.

Building on Freud, Pavlov, and Darwin, as well as his own research, Kandel noted that the ability to anticipate threats is biologically adaptive. Through learning and memory we can acquire information that allows us to begin to respond to a threatening stimulus prior to the arrival of the actual harm. If such learning capacities are conserved across species, it might be possible to extrapolate findings in simple organisms, like snails, in the effort to understand and treat anxiety.

To some, Kandel's idea that research on snails could have any relevance to one of the most profound and troubling human conditions, one that has preoccupied poets and philosophers for centuries, if not millennia, might seem absurd. After all, anxiety is a problem of subjective experience. A snail does not have a brain, and whatever mind it has, if any, is not likely to have much in common with human consciousness. The key to understanding why Kandel was correct in principle rests not on any similarity between the mental life of a snail and the subjective or conscious manifestation of anxiety in humans. Instead, the key is that the mechanism underlying learning and information storage, including about threats, is likely to be conserved at the level of molecules and possibly genes. In 1983, this was still somewhat of a long shot. Today, it is essentially taken for granted by neuroscientists.

For example, certain gene products (proteins) that have been implicated in learning and memory storage in the snail have also been shown to be involved in learning and information storage in flies, worms, and bees. If that's where it stopped, no one would have been surprised. But through the work of Kandel and others, similar molecules have also been shown to play a role in memory formation in mammals, especially mice and rats.

Let's look a little more closely at the specific proposals in Kandel's 1983 paper. He divides Freud's notion of acquired or signal anxiety into several categories and focuses on two of these: anticipatory anxiety and chronic anxiety. He then argues that anticipatory anxiety can be modeled in the laboratory using Pavlovian fear conditioning, a procedure in which a neutral stimulus comes to elicit fear reactions after pairing with an aversive event (such as electric shock), while chronic anxiety can be modeled by the procedure called sensitization, in which an aversive stimulus primes or sensitizes the organism to respond to any neutral stimulus that later appears. The difference is that in conditioned fear the neutral stimulus is specifically paired with the aversive event and only it is conditioned, whereas in sensitization the response to any neutral stimulus is facilitated. A very important point is that fear and anxiety, in these models, do not refer to subjective manifestations but only to the role of the nervous system in controlling protective or defensive responses, what we would call the fight-flight response in humans.

Over the years, Kandel's work has elucidated in exquisite detail how these two forms of learning work in the snail at the levels of behavior, neural circuits, cells and synapses, and molecules. More recently, building on Kandel's work, and especially following the cellular-connectionist strategy he pioneered in the *Aplysia,* I and some others have made progress at exploring the neural basis of fear conditioning at the level of the system, cells, synapses, and molecules in mammals (Kandel himself has also done important research on this topic). Although the neural system of fear conditioning is different in mammals and invertebrates, there are fundamental similarities in the rules that govern learning at the level of behavior and, as already noted, in the molecules that allow the cells and synapses to change with experience and to store the results in memory.

In the end, Kandel's 1983 paper is important not because it solved the problems of pathological fear and anxiety, but because it suggested a strategy about how we might go about using what we know about the neural basis of learning and memory to gain insights into acquired fear and anxiety. Kandel's groundbreaking research on learning and memory in the *Aplysia* continues today—so does his effort to bring modern neuroscience into psychiatry, as the various other articles in this volume illustrate. Sigmund Freud started his career studying the nervous system before he turned his efforts to the human mind. Were he alive today, he would likely be a very big fan of Eric Kandel's research and writings.

CHAPTER 4

FROM METAPSYCHOLOGY TO MOLECULAR BIOLOGY

Explorations Into the Nature of Anxiety

Eric R. Kandel, M.D.

Despite important progress during the last decade, the cell-biological mechanisms of mentation have until very recently eluded analysis. However, growth in the conceptual and technical power of cognitive psychology on the one hand and neurobiology on the other now makes it possible to confront one of the last frontiers of science. Within a theoretical framework

This article was originally published by the *American Journal of Psychiatry,* Volume 40, Number 10, 1983, pp. 1277–1293.

Expanded version of the first John Flynn Memorial Lecture, presented to the Department of Psychiatry, Yale University Medical School, New Haven, Connecticut, April 1982. Received November 4, 1982; revised April 7, 1983; accepted May 13, 1983. From the Center for Neurobiology and Behavior, College of Physicians and Surgeons, Columbia University; and the New York State Psychiatric Institute, New York, New York.

The original work described in this paper was supported by Career Scientist Award MH-18558 from NIMH, grant MH-26212 from NIMH, and a grant from the McKnight Foundation.

The author thanks Morton Rieser, Ethel Person, E. Terrell Walters, Tom Carew, Sally Muir, and James H. Schwartz for comments on an earlier draft of this paper.

117

based largely on insights from experimental psychology and psychiatry, biologists are beginning to study successfully elementary aspects of mentation and to address a number of central questions: What functional changes must take place in nerve cells for learning and memory to occur? Are there unifying cellular and molecular principles that relate one form of learning to another? That relate short-term to long-term memory? Can experience lead to enduring structural changes in the nervous system? Do these structural changes involve alteration of gene expression, and, if so, is psychotherapy successful only when it induces such changes? These questions, although originating from different behavioral and neurobiological perspectives, are increasingly converging on a common ground. In this essay, I shall try to illustrate how the independent contributions of psychiatry, psychology, and neurobiology can be combined in animal models to yield insights into mentation that extend from the behavioral to the molecular level and that promise to apply to experimental animals and humans alike.

To emphasize the relevance of these new developments for psychiatry, I shall concentrate on two learned abnormalities of behavior: anticipatory anxiety and chronic anxiety. I hope to document how cognitive psychology, which has shown that the brain stores an internal representation of experiential events, converges with neurobiology, which has shown that this representation can be understood in terms of individual nerve cells, so as to yield a new perspective in the study of learned anxiety. In a larger sense, I shall try to illustrate that mentation loses none of its power or its beauty when the approach is moved from the domain of metapsychology into the range of molecular biology. On the contrary, the combined developments in cognitive psychology and in neurobiology promise to renew interest in aspects of mentation that until now have been out of experimental reach. Although behaviorist psychology has been content to explore observable aspects of behavior, advances in cognitive psychology indicate that investigations which fail to consider internal representations of mental events are inadequate to account for behavior, not only in humans but—perhaps more surprisingly—also in simple experimental animals. This recognition of the importance of internal representations, a conclusion intrinsic to psychoanalytic thought, might have been scientifically disappointing as recently as 10 years ago, when internal mental processes were essentially inaccessible to experimental analysis. However, subsequent developments in cell and molecular biology have made it feasible to explore elementary aspects of internal mental processes. Thus, contrary to some expectations, biological analysis is unlikely to diminish the interest in mentation or to make mentation trivial by reduction. Rather, cell and molecular biology have merely expanded our vision, allowing us to perceive previously unanticipated interrelationships between biological and psychological phenomena.

The boundary between behavior and biology is arbitrary and changing. It has been imposed not by the natural contours of the disciplines but by lack of knowledge. As knowledge expands, the biological and behavioral disciplines will begin to merge at certain points, and it is at these points that the ground on which modern psychiatry is based will become particularly secure.

The Clinical Syndromes of Anxiety

Anxiety is a normal inborn response either to threat—to one's person, attitudes, or self-esteem—or to the absence of people or objects that assure and signify safety (Bowlby 1969; Freud 1925–1926/1959; Nemiah 1975). Anxiety has subjective as well as objective manifestations. Subjective manifestations range from a heightened sense of awareness to deep fear of impending disaster. The objective manifestations of anxiety consist of increased responsiveness, restlessness, and autonomic changes (for example, changes in heart rate and blood pressure). Anxiety can be adaptive: it prepares us for potential danger and can contribute to the mastery of difficult circumstances and thus to personal growth. On the other hand, anxiety can become dysfunctional, by either being inappropriately intense or being displaced by association with neutral events that neither are dangerous themselves nor indicate danger. Thus, anxiety is pathological when it becomes inappropriately severe and persistent or no longer serves only to signal danger.

The biological mechanisms that give rise to feelings of anxiety represent a central problem in the neurobiology of normal affective behavior. Anxiety is also an important component of neurotic and psychotic illnesses. Yet despite the importance of anxiety, little is known of its underlying cellular and molecular mechanisms.

As in other areas of behavior, most of the initial insights into anxiety have come from clinical observation. One key insight was the appreciation, first by Freud and then by others, that anxiety is not unitary but is manifested in a variety of forms. Thus, in his later writings, Freud distinguished *actual (automatic) anxiety*, an automatic, inborn response to external or internal danger, from *signal anxiety*, an acquired fear response in anticipation of danger, either internal (unconscious) or external (Freud 1925–1926/1959). (Actual anxiety is referred to as fear by some investigators [Kimmel and Burns 1977; Mowrer 1939]. For a clear discussion of the evolution of Freud's writings on anxiety, see Strachey's editorial introduction to Freud's essay in the *Standard Edition* [Freud 1925–1926/1959] and Brenner 1973.) Subsequent work (for reviews, see Goodwin and Guze 1979; Klein 1981; Sheehan 1982) has shown that acquired anxiety can further be subdivided into three forms on the basis of clinical characteristics and response to psychopharmacological agents. These forms are panic attacks, anticipatory anxiety, and chronic anxiety.

Panic attacks are brief, spontaneous episodes of terror without manifest or clearly identifiable precipitating cause. The attacks are characterized by a sense of impending disaster accompanied by a sympathetic crisis: the heart races; breath is short and unsteady. This form of anxiety often responds to tricyclic antidepressants and to MAOIs (Klein 1962, 1964; Sheehan 1982). *Anticipatory* (or signaled) *anxiety* also is typically brief. Unlike panic attacks, anticipatory anxiety is triggered by an identifiable signal, real or imagined, that has come to be associated with danger. This form of anxiety tends to respond to benzodiazepines and β-receptor–blocking agents such as propranolol (Goodwin and Guze 1979; Klein 1981; Sheehan 1982). *Chronic anxiety* is a persistent feeling of tension that cannot be related to obvious external threats; it may or may not be reduced by benzodiazepines (Mayer-Gross 1969).

Panic attacks, occurring suddenly and without an apparent trigger, are not under obvious stimulus control. In contrast, anticipatory and chronic anxiety are to some degree under stimulus control. This feature suggests that both forms are at least partly learned. That is, each form involves learning a relationship (or the absence of a relationship) between a neutral and a threatening stimulus.

The idea that anxiety is inborn and that a neutral stimulus can be associated with it through learning has come from two sources. First, work in comparative and evolutionary biology beginning with Darwin (1873) and Romanes (1883, 1888) has shown that most animals, like humans, have a repertoire of inborn defensive behaviors. Aware of the contributions of Darwin and Romanes, William James proposed in 1893 that in animals and in humans these built-in defensive behaviors are triggered by anxiety, an inborn tendency to react with fear to dangerous situations. Experimental support for the notion that anxiety can be learned came from Pavlov's discovery at the turn of the century that defensive reflexes can be modified by experience and can be elicited by a previously neutral stimulus. Thus, in 1927 Pavlov noted the utility of such associative learning for an animal's survival:

> It is pretty evident that under natural conditions the normal animal must respond not only to stimuli which themselves bring immediate benefit or harm, but also to other physical or chemical agencies....which in themselves only *signal* the approach of these stimuli; though it is not the sight or the sound of the beast of prey which is itself harmful to smaller animals, but its teeth and claws. (Pavlov 1927, p. 14)

A similar proposal was made independently by Freud. Because painful stimuli are often associated with neutral stimuli, symbolic or real, Freud postulated that repeated pairing of a neutral and a noxious stimulus can cause the neutral stimulus to be perceived as dangerous and to elicit by itself the anxiety response. Placing this argument in a biological context, Freud wrote in 1926:

> The individual will have made an important advance in his capacity for self-preservation if he can foresee and expect a traumatic situation of this kind which entails helplessness, instead of simply waiting for it to happen. Let us call a situation which contains the determinant for such expectation a *danger situation*. It is in this situation that the *signal of anxiety* is given [italics added]. (Freud 1925–1926/1959, p. 166)

Pavlov and Freud not only appreciated that anxiety can be learned, but each also had the important insight that the ability to manifest anticipatory defensive responses to danger signals is biologically adaptive. Anxiety as a signal prepares the individual for fight or flight if the danger is external. For internal danger, Freud suggested that defensive mental mechanisms substitute for actual flight or withdrawal. I would only make a cautionary comment here: simply because aspects of anxiety may be learned and thus acquired does not exclude the possible contribution of a genetic predisposition to anxiety. In fact, what might be inherited is the predisposition to learn certain stimulus relationships (Cohen et al. 1951; Crowe et al. 1980; Goodwin and Guze 1979; Pauls et al. 1980; Sargant and Slater 1963; Sheehan 1982; Slater and Shields 1969).

Anxiety Can Be Studied in Animal Models

In people suffering from anticipatory anxiety, a cue stimulus is thought to predict the occurrence of an aversive stimulus (Estes and Skinner 1941; Miller 1948; Mowrer 1939; Pavlov 1927). By contrast, chronic or unsignaled anxiety is thought to occur when people learn either that danger is associated with a wide range of ever-present environmental cues or that danger is always present and not signaled by any cues (Kandel 1976; Seligman 1975). As a result, chronic anxiety is triggered in a less discriminating way.

With the recognition of distinct types of acquired anxiety, experimental interest turned to the development of animal models for studying each type. From work with experimental models it soon became clear that animals can learn to manifest aspects of anticipatory and chronic anxiety. This evidence has strengthened the initial clinical distinction between anticipatory and chronic anxiety and supports the belief that aspects of these forms of anxiety are also learned in humans. (For earlier discussions of animal models, see Dollard and Miller 1950; Estes and Skinner 1941; Hammond 1970; Miller 1948; Mowrer 1939.)

Anticipatory (signaled) anxiety	Chronic (unsignaled) anxiety
Neutral signal predicts danger (anxiety)	No signal to predict danger therefore No signal to predict safety (chronic anxiety)
Lack of neutral signal predicts safety (no anxiety)	
Aversive conditioning	**Sensitization**
Neutral signal (CS) predicts aversive stimulus (US)	No neutral signal (CS) to predict aversive stimulus (US)
Lack of CS predicts no US	

FIGURE 4–1.   Two forms of learning that give rise to two forms of anxiety.

Comparison of anticipatory anxiety and its animal model, aversive conditioning, to chronic anxiety and its animal model, long-term sensitization.

Since we think of anxiety as characteristically human, it is important to review the evidence that simple animals can learn anxiety and that conditioned fear in these animals approximates certain forms of anxiety in humans. I shall argue that classically conditioned fear and long-term sensitization provide models of anticipatory and chronic anxiety, respectively (see Figure 4–1).

In classical (Pavlovian) conditioning, an animal learns to associate two stimuli, a *conditioned* (or cue) stimulus (CS) and an *unconditioned* (or reinforcing) stimulus (US). The US, by definition, elicits from the animal an effective reflex, or instinctive response, that is called the unconditioned response because it is present before conditioning. In contrast, the CS need not elicit a reflex response before conditioning takes place. After repeated pairing, the animal learns to associate the CS with the US and as a consequence will show reliable conditioned responses to the CS that often resemble the inborn unconditioned response to the US. In Pavlov's classic experiment, food (meat powder) served as a US, eliciting an inborn response, reflex salivation. After several trials in which the food was paired with a neutral stimulus, a tone, the tone reliably elicited salivation. For pairing to be effective, Pavlov found that he had to present the tone and the food in a precise sequence: the tone had to precede the food on each training trial. This, as we shall see later, is because what the animal actually learned during classical conditioning is that tone *predicts* food. The animal salivates after the tone to prepare for food. Thus, if the pairing sequence is reversed (backward conditioning), the animal does not respond to the tone by salivating.

Pavlov further found that by varying the nature of the US he could produce different types of learning. Unconditioned stimuli that satisfied the animal's needs or that enhanced survival gave rise to *appetitive learning,* leading to satisfaction and ultimately satiation. Unconditioned stimuli that threatened survival, such as a painful shock, produced *aversive learning,* leading to conditioned fear (Estes and Skinner 1941; Mowrer 1939; Watson and Rayner 1920).

Is there a relationship between aversive conditioning in animals and specific anxiety in humans? One of the first experiments to apply aversive conditioning to humans illustrated how classical conditioning could give rise to anticipatory anxiety. In 1920, Watson and Rayner found that an infant they were studying cried readily (Watson and Rayner 1920). Loud and sudden noise proved a particularly effective US for eliciting the unconditioned response, crying. They then added a neutral CS, a white rat that initially did not elicit crying. After several pairings of CS and US, the infant started to cry the instant the white rat was presented: the previously neutral CS produced the conditioned response.

Aversive conditioning can lead to one of two forms of anxiety, depending on whether the aversive stimulus is presented in a signaled or an unsignaled manner (Figure 4–1). As emphasized by Mowrer (1939) and Miller (1948), the presence of a CS, as a cue that predicts the occurrence of the aversive stimulus (US), allows the animal to learn to focus its anxiety on a particular event in time (see also Estes and Skinner 1941; Pavlov 1927). By contrast, repeated exposure to the aversive stimulus *without* a cue stimulus produces chronic anxiety (long-term sensitization), as pointed out by Seligman (1975) and by Pinsker, Hening, Carew, and me (Pinsker et al. 1973) (for a review, see Kandel 1975).

Seligman (1975) has used the following analogy to illustrate the distinctions between the two forms of anxiety in terms of biological adaptation. Imagine a world in which each aversive stimulus capable of causing pain, and therefore fear, is predicted accurately and invariably by a brief neutral stimulus so that the presence of this neutral stimulus comes to produce a brief episode of anxiety. As long as the cue stimulus is not present the animal can relax and do what it wants. A consequence of traumatic events being predictable is that the absence of traumatic events is also predictable. When, however, aversive events are unpredictable, safety also is unpredictable: no reliable event exists to indicate that the trauma will not occur. Lacking a safety signal, organisms remain in a state of chronic anxiety. According to this view, people and animals seek safety signals (Seligman 1975). They look for predictors of danger because these also provide information about safety (Badia and Culbertson 1970; Badia et al. 1967; Weiss 1970).

Thus, both in people and in experimental animals, what distinguishes signaled (anticipatory) from unsignaled (chronic) anxiety is that signaled anxiety is predictive with respect to its cause, whereas unsignaled anxiety is completely unpredictive. The two variants of aversive training that we have already considered therefore model the essential difference between anticipatory and chronic anxiety.

Mechanisms Underlying Anxiety Are Likely to Be General Throughout Phylogeny

An implication of these observations is that the development of anticipatory and chronic anxiety in animals depends on mechanisms by which animals process information about the predictive interrelationships among various environmental events. A useful perspective on these problems has been provided by cognitive psychologists, who have shown that learning involves considerable mental processing and elaboration of sensory information, with the result that humans and animals develop internal representations ("cognitive structures") of environmental events that allow flexible behavioral decisions (for reviews, see Bindra 1978; Bolles and Fanselow 1980; Dickinson 1980; Rescorla 1978).

If learning in humans and other higher animals involves the establishment of certain cognitive structures, why are aspects of such cognitive mechanisms likely to be similar in humans and in simple animals like the snail *Aplysia?* One good reason for believing that this would be the case lies in the consequences of adaptation to evolutionary pressure. Animals that differ greatly in habitat and heritage nonetheless face common problems of adaptation and survival, problems for which learning and flexible decision making are useful. When different species face a common environmental constraint, they often manifest homologous patterns of adaptation because a successful solution to an environmental challenge, first evolved in a common primitive ancestor, will continue to be inherited as long as it remains useful and the selective pressure is present. As the physicist Weisskopf (1981) put it, "nature likes to use the same old tricks again." In addition, common environmental pressures often lead to the independent evolution of functionally analogous processes in distantly related species.

What constraint might have shaped or maintained a common cognitive learning mechanism in a wide variety of species? Testa (1974) and Dickinson (1977) have argued that to function effectively, animals need to recognize certain key relationships between events in their environment. They must be able to recognize and mate with their own species and to avoid even closely related species; they must distinguish animals that are prey and learn to avoid those that are predators; they must search out food that is nutritious and avoid food that is poisonous. There are two ways in which an animal arrives at such knowledge: the correct information for every choice can be pre-programmed in the animal's nervous system, or the ability to choose correctly among alternatives can be acquired through learning. Genetic and developmental programming may suffice for all of the behavior of simple parasites and certain free-living forms such as the nematode worm *C. elegans,* which exists in a limited and relatively invariant environment in the soil. But for more complex animals, extensive learning is probably required to cope efficiently with varied or novel situations. Complex animals need to maximize their ability to order the world. An effective way to do this is to be able to learn about predictive relationships between related events.

Given that humans and experimental animals should be capable of learning predictive relationships, do such relationships have properties in common that could constitute a universally selective evolutionary pressure? They do. First, predictors always precede the signaled event, and second, they are highly correlated with it; they provide optimal information about the probability of its occurrence. As Dickinson (1980, 1981) and Testa (1974) have argued, these distinctive properties of predictive relationships are of such importance that they probably form the basis of widespread adaptational and evolutionary pressures that have acted on all animals and enhanced the survival of those species capable of taking them into account (for a related discussion, see also Staddon and Simelhag 1971). Some psychologists therefore believe that common associative mechanisms of learning exist in all species capable of learning and that these common mechanisms are designed to recognize and store information about predictive relationships in the environment (Dickinson 1980; Rescorla 1968, 1973, 1978, 1979). As we have already seen, this issue is not new but was first raised by William James in 1892, when, following Darwin, he argued with his usual prescience that mental processes evolved to serve adaptive functions for animals in their struggle with a complex environment: "Mental facts cannot be properly studied apart from the physical environment of which they take cognizance....Our inner faculties are *adapted in advance to the features of the world in which we dwell,* adapted, I mean, so as to secure our safety and prosperity in its midst....Mind and world in short have evolved together, and in consequence are something of a mutual fit" (James 1892, p. 4).

What is the evidence that animals are particularly adept at learning predictive relationships? Actually, until quite recently, animal psychologists thought that classical conditioning depended only on temporal contiguity: A conditioned stimulus had only to precede the reinforcing unconditioned stimulus by a certain critical period to be effectively conditioned (Gormezano and Kehoe 1975; Guthrie 1935). However, this simple idea appears to be inadequate. If animals learned to derive predictive information simply from the occurrence of two events in close temporal contiguity, they might acquire a variety of erroneous notions about signals in the environment and begin to act maladaptively. The world is full of chance, and events sometimes occur together without being highly correlated or causally related.

Analyses of learning by Prokasy (1965), Rescorla (1968), Kamin (1969), Mackintosh (1974), Wagner (1976), and their colleagues have shown that classical conditioning develops best when in addition to *contiguity* there is also a *contingency*—a truly predictive relationship—between the conditioned (or cue) stimulus and the unconditioned stimulus (Dickinson 1980, 1981; Rescorla 1973, 1978; Rescorla and Wagner 1972). Classical conditioning works not simply because the CS and US are temporally paired but also because there are time intervals *between* successive pairings *within which the US does not occur*. Thus, in addition to being paired in time, the signal and reinforcer need to be positively correlated; the signal must indicate an increased probability that the reinforcer will occur. It therefore appears likely that animals learn classical conditioning, and perhaps all forms of associative learning, so readily because *the brain has evolved to enable animals to distinguish events that reliably and predictively occur together from those which are unrelated.*

Behaviorism, Cognitive Psychology, and Renascence of Psychoanalytic Perspective

These several arguments indicate that explanations of learning based solely on temporal contiguity are limited. The behaviorist position, which has emphasized temporal contiguity (Gormezano and Kehoe 1975; Guthrie 1959; Skinner 1957), has also run into difficulty in addressing questions central to other areas of behavior and learning (for critiques of the behaviorist position, see Chomsky 1959; Kandel 1976; Neisser 1967; Rescorla 1978; Tolman 1932). Objective measurements of behavior through analysis of stimuli and responses are clearly important for the study of behavior. They are the only indices of behavior that can be manipulated experimentally and the only ones that can be measured objectively. Indeed, the most useful definition of behavior—*observable movement*—derives from traditional behaviorism. Nonetheless, despite the great technical and conceptual debt that psychology owes to behaviorism, there is a substantial difference between the view of mental life held by behaviorists such as Watson and Skinner and that found useful by most current students of behavior (Klatzky 1980; Neisser 1967; Posner 1973; Seligman and Meyer 1970). The extreme behaviorist view (Watson 1913, 1925) is that observable behavior is synonymous with mental life. This view narrowly defines a larger reality, psychic life, in terms of the scientific techniques available for studying it. By so doing, this approach denies the existence both of consciousness and of unconscious mentation, feelings, and motivation merely because they cannot be studied objectively. Broader perspectives such as those used by cognitive psychology are necessary to account for the behavioral capabilities both of people and of animals (Dickinson 1980; Dickinson and Macintosh 1978; Griffin 1982; Kandel 1976). Cognitive psychologists have emphasized the richness of the internal representations that intervene between stimuli and response. Even the acquisition of simple associative tasks by invertebrates involves the learning of surprisingly complex predictive relationships, which suggests that many animals may form "cognitive" representations of relationships among events in their environment (Sahley et al. 1981).

In the past, ascribing a particular behavioral feature to an unobservable mental process essentially excluded the problem from direct biological study because the complexity of the brain posed a barrier to any complementary biological analysis. But the nervous systems of invertebrates are quite accessible to a cellular analysis of behavior, including certain internal representations of environmental experiences that can now be explored in detail. This encourages the belief that elements of cognitive mentation relevant to humans and related to psychoanalytic theory can be explored directly and need no longer be merely inferred.

The psychoanalytic perspective has been devalued recently and, in the United States, is in decline. Its propositions have relied heavily on intervening constructs: on a mental apparatus for unconscious and conscious mental activity and on postulated libidinal and aggressive drives. Some of these ideas are vague; all are difficult to quantify. As a result, the exploration of psychoanalytic theories has been hampered by a lack of opportunities for experimental verification. Nevertheless, psychoanalytic thought has been particularly valuable for its recognition of the diversity and complexity of human mental experience, for discerning the importance both of genetic and learned (social) factors in determining the mental representation of the world, and for its view of behavior as being based on that representation. By emphasizing mental structure and internal representation, psychoanalysis served as a source of modern cognitive psychology, a psychology that has stressed the importance of the logic of mental operations and of internal representations. Just as I believe that the vigor now evident in cognitive psychology will be strengthened by its contact with the cellular neurobiology of behavior through work on simple systems such as invertebrates, I also think that the emergence of an empirical neuropsychology of cognition based on cellular neurobiology can produce a renascence of scientific psychoanalysis. This form of psychoanalysis may be founded on theoretical hypotheses that are more modest than those applied previously but that are more testable because they will be closer to experimental inquiry.

In the remainder of this paper I shall try to document these ideas by describing studies directed at developing an animal model for both anticipatory anxiety and chronic anxiety in *Aplysia* that makes it possible to explore their cellular and molecular mechanisms.

Aplysia Shows Aspects of Anticipatory and Chronic Anxiety

Because of language, we know that in humans anxiety is to an important extent subjective. In assessing anxiety in animals, we must rely exclusively on inferences derived from objective manifestations. Although the correspondence of the two types of anxiety to the two objectively defined laboratory forms of aversive learning is imperfect, the analogy is useful because it allows aspects of acquired anxiety to be explored experimentally. Being able to study aspects of anxiety in animals like rats and monkeys is not surprising. But it is surprising—at least it surprised my colleagues Terrell Walters and Thomas Carew and me—to find that even simpler animals, such as the marine snail *Aplysia,* manifest behavioral changes that by inference resemble anxiety and that can be used as a model of anxiety in higher forms. The advantage of invertebrate animals over monkeys and rats is that their nervous systems are much simpler, which offers a chance to explore the cellular and molecular mechanisms that contribute to anxiety.

I emphasize at the outset that I do not, even in my most optimistic moments, believe that the mechanisms for anxiety in simple animals are likely to be identical to those in humans. However, I would argue that, at this early stage in our understanding of the biological mechanisms of anxiety, the precision of the fit is not critically important. What is important initially is that we learn something meaningful about the mechanisms by which *any* form of anxiety arises in *any* animal, no matter how simple the animal (or the anxiety). In view of the biological adaptiveness of anxiety and its apparent conservation across diverse species, it seems likely that a rigorous analysis of the mechanisms of anxiety in any animal will prove instructive for understanding human anxiety. An analogy from recent developments in molecular biology underscores this point. The regulation of gene expression in eukaryotes (animal cells) has recently been shown to be more complex than in prokaryotes (bacteria) (for reviews, see Crick 1979; Darnell 1979; Gilbert 1978; Lewin 1980; Stryer 1981). Nevertheless, almost everything that has been learned from bacteria applies to animal cells, and without the foundation provided by the earlier work with bacteria, our understanding of gene regulation in animals would still be fairly primitive.

As with humans, *Aplysia* manifests behavioral states resembling anticipatory anxiety (or fear) in response to a classical aversive conditioning paradigm and chronic anxiety in response to a long-term sensitization paradigm

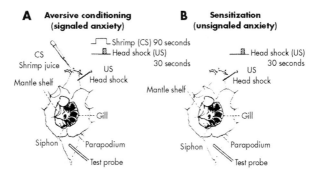

FIGURE 4-2. Experimental arrangements for studying aversive conditioning and long-term sensitization in *Aplysia*.

In both cases a strong noxious stimulus to the head serves as the unconditioned stimulus (US), and a weak test probe to the tail elicits escape locomotion and other defensive reflexes.

(Walters et al. 1979, 1981) (Figures 4–2 and 4–3). This is most clearly demonstrated by using the same aversive, unconditioned stimulus (strong shock to the head) during training and the same test pathway (escape locomotion in response to a weak tail shock) to assess learning in both paradigms. The degree of learned anxiety is assayed by measuring the amount of escape locomotion an animal displays following training. The only difference between sensitization training and classical conditioning is the presence in the latter of a cue, the conditioned or cue stimulus, specifically paired with the head shock (Figure 4–3). A particularly effective cue is a neutral chemical stimulus, extract of shrimp. *Aplysia* is a herbivore; it eats only seaweed. Although it normally ignores shrimp, its chemosensitivity can readily detect the presence of shrimp. Thus, shrimp can be an effective signal.

The two training procedures produce two forms of conditioned anxiety that differ primarily in their specificity to the neutral signal. To demonstrate this difference, Walters and associates (1979) tested the conditioned and sensitized groups twice (Figure 4–4), first in the absence and then in the presence of the CS. The animals trained with a paired stimulus showed no increase in escape locomotion when tested in the absence of the CS; when the signal was present, however, this group exhibited significantly more escape locomotion than it had either to the same test stimulus in the absence of the signal stimulus or before training. For animals trained with a cue signal, the neutral signal is required for anxiety to be expressed; they thus show a form of anticipatory anxiety. In the absence of the cue, the animals show no apprehension. In contrast, the chronically sensitized animals, trained

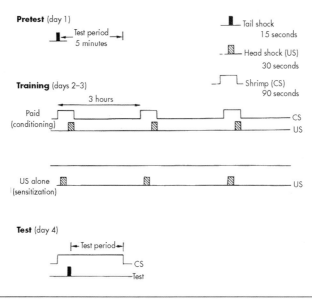

FIGURE 4–3.   Experimental protocols used on *Aplysia* for conditioning and sensitization.

The test stimulus was a train of electrical pulses to the tail. The US was a shock to the head. The CS was 1.5 mL of crude shrimp extract. These procedures were identical for paired and sensitized groups except that the latter was not exposed to the CS during training.

without a signal stimulus, show a generally heightened responsiveness that is unaffected by the presence or absence of the cue and thus show a form of chronic anxiety.

Human anticipatory anxiety and anxiety in *Aplysia* are also comparable in the pattern of effects produced by the anxiety. Some behaviorists, including Pavlov (1927), assumed that anticipatory anxiety results from stimulus substitution: the learned response to the previously neutral stimulus comes to produce the same overt response as does the painful stimulus. In contrast, Freud, in some ways the founder of modern cognitive psychology, assumed that actual danger produces an internal state, which, as we have seen, he called "actual anxiety." A neutral signal, he argued, that comes to be associated with danger may elicit any of a variety of responses, which may differ dramatically from the response to the actual trauma that the signal predicts. Indeed, the immediate reaction to the danger signal is not an overt response but an internal state of tension, an augmented preparedness for action, which Freud called "signal anxiety." By signaling fear, anxiety motivates behavioral response systems designed to reduce danger. In Freud's view, anxi-

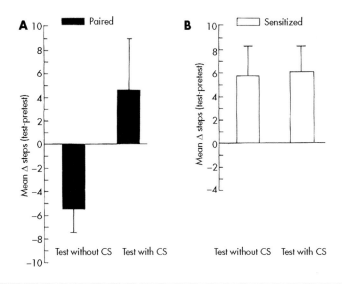

FIGURE 4–4. Comparison of responses of conditioned (paired) and sensitized animals after training.

Differences between test and pretest scores are shown for two administrations of the test stimulus: one in the absence and the other (3 hours later) in the presence of the CS. Zero on the scale represents the mean number of steps of escape locomotion taken in the pretest. Paired animals showed significantly more escape locomotion after training than before when tested in the presence of the CS but not when tested in the absence of the CS. Sensitized animals took significantly more steps to escape than they did before training in both the presence and absence of the CS.

Source.   Adapted from Walters et al. 1979.

ety is a motivational state, a defensive arousal, similar to other motivational states stemming from hunger, thirst, or the need for sex (Figure 4–5).

In humans and other mammals, Freud's view appears accurate. Both chronic and anticipatory anxiety represent a motivational (defensive) state in preparation for expected danger, a preparation that is not necessarily expressed in motor activity. As such, these forms of anxiety have two characteristics. They act not just on a single response but are likely to engage a repertory of responses, some components of which are enhanced while others are suppressed, and the effects are motivationally consistent: defensive responses are enhanced and appetitive responses suppressed (Figure 4–5).

To test the possibility that simple invertebrates can also learn to associate a neutral stimulus with a motivational state analogous to anxiety during aversive conditioning, Walters and associates (1981) examined the effect of the CS, after aversive conditioning (pairing of shrimp extract and head shock), on

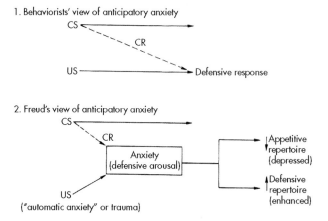

FIGURE 4–5. Comparison of behaviorists' and Freud's views of anticipatory anxiety.

The behaviorists view anxiety as a form of stimulus substitution. Initially, defensive responses are produced only in response to the US. After pairing of the CS and US, the CS produces the same defensive responses elicited by the US (a conditioned response, or CR). According to Freud's view (and that of modern cognitive psychology), the learning of anticipatory anxiety involves the endowment of a neutral stimulus with the ability to trigger an internal state—anxiety or defensive arousal—which then modulates in a motivationally consistent way not just a single response but an entire family of responses or a behavioral repertoire.

three defensive responses in addition to locomotion: two graded reflex acts (head and siphon withdrawal) and an all-or-none fixed act (inking). They also examined the suppressive effects of the CS on an appetite behavior, feeding. They found that conditioning modulates these responses in a manner that is motivationally consistent with anxiety in mammals: defensive responses are enhanced and appetitive responses are inhibited (Figure 4–6).

A Simple Form of Chronic Anxiety Can Now Be Understood in Terms of Its Cell-Biological Mechanisms

Knowing that *Aplysia* shows elementary forms of both anticipatory and chronic anxiety, we are in a position to explore the cellular mechanisms of each and to examine the relationships between them. I shall begin by considering sensitization, the animal model of chronic anxiety, because this form of anxiety has been analyzed in detail at both the cellular and molecu-

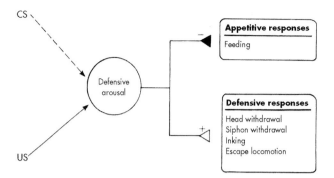

FIGURE 4-6. Conditioning of anticipatory anxiety in *Aplysia*.

The conditioning is consistent with a cognitive interpretation whereby the US elicits a motivational state (defensive arousal), which then becomes associated with the CS after repeated pairing. This motivational state suppresses appetitive behavior and augments defensive responses.

Source. Reprinted from Walters ET, Carew TJ, Kandel ER: "Associative Learning in *Aplysia*: Evidence for Conditioned Fear in an Invertebrate." *Science* 211:504–506, 1981. Used with permission from the American Association for the Advancement of Science.

lar levels. I shall focus on a very simple defensive system modulated by anxiety—siphon and gill withdrawal—because it is understood most fully (Figure 4–7).

As is the case with other snails, *Aplysia* has a respiratory chamber called the mantle cavity that houses the *gill*. This cavity is covered by a protective sheet, the mantle shelf, which terminates in a fleshy spout, the *siphon*. When the siphon or mantle shelf is touched, the siphon and gill contract vigorously, withdrawing into the mantle cavity. The gill-withdrawal reflex to stimulation of the siphon is analogous to simple defensive responses in humans such as withdrawing a hand from a hot object.

Most of the nerve cells making up the neural circuit (or wiring diagram) of the gill-withdrawal reflex have now been identified (Figure 4–7). The siphon skin is innervated by 24 sensory neurons, and there are six motor neurons. There are also several interneurons, one of which produces inhibition and five others which produce excitation. The sensory neurons that carry tactile input from the siphon skin connect to the interneurons and to the motor neurons; the motor neurons connect directly to the muscles of the gill that effect the behavior. By examining this neural circuit during sensitization, Castellucci and associates (Castellucci and Kandel 1976; Castellucci et al. 1970) found that a stimulus which produces chronic anxiety in *Aplysia*

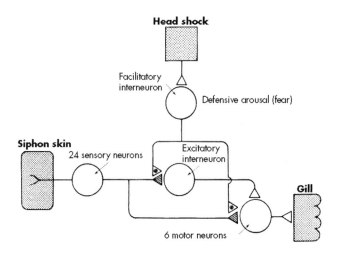

FIGURE 4-7. Neural circuit mediating sensitization of the gill-withdrawal reflex in *Aplysia*.

The interneurons and the motor cells are all unique, identified cells.

leads to an enhancement of the connections made by the sensory neurons on their target cells: the interneurons and the motor neurons. This enhancement, called presynaptic facilitation, works as follows: a noxious stimulus to the head activates a group of cells—the L29 cells—that are thought to use a transmitter closely related to serotonin as their chemical transmitter (Bailey et al. 1981; Hawkins et al. 1981a, 1981b). This group of facilitating cells acts as a defensive arousal system. The cells impinge on the synaptic terminals of the sensory neurons of the reflex system for gill withdrawal, and they amplify the strength of the connections that these sensory synapses make onto the motor neurons and interneurons (Hawkins 1981; Hawkins et al. 1981a, 1981b). Serotonin simulates all the actions of this defensive arousal and produces its amplifying action by increasing the intracellular messenger cyclic AMP in the sensory neurons (Bernier et al. 1982). The increase in intracellular cyclic AMP in turn strengthens the connections of the sensory neurons by facilitating transmitter release from their terminals (Brunelli et al. 1976; Castellucci and Kandel 1976) (Figure 4–8).

A Molecular Explanation for Chronic Anxiety

On the basis of pharmacological and biochemical studies, we have been able to piece together a coherent sequence of biochemical steps that take place in the sensory neurons when the behavior is altered by anxiety (Kandel and Schwartz 1982; M. Klein and Kandel 1980). As an action potential propa-

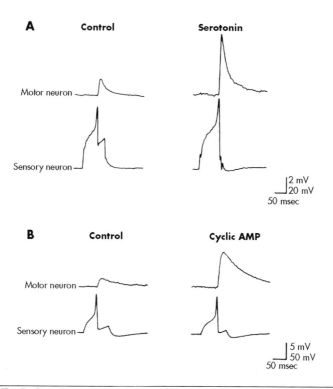

FIGURE 4-8. Experimental application of serotonin and cyclic AMP to simulate presynaptic facilitation in *Aplysia*.

In part A, serotonin, a sensory neuron was stimulated 15 times, at the rate of once every 10 seconds, and produced a monosynaptic excitatory postsynaptic potential in a gill or siphon motor neuron, as seen in the set of electrical recordings on the left. Between the fifteenth and sixteenth action potentials, the ganglion was bathed with 10^4 M serotonin, which produced presynaptic facilitation (right). In part B, cyclic AMP, as in part A, a sensory neuron was stimulated once every 10 seconds for 15 stimuli (left). Between the fifteenth and sixteenth action potentials, cyclic AMP was injected into the cell body of the sensory neuron and produced presynaptic facilitation (right).

Source.   Adapted from Brunelli et al. 1976.

gates toward the synaptic terminals of the sensory neurons, it begins to depolarize the terminals and open up the sodium channels, thereby producing more depolarization and generating an action potential in the terminal. The depolarizing component of the action potential in the terminal then opens up calcium channels and allows a certain amount of calcium to come into the cell. The depolarizing component of the action potential also opens up potassium channels; the resulting influx of potassium repolarizes the action

potential and turns the calcium channels off. Thus the activation of the sodium and potassium channels not only generates the action potential and
determines its duration but also activates the calcium channels and determines how long they remain open. Entry of calcium into the terminals is
critical for transmitter release. Calcium is thought to allow the vesicles that
contain the transmitter to bind to discharge sites—a necessary step for transmitter release. Serotonin and cyclic AMP work to prolong the action potential and thus enhance calcium influx into the sensory neuron terminals.
When the action potential is prolonged, the calcium channels stay open
longer, and more calcium is available to allow more transmitter-containing
vesicles to bind to release sites.

My colleagues Schwartz, Castellucci, Hawkins, and Klein and I have
Klein and I have outlined a molecular model for sensitization based on a
series of biophysical and biochemical experiments (Klein and Kandel 1980)
(Figure 4–9). According to this model, serotonin released by the facilitating
neurons acts on a serotonin receptor in the membrane of the presynaptic
terminals of the sensory neuron; the receptor then activates a serotonin-
sensitive adenylate cyclase. The adenylate cyclase increases cyclic AMP
within the terminals, which activates a protein kinase—the enzyme thought
to be the common site of action for cyclic AMP in all eukaryotic cells. Protein kinases are enzymes that phosphorylate proteins; that is, they add phosphoryl groups to certain amino acid residues in the protein. The addition of
a phosphoryl group changes the charge of the protein, making it more negative. This in turn changes the three-dimensional shape and therefore the
functional state of the protein. We have found that the activated protein kinase phosphorylates a certain species of potassium-channel protein or a protein that is associated with the potassium channel. Phosphorylation in effect
closes this species of potassium channel and thereby reduces the potassium
currents that normally repolarize the action potential. Reduction of these
currents prolongs the action potential, allowing more calcium to flow into
the terminals. Consequently, more vesicles bind to release sites, more transmitter is released, the functional output of the cell increases, and the animal
shows the enhanced responsiveness that characterizes chronic anxiety in
Aplysia.

My colleagues Schwartz, Castellucci, Hawkins, and Klein and I have
tested this model in a variety of ways, and we have found that we can either
trigger or block the enhancement of transmitter release by perturbing any
one of the several steps in the biochemical cascade (Castellucci et al. 1980,
1982). Thus, there now is compelling evidence that serotonin increases the
level of cyclic AMP in individual sensory cells, that cyclic AMP activates a
protein kinase, and that kinase activation leads to closing of a certain species
of potassium channel (Figure 4–9). Indeed, recently Siegelbaum, Camardo,
and I have been able to record the activity of a single potassium channel and

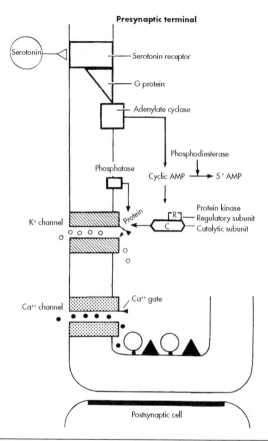

FIGURE 4-9. Molecular model of presynaptic facilitation underlying sensitization in *Aplysia*.

Serotonin or related amine released from the facilitating interneurons reaches a serotonin receptor in the presynaptic terminal of the sensory neuron, where it activates an adenylate cyclase. The receptor does not activate the cyclase directly but, as indicated in the diagram, through another membrane protein called the G protein. Once activated, the adenylate cyclase causes an increase in cyclic AMP, which in turn activates a protein kinase (an enzyme that is composed of separate regulatory and catalytic subunits). This kinase phosphorylates a protein associated with the potassium channel so that the channel closes. With the potassium channel closed, the inflow of potassium ions that would normally repolarize the action potential is reduced. Consequently, the action potential is prolonged and calcium remains free to continue entering the cell, where it binds vesicles of transmitter (large circles at bottom) to their release sites. Thus, in presynaptic facilitation the amount of time available for transmitter release from the sensory neuron terminal is extended. Eventually, the enzyme phosphatase dephosphorylates the potassium channel, which causes it to reopen, thereby terminating the action of the cascade activated by serotonin.

Source. Adapted from M. Klein and Kandel 1980.

the conformational changes in a *single protein molecule,* and we have shown that phosphorylation of this species of potassium channel either directly or indirectly (by means of a regulatory protein that affects the channel) decreases the probability that the channel will open (Siegelbaum et al. 1982).

Thus, we have been able to take the analysis of a form of anxiety from the behavior of the intact animal to the neural circuit of the behavior and to some of the critical cells involved. Within these critical cells (the sensory neurons of the reflex), we localized the change to a particular component of the neuron, the presynaptic terminals, and demonstrated that the expression of anxiety involves enhancement of transmitter release. We found that the molecular mechanism of this enhancement is protein phosphorylation, which leads to a broadening of the action potential and a greater influx of calcium. We are now able to focus on the individual protein molecules modulated by learning and explore them in a behavioral as well as a biochemical context.

The Maintenance of Chronic Anxiety Involves Structural Changes

Sensitization is a form of chronic anxiety whereby a defensive arousal system is activated and increases the release of transmitter from particular identified synapses. We can therefore ask: Does the maintenance of this learned anxiety involve a morphological change? To answer this question, Bailey and Chen (1983) have visualized the synaptic terminals of the sensory neuron electron-microscopically using the electron-dense marker horseradish peroxidase. Their evidence suggests that, as in other neurons, synaptic vesicles, the likely storage sites for transmitter, are released at varicose expansions of the presynaptic terminal of the axon. The varicosities contain specialized regions called *active zones,* where the vesicles are loaded into release sites from which they subsequently discharge their contents. Comparing sensory neurons from chronically sensitized and control animals, these researchers have analyzed the changes in the number and distribution of the synaptic vesicles and in the size and extent of the active zones. They found that in normal animals not all varicosities contain active zones. Rather, the incidence of active zones and the average size of each active zone can be modified by anxiety. In sensory neurons from naive animals, only 41% of varicosities have active zones; the rest do not. In contrast, in animals that have been sensitized, the incidence of active zones is increased to 65%. In addition, the average size of each zone is larger in sensitized than in naive animals (Figure 4–10).

Thus, simple forms of anxiety produce profound morphological as well

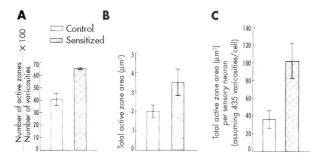

FIGURE 4-10. Morphological correlates of long-term sensitization in *Aplysia*.

Long-term sensitization produces an increase in both the number (A) and size (B) of sensory neuron active zones. These complementary changes are even more apparent when viewed together, as illustrated in part C. The value for the average number of varicosities per sensory neuron has been taken from total reconstructions of simple horseradish peroxidase-injected sensory neurons in untrained animals (N=2, mean±SE). *Source.* Based on data reported in Bailey and Chen 1983.

as functional changes (Figure 4–11). The normal set of varicosities serves as a mere scaffolding for behavior. Learning experiences, such as the acquisition of chronic anxiety, can build upon this scaffolding by altering the functional expression of neural connections.

Chronic Anxiety Might Involve Alterations in Gene Expression

How is this structural change achieved? We do not yet know the answer to this question. But recent progress in the molecular genetics of animal cells suggests a possible mechanism. Each somatic cell in the body contains all the genes present in every other cell. What makes a liver cell a liver cell and a brain cell a brain cell is that during development from a single fertilized egg cell, the various cells of the body differentiate by shutting off the activity of certain genes while allowing others to be expressed. This developmentally determined repression and activation of organ- and tissue-specific genes occurs during certain critical periods in development and is then self-maintained throughout the life of the differentiated cell (for review, see Davidson 1976; Gurdon 1974; Lewin 1980; Stryer 1981; J. D. Watson 1976). As a result, in any given cell most genes are closed; only some are open and available for transcription.

    In addition to these relatively permanent changes in gene expression

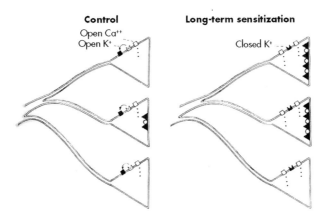

FIGURE 4–11.   Ionic and morphological mechanisms of long-term sensitization in *Aplysia*.

The ionic mechanism leads to the closing of a species of potassium channel, which results in an increase in calcium influx due to a broadening of the action potential. This contrasts with the situation in control, in which potassium flowing in through open channels leads more quickly to a closure of the calcium channels (filled squares). In addition, a morphological stabilization of the varicosities occurs during long-term sensitization, whereby there are increases in both the number of varicosities containing active zones (filled triangles) and the number of active zones per active varicosity. Consequently, more synaptic vesicles can be released.

produced by differentiation, the genes that are available for expression within a given cell type can be regulated. For example, a gene's rate of activity (its rate of transcribing mRNA) can be transiently enhanced or depressed (Brown 1981; Darnell 1982) by a variety of molecules, such as hormones that act directly on the genes or on proteins that regulate the genes. In contrast to differentiation, these forms of gene modulation can be either rapid and readily reversible or self-maintained and enduring. Because learning produces enduring changes in the structure and function of synapses, Schwartz and I have proposed that learning is likely to involve enduring, self-maintained alterations in gene expression (Kandel and Schwartz 1982). This idea is consistent with the suggestion that new protein synthesis is required for long-term memory (Agranoff et al. 1978; Barondes 1970). If this speculation proves correct, it would provide a new perspective on the nature of normal learning and thereby on the nature of certain learned neurotic illnesses such as chronic anxiety. Specifically, the possibility of gene regulation by experience suggests a class of molecular regulatory defects that might be caused by learning.

To put this view into perspective, let me illustrate one way of looking at the relationship between psychotic and neurotic illness. There is substantial evidence that the major psychotic illnesses, such as schizophrenia and depression, are heritable. The illnesses presumably represent mutations—alterations in the nucleotide sequence of the DNA—leading to abnormal messenger RNA and abnormal protein. The hereditary information of a cell is carried in its nucleic acid, DNA. The strands of DNA contain one of four characteristic bases: adenine, guanine, thymine, and cytosine. The information carried by a gene is defined by the sequence of bases along the strand. Consecutive triplets of bases serve as code words called *codons*; with some exceptions, each codon specifies an amino acid, and a string of 100 or more codons provides the genetic code for the assembly of a protein chain. The sequence of amino acids in a protein chain determines how the chain folds and therefore how it assumes the three-dimensional structure necessary for its biological activity. Alteration by mutation in only one nucleotide subunit—one base—of one codon will be sufficient to alter the amino acid sequence of the protein and thereby alter the protein properties, possibly even making it inactive.

The information of DNA is not translated directly into a protein. Rather, the sequence of bases that codes for a protein is transcribed into a complementary strand of RNA called messenger RNA (mRNA) because it carries the information for the sequence of amino acids necessary to construct the protein. The mRNA in turn is translated into protein. Thus, altered genes give rise to altered mRNAs, which produce altered proteins.

How the genetic abnormalities of schizophrenia and depression are manifest in the brain is still not known, but it is thought that they lead to changes in synaptic function either by altering the release process of the biogenic amine that serves as a transmitter in the presynaptic neuron or by affecting the expression or onset of receptors on the postsynaptic cell (for a review, see Kety 1979; Sachar 1981a, 1981b). I would now suggest that whereas the major psychotic illnesses (that do not respond to psychotherapy because the disease is not fundamentally acquired or altered by learning) may involve *alteration in the structure of specific genes*, certain neurotic illnesses such as chronic anxiety (that are acquired by learning and that can respond to psychotherapy) might involve *alterations in the regulation of gene expression*. According to this speculative view, schizophrenia and depression would be due primarily to heritable genetic changes in synaptic function in a substrain or mutant population; neurotic illnesses would not be. Rather, neurotic illnesses might represent alterations in synaptic function produced by environmentally induced modulation of gene expression (Figure 4–12). Even though the learning mechanisms are inherited, neurotic individuals would be neurotic only if experience taught their genes to be pathologically expressed. A corollary to this argument is that insofar as psychotherapy works

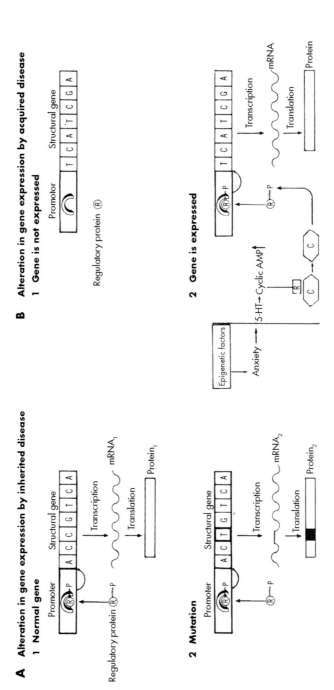

FIGURE 4–12. Comparison of mutation of the DNA sequence by disease, leading to the expression of an altered gene, and modulation of gene expression by environmental stimuli, leading to the transcription of a previously inactive gene *(opposite page)*.

For simplicity, a specific example is illustrated. The gene is illustrated as having two segments, a *structural gene* (that is transcribed by a mRNA and in turn is then translated into a specific protein) and a *regulatory or promoter* segment. The promoter is located upstream from the structural gene and regulates the initiation of the transcription of the structural gene. In this example, the promoter segment can be activated (and made accessible to transcription) only when a regulatory protein binds to the promoter. To bind, the regulatory protein must first be phosphorylated. Thus, in part A, the phosphorylated regulatory protein binds to the promoter, thereby activating the transcription of the structural gene leading to the production of a gene product: protein 1. In part A(2), a mutant form of the structural gene is illustrated in which a single base change has occurred; a thymidine (T) has been substituted for cytosine (C). As a result, an altered mRNA is transcribed and an abnormal protein (protein 2) is produced, giving rise to the disease state. This alteration in gene structure is present in the germ line and is inherited. Part B illustrates a specific example of alteration in expression of a normal structural gene that is not heritable. The regulatory protein is indicated in its dephosphorylated state; it therefore cannot bind to the promoter site and gene translation cannot be initiated. An (epigenetic) learning experience, such as learned anxiety, acting through serotonin and cyclic AMP, activates a protein kinase enzyme. This enzyme has both a regulatory unit component (R) and a catalytic unit component (C). The increase in cyclic AMP removes the regulatory unit, thereby activating the catalytic unit. The catalytic unit phosphorylates the regulatory protein, which can now bind to the promoter and consequently initiate gene transcription.

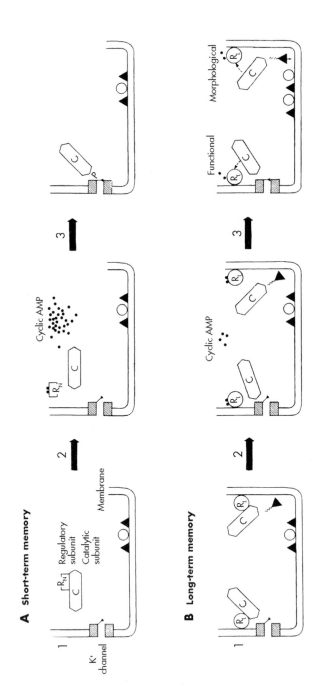

FIGURE 4–13. A model for the biochemical basis of long-term memory *(opposite page).*

In short-term sensitization (part A1), the cyclic AMP–dependent protein kinase is proposed to have a normal regulatory subunit (R_N) and no particular orientation with respect to a substrate membrane protein associated with the K^+ channel. In naive terminals, relatively high concentrations of cyclic AMP are needed to activate the catalytic subunit (C) (part A2) to phosphorylate the membrane protein (part A3). This phosphorylation brings about enhanced release of transmitter; the neurophysiological event underlying sensitization. The memory is brief because the concentration of cyclic AMP diminishes soon after stimulation with serotonin. In trained neurons (part B1), a new class of regulatory subunit (R_L) has been induced. As a result, the protein kinase differs from the naive enzyme in being site specific and thus being advantageously oriented both to the channel and to the mechanism that governs the organization of dense projections at the active zone (filled triangles), where synaptic vesicles line up to release transmitter. In addition, this new kinase has higher affinity for cyclic AMP. Consequently, lower concentrations of cyclic AMP are required to phosphorylate these target proteins (part B2). In part B3, functionally, as in part A3, the K^+ channel is inhibited as long as the channel protein remains phosphorylated. Morphologically, protein phosphorylation leads to the stable enlargement of the synapse. In this form of sensitization, the memory persists for longer periods of time because it is embodied in R_L, a protein molecule.

Source. Reprinted from Kandel ER, Schwartz JH: "Molecular Biology of an Elementary Form of Learning: Modulation of Transmitter Release by Cyclic AMP." *Science* 218:433–443, 1982. Used with permission from the American Association for the Advancement of Science.

and produces long-term learned changes in behavior, it may do so by producing alterations in gene expression. Needless to say, psychotic illness, in addition to partaking obligatorily of alterations in gene structure, may also involve a secondary disturbance in gene expression.

A Molecular Genetic Model for the Maintenance of Anxiety

How might one envision the alteration of gene expression in learning? The model in Figure 4–9 accounts only for the immediate (short-term) effects of sensitization, that is, for the acquisition of anxiety. It is attractive to think, however, that this model might be more general and might account for the long-term maintenance of anxiety, including the structural changes. Indeed, Schwartz and I recently extended this model to account for the long-term maintenance of anxiety by positing a specific kind of alteration in gene expression (Kandel and Schwartz 1982). According to this theory, a gene is induced to produce a new protein kinase that ensures prolonged phosphorylation of the potassium channel (Figure 4–13).

The cyclic AMP–dependent kinase is a protein consisting of two classes of subunits, a catalytic subunit and a regulatory subunit. The free catalytic subunit carries out the phosphorylation. The regulatory subunit binds to the catalytic subunit and prevents it from acting. The function of cyclic AMP is to cause the regulatory unit to dissociate from the catalytic unit and free it for action.

Serotonin, acting repeatedly on the terminals of the sensory neuron (as a result of repeated aversive stimulation), might activate a gene able to produce a novel class of regulatory subunit for the protein kinase. The specific inducer that activates the gene for the regulatory subunit might be cyclic AMP. The prolonged elevation of cyclic AMP that occurs in short-term sensitization may allow cyclic AMP to enter the nucleus of the cell and there to cause the gene for a new regulatory subunit to be expressed by cyclic AMP–dependent phosphorylation (of perhaps one or more proteins). Activation of a gene for a new class of subunit could be permanent or it could slowly decay if not reinforced by subsequent aversive training. Schwartz and I posited that this regulatory subunit would have two novel features: it would have greater sensitivity to cyclic AMP, thereby allowing the catalytic subunit to dissociate more readily; and it would be site specific, allowing the kinase to be bound to the presynaptic membrane near the potassium channels that are to be modulated (Kandel and Schwartz 1982) (Figure 4–13B). The synthesis of a regulatory subunit with greater affinity for cyclic AMP would allow the cyclic AMP–dependent protein kinase to work at relatively normal concentrations of the cyclic nucleotide. Slight elevations above the normal concentrations of cyclic AMP of the sort that accompany the arousing stimuli of everyday life are inadequate to evoke sensitization in the untrained terminal. However, with the new regulatory subunit facilitating the work of the protein kinase, such slight elevations in cyclic AMP would now be sufficient to provide, by modification of ion channels, the enhanced influx of calcium required to increase transmitter release and thus sustain the learned anxiety reaction (Figure 4–13B, 2 and 3).

A subunit that would position the protein kinase optimally could allow it to trigger a family of parallel cyclic AMP–dependent changes in the sensory neuron. For example, it could 1) produce the functional change in the potassium-channel protein and 2) alter the assembly of the protein components that constitute the active zone (Figure 4–13B) and thereby initiate the striking change in morphology of sensory terminals observed by Bailey and Chen (1983). These two molecular changes, both caused by the same cyclic AMP–dependent kinase, could operate together to bring about enhanced transmitter release from the long-term sensitized neuron.

Although obviously premature because of lack of experimental support, this speculative explanation for the maintenance of anxiety can be tested. The large nerve cells of *Aplysia* offer special experimental advantages for molecular genetic studies of the nervous system, since nuclei of individual cells can be isolated by hand dissection (Lasek and Dower 1971; Schwartz et al. 1971). In addition, recombinant DNA techniques have recently been successfully applied to *Aplysia*, allowing genes of known function to be isolated (Scheller et al. 1982). But the primary reason I have engaged in this speculative digression is to illustrate that the molecular regulatory processes are likely to prove important for understanding long-term modification in behavior produced by natural experience and by psychotherapeutic intervention.

Anticipatory Anxiety Shares Molecular Components With Chronic Anxiety

We do not yet know in cellular detail the mechanisms underlying the aversive conditioning that is used as a model for anticipatory anxiety in *Aplysia*, but there is already good evidence that the cellular mechanisms of aversive conditioning are related to those of long-term sensitization (Duerr and Quinn 1982; Hawkins et al. 1983; Walters and Byrne 1983). Analyses of classical conditioning of simple reflexes in *Aplysia* suggest that the learning of signaled anxiety involves a modified form of the same cellular and molecular mechanisms—those of presynaptic facilitation—that underlie chronic anxiety. The mechanism for associative specificity is an augmented form of presynaptic facilitation called activity-dependent enhancement of presynaptic facilitation. The invasion of the sensory neuron terminals by action potentials resulting from activation of the CS pathway makes these terminals more responsive to the effects of serotonin released by the facilitating neurons in the US pathway. Thus, classical conditioning uses an amplification of the molecular machinery used by sensitization, suggesting that there may be a molecular alphabet to learning, whereby complex forms of learning use components found in simple forms. In signaled anxiety, these presynaptic facilitating mechanisms appear to be used for two components of the learning at two points in the neural circuit: an associative component to provide for the temporal specificity of the modulation and a modulatory component to enhance defensive reflexes.

The first component, enhancement, is a modulatory component identical to the presynaptic facilitation that accounts for sensitization. As is the case with sensitization, the *modulatory component* enhances defensive reflex responses (such as gill withdrawal) through the serotonergic defensive arousal cells (Figure 4–14).

The second component, called the *associative component,* consists of a modified and augmented form of the same mechanism and gives classical conditioning its temporal specificity. Using a simple reflex pathway that can be associatively conditioned, Hawkins and associates (1983) and Walters and Byrne (1983) found that after a series of pairing trials in which action potentials in the sensory neuron of a CS pathway immediately precede activity in the US pathway the sensory neuron releases more transmitter than when action potentials in the sensory neuron are not paired with the US. This evidence suggests that classical conditioning is essentially an amplified form of the mechanism of presynaptic facilitation. It produces a more profound depression of the potassium channels and a larger increase in the duration of the action potential than does conventional presynaptic facilitation.

According to this model for anticipatory anxiety, when head shock is paired with shrimp, serotonergic cells in the head ganglia produce a highly robust presynaptic facilitation of the sensory neurons that respond to shrimp. In this augmented form of facilitation, the ability of the serotonergic neurons to produce presynaptic facilitation is substantially enhanced because an action potential in the neurons of the CS pathway that responds to shrimp immediately precedes the action of the serotonergic cell (which is activated by the aversive US). Activity dependence of classical conditioning explains why the CS must precede the US during pairing for anticipatory anxiety to be acquired.

A Molecular Model for Anticipatory Anxiety

How does the action potential in the neurons of the CS pathway lead to the enhanced presynaptic facilitation that underlies the association of the anxiety

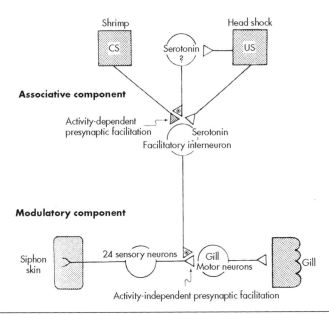

FIGURE 4-14. Relationship of classical conditioning to sensitization in *Aplysia*.

According to the model this form of learning involves two components: an associative component that accounts for the timing or temporal specificity of the modulation and a modulatory component that enhances or reduces a number of behavioral responses. It is attractive to think that presynaptic facilitation, the mechanism underlying sensitization, could be used in each component, although in different ways, conventional and amplified, to modulate the strength of a particular connection. The key point of the model is that the two components represent distinct neuronal processes: 1) Activity-dependent presynaptic facilitation could yield the temporal discrimination necessary to achieve associative specificity, and 2) conventional presynaptic facilitation could provide the modulatory component responsible for enhancing the strength of the response.

state with a specific environmental signal? Clearly, one or more aspects of the action potential produce the amplification. Four events occur during the action potential: depolarization, sodium influx, calcium influx, and potassium efflux. One attractive possibility is that calcium influx is the signal, since calcium affects the activity of cyclic AMP in a number of ways (Cheung 1980). The activity-dependent enhancement produced by calcium could be achieved by modulating one or more steps in the cyclic AMP cascade. For example, calcium might modulate the activation of the adenylate cyclase in the sensory cells by serotonin. According to this idea, the action of the serotonergic facilitating neuron would be more effective in classical conditioning than in sensitization because activation of the cyclase by serotonin is preceded by an influx of calcium within the sensory neuron. Calcium, perhaps in association with calmodulin, could bind to a site on the adenylate cyclase, producing a conformational change that leads to its greater activation by serotonin.

Thus, a basic presynaptic regulatory mechanism involving enhancement of transmitter release by cyclic AMP–dependent phosphorylation could be used in different ways (conventional and amplified) to achieve chronic (unsignaled) anxiety or anticipatory (signaled) anxiety. This model of classical conditioning further suggests that we are on the threshold of understanding our ability to learn predictive relationships. I would argue that the ability to learn predictive relationships—an ability critical to our mental life—lies in the precise temporal requirement of a signal produced by spike activity (perhaps calcium) for the adenylate cyclase system of certain neurons.

An Overall View

The models that I have considered emphasize the cellular interrelationship of chronic and anticipatory anxiety in *Aplysia*. Both forms of anxiety involve the strengthening of connections by modulating synaptic transmission. Both lead to enhancement of transmitter release by depressing a potassium channel and thereby increasing influx of calcium. Thus, these studies suggest that there is a basic molecular grammar underlying the various forms of anxiety, a set of mechanistic building blocks that can be used in different combinations and permutations. They further suggest that a variety of mental processes that appear phenotypically unrelated may share a fundamental unity on the cellular and molecular levels.

I have in this discussion purposely gone beyond the facts in pointing arguments based on animal studies toward human behavior because I have wanted to emphasize two conceptual points that I believe will be fundamental for the future study of the cellular mechanisms of anxiety. The first is the power of experience in modifying brain function through altering synaptic strength and regulating gene expression. The second is the utility and promise of animal models for the study of anxiety. Unlike schizophrenia, which does not exaggerate a normal adaptive process and is therefore a characteristically human mental illness, fear or anxiety is a general adaptive mechanism found in simple as well as complex animals. There is good reason to believe that some of the cellular mechanisms of anxiety may also be general.

Moreover, I have suggested that normal learning, the learning of anxiety and unlearning it through psychotherapeutic intervention, might involve long-term functional and structural changes in the brain that result from alterations in gene expression. Thus, we can look forward, in the next decade of research into learning, to a merger between aspects of molecular genetics and cellular neurobiology. This merger, in turn, will have important consequences for psychiatry—for psychotherapy on the one hand and for psychopharmacology on the other.

References

Agranoff BW, Burrell HR, Dokas LA, et al: Progress in biochemical approaches to learning and memory, in Psychopharmacology: A Generation of Progress. Edited by Lipton MA, DiMascio A, Killam KF. New York, Raven, 1978

Badia P, Culbertson S: Behavioral effects of signaled versus unsignaled shock during escape training in the rat. J Comp Physiol Psychol 72:216–222, 1970

Badia P, Suter S, Lewis P: Preference for warned shock: information and/or preparation. Psychol Rep 20:271–274, 1967

Bailey CH, Chen MC: Morphological basis of long-term habituation and sensitization in Aplysia. Science 220:91–93, 1983

Bailey CH, Hawkins RD, Chen MC, et al: Interneurons involved in mediation and modulation of the gill-withdrawal reflex in Aplysia, IV: morphological basis of presynaptic facilitation. J Neurophysiol 45:340–360, 1981

Barondes SH: Multiple steps in the biology of memory, in The Neurosciences: Second Study Program. Edited by Schmitt FO, Quarton GC, Melnechuk T, et al. New York, Rockefeller University Press, 1970

Bernier L, Castellucci VF, Kandel ER, et al: Facilitatory transmitter causes a selective and prolonged increase in cAMP in sensory neurons mediating the gill- and siphon-withdrawal reflex in Aplysia. J Neurosci 2:1682–1691, 1982

Bindra D: How adaptive behavior is produced: a perceptual motivational alternative to response reinforcement. Behav and Brain Sci 1:41–91, 1978

Bolles RC, Fanselow MS: A perceptual, defensive, recuperative model of fear and pain. Behavioral and Brain Sciences 3:291–323, 1980

Bowlby J: Attachment and Loss, Vol 1: Attachment. London, Hogarth Press, 1969

Brenner C: An Elementary Textbook of Psychoanalysis, Revised Edition. New York, International Universities Press, 1973

Brown DD: Gene expression in eukaryotes. Science 211:667–674, 1981

Brunelli M, Castellucci VF, Kandel ER: Synaptic facilitation and behavioral sensitization in Aplysia: possible role of serotonin and cyclic AMP. Science 194:1178–1181, 1976

Castellucci VF, Kandel ER: Presynaptic facilitation as a mechanism for behavioral sensitization in Aplysia. Science 194:1176–1178, 1976

Castellucci VF, Pinsker H, Kupfermann I, et al: Neuronal mechanisms of habituation and sensitization of the gill-withdrawal reflex in Aplysia. Science 167:1745–1748, 1970

Castellucci VF, Kandel ER, Schwartz JH, et al: Intracellular injection of the catalytic subunit of cyclic AMP–dependent protein kinase simulates facilitation of transmitter release underlying behavioral sensitization in Aplysia. Proc Natl Acad Sci USA 77:7492–7496, 1980

Castellucci VF, Nairn A, Greengard P, et al: Inhibitor of adenosine 3':5'-monophosphate–dependent protein kinase blocks presynaptic facilitation in Aplysia. J Neurosci 12:1673–1681, 1982

Cheung WY: Calmodulin plays a pivotal role in cellular regulation. Science 207:19–27, 1980

Chomsky N: A review of BF Skinner's Verbal Behavior. Language 35:26–58, 1959

Cohen ME, Badal DW, Kilpatrick A, et al: The high familial prevalence of neurocirculatory asthenia (anxiety neurosis, effort syndrome). Am J Hum Genet 3:126–158, 1951

Crick F: Split genes and RNA splicing. Science 204:264–271, 1979

Crowe RR, Pauls DL, Slymen DJ, et al: A family study of anxiety neurosis. Arch Gen Psychiatry 37:77–79, 1980

Darnell JE Jr: Transcription units for mRNA production in eukaryotic cells and their DNA viruses. Prog Nucleic Acid Res Mol Biol 22:327–353, 1979

Darnell JE Jr: Variety in the level of gene control in eukaryotic cells. Nature 297:365–371, 1982

Darwin C: The Expression of the Emotions in Man and Animals. New York, D Appleton, 1873

Davidson EH: Gene Activity in Early Development, 2nd Edition. New York, Academic Press, 1976

Dickinson A: Appetitive-aversive interactions: superconditioning of fear by an appetitive CS. Q J Exp Psychol 29:71–83, 1977

Dickinson A: Contemporary Animal Learning Theory. Cambridge, England, Cambridge University Press, 1980

Dickinson A: Conditioning and associative learning. Br Med Bull 37:165–168, 1981

Dickinson A, Mackintosh NJ: Classical conditioning in animals. Annu Rev Psychol 29:587–612, 1978

Dollard J, Miller NE: Personality and Psychotherapy. New York, McGraw-Hill, 1950

Duerr JS, Quinn WG: Three Drosophila mutations that block associative learning also affect habituation and sensitization. Proc Natl Acad Sci USA 79:3646–3650, 1982

Estes WK, Skinner BF: Some quantitative properties of anxiety. J Exp Psychol 29:390–400, 1941

Freud S: Inhibitions, symptoms and anxiety (1925–1926), in Complete Psychological Works, Standard Edition, Vol 20. London, Hogarth Press, 1959, pp 75–175

Gilbert W: Why genes in pieces? Nature 271:501, 1978

Goodwin DW, Guze SB: Psychiatric Diagnosis, 2nd Edition. Oxford, Oxford University Press, 1979

Gormezano I, Kehoe EJ: Classical conditioning: some methodological conceptual issues, in Handbook of Learning and Cognitive Processes: Conditioning and Behavior Theory, Vol 2. Edited by Estes WK. Hillsdale, NJ, Erlbaum, 1975

Griffin DR (ed): Animal Mind-Human Mind: Report on the Dahlem Workshop. New York, Springer-Verlag, 1982

Gurdon JB: The Control of Gene Expression in Animal Development. Cambridge, MA, Harvard University Press, 1974

Guthrie ER: The Psychology of Learning. New York, Harper & Brothers, 1935

Guthrie ER: Association by contiguity, in Psychology: A Study of a Science, Vol 2. Edited by Koch S. New York, McGraw-Hill, 1959

Hammond LJ: Conditioned emotional states, in Physiological Correlates of Emotion. Edited by Black P. New York, Academic Press, 1970

Hawkins RD: Interneurons involved in mediation and modulation of gill-withdrawal reflex in Aplysia, III: identified facilitating neurons increase Ca^{2+} current in sensory neurons. J Neurophysiol 45:327–339, 1981

Hawkins RD, Castellucci VF, Kandel ER: Interneurons involved in mediation and modulation of gill-withdrawal reflex in Aplysia, I: identification and characterization. J Neurophysiol 45:304–314, 1981a

Hawkins RD, Castellucci VF, Kandel ER: Interneurons involved in mediation and modulation of gill-withdrawal reflex in Aplysia, II: identified neurons produce heterosynaptic facilitation contributing to behavioral sensitization. J Neurophysiol 45:315–326, 1981b

Hawkins RD, Abrams TW, Carew TJ, et al: A cellular mechanism of classical conditioning in Aplysia: activity–dependent amplification of presynaptic facilitation. Science 219:400–405, 1983

James W: Psychology: Briefer Course. New York, Holt, 1892

James W: The Principles of Psychology, Vols 1, 2. New York, Holt, 1893

Kamin LJ: Predictability, surprise, attention, and conditioning, in Punishment and Aversive Behavior. Edited by Campbell BA, Church RM. New York, Appleton-Century-Crofts, 1969

Kandel ER: Perspectives in the neurophysiological study of behavior and its abnormalities, in New Psychiatric Frontiers: American Handbook of Psychiatry, 2nd Edition, Vol 6. Edited by Hamburg DA, Brodie HKH. New York, Basic Books, 1975

Kandel ER: Cellular Basis of Behavior: An Introduction to Behavioral Neurobiology. San Francisco, CA, WH Freeman, 1976

Kandel ER, Schwartz JH: Molecular biology of an elementary form of learning: modulation of transmitter release by cyclic AMP. Science 218:433–443, 1982

Kety SS: Disorders of the human brain. Sci Am 241:202–214, 1979

Kimmel HD, Burns RA: The difference between conditioned tonic anxiety and conditioned phasic fear: implications for behavior therapy, in Stress and Anxiety, Vol 4. Edited by Spielberger CD, Sarason IG. Washington, DC, Hemisphere, 1977

Klatzky RL: Human Memory: Structures and Processes, 2nd Edition. San Francisco, CA, WH Freeman, 1980

Klein DF: Delineation of two drug-responsive anxiety syndromes. Psychopharmacologia 5:397–408, 1964

Klein DF: Anxiety reconceptualized, in Anxiety: New Research and Changing Concepts. Edited by Klein DF, Rabkin JG. New York, Raven, 1981

Klein DF, Fink M: Psychiatric reaction patterns to imipramine. Am J Psychiatry 119:432–438, 1962

Klein M, Kandel ER: Mechanism of calcium current modulation underlying presynaptic facilitation and behavioral sensitization in Aplysia. Proc Natl Acad Sci USA 77:6912–6916, 1980

Lasek RJ, Dower WJ: Aplysia californica: analysis of nuclear DNA in individual nuclei of giant neurons. Science 172:278–280,1971

Lewin B: Gene Expression: Eucaryotic Chromosomes, 2nd Edition, Vol 2. New York, Wiley, 1980

Mackintosh NJ: The Psychology of Animal Learning. London, Academic Press, 1974

Mayer-Gross W: Clinical Psychiatry, 3rd Edition. Edited by Slater E, Roth M. Baltimore, MD, Williams & Wilkins, 1969

Miller NE: Studies of fear as an acquirable drive: fear as motivation and fear-reduction as reinforcement in the learning of new responses. J Exp Psychol 38:89–101, 1948

Mowrer OH: A stimulus-response analysis of anxiety and its role as a reinforcing agent. Psychol Rev 46:553–565, 1939

Neisser U: Cognitive Psychology. Englewood Cliffs, NJ, Prentice-Hall, 1967

Nemiah JC: Anxiety neurosis, in Comprehensive Textbook of Psychiatry, 2nd Edition, Vol 1. Edited by Freedman AM, Kaplan HI, Sadock BJ. Baltimore, MD, Williams & Wilkins, 1975

Pauls DL, Bucher KD, Crowe RR, et al: A genetic study of panic disorder pedigrees. Am J Hum Genet 32:639–644, 1980

Pavlov IP: Conditioned Reflexes: An Investigation of the Physiological Activity of the Cerebral Cortex. Translated and edited by Anrep GV. London, Oxford University Press, 1927

Pinsker HM, Hening WA, Carew TJ, et al: Long-term sensitization of a defensive withdrawal reflex in Aplysia. Science 182:1039–1042, 1973

Posner MI: Cognition: An Introduction. Chicago, Scott, Foresman, 1973

Prokasy WF (ed): Classical Conditioning: A Symposium. New York, Appleton-Century-Crofts, 1965

Rescorla RA: Probability of shock in the presence and absence of CS in fear conditioning. J Comp Physiol Psychol 66:1–5, 1968

Rescorla RA: Second order conditioning: implications for theories of learning, in Contemporary Approaches to Conditioning and Learning. Edited by McGuigan FJ, Lumsden DB. Washington, DC, VH Winston, 1973

Rescorla RA: Some implications of a cognitive perspective on Pavlovian conditioning, in Cognitive Processes in Animal Behavior. Edited by Hulse SH, Fowler H, Honig WK. Hillsdale, NJ, Erlbaum, 1978

Rescorla RA: Conditioned inhibition and extinction in mechanisms of learning and motivation, in Mechanisms of Learning and Motivation: A Memorial Volume of Jerzy Konorski. Edited by Dickinson A, Boakes RA. Hillsdale, NJ, Erlbaum, 1979

Rescorla RA, Wagner AR: A theory of Pavlovian conditioning: variations in the effectiveness of reinforcement and nonreinforcement, in Classical Conditioning II: Current Research and Theory. Edited by Black AH, Prokasy WF. New York, Appleton-Century-Crofts, 1972

Romanes GJ: Animal Intelligence. New York, D Appleton, 1883

Romanes GJ: Mental Evolution in Man: Origin of Human Faculty. London, Paul, 1888

Sachar EJ: Psychobiology of affective disorders, in Principles of Neural Science. Edited by Kandel ER, Schwartz JH. New York, Elsevier-North Holland, 1981a

Sachar EJ: Psychobiology of schizophrenia, in Principles of Neural Science. Edited by Kandel ER, Schwartz JH. New York, Elsevier-North Holland, 1981b

Sahley CL, Rudy JW, Gelperin A: An analysis of associative learning in a terrestrial mollusc, I: higher-order conditioning, blocking and a transient US-pre-exposure effect. Comp Physiol [A] 144:1–8, 1981

Sargant W, Slater E: An Introduction to Physical Methods of Treatment in Psychiatry, 4th Edition. Edinburgh, Livingstone, 1963

Scheller RH, Jackson JF, McAllister LB, et al: A family of genes that codes for ELH, a neuropeptide eliciting a stereotyped pattern of behavior in Aplysia. Cell 28:707–719, 1982

Schwartz JH, Castellucci VF, Kandel ER: Functioning of identified neurons and synapses in abdominal ganglion of Aplysia in absence of protein synthesis. J Neurophysiol 34:939–953, 1971

Seligman MEP: Helplessness: On Depression, Development, and Death. San Francisco, CA, WH Freeman and Co, 1975

Seligman MEP, Meyer B: Chronic fear and ulcers in rats as a function of the unpredictability of safety. J Comp Physiol Psychol 73:202–207, 1970

Sheehan DV: Current concepts in psychiatry: panic attacks and phobias. N Engl J Med 307:156–158, 1982

Siegelbaum SA, Camardo JS, Kandel ER: Serotonin and cyclic AMP close single K^+ channels in Aplysia sensory neurones. Nature 299:413–417, 1982

Skinner BF: Verbal Behavior. New York, Appleton-Century-Crofts, 1957

Slater E, Shields J: Genetical aspects of anxiety. Br J Psychiatry 3:62–71, 1969

Staddon JER, Simelhag VL: The "superstition" experiment: a reexamination of its implications for the principles of adaptive behavior. Psychol Rev 78:16–43, 1971

Stryer L: Biochemistry, 2nd Edition. San Francisco, CA, WH Freeman, 1981

Testa TJ: Causal relationships and the acquisition of avoidance responses. Psychol Rev 81:491–505, 1974

Tolman EC: Purposive Behavior in Animals and Men. New York, Appleton-Century-Crofts, 1932

Wagner AR: Priming in STM: an information-processing mechanism for self-generated or retrieval-generated depression in performance, in Habituation: Perspectives From Child Development, Animal Behavior, and Neurophysiology. Edited by Tighe TJ, Leaton RN. Hillsdale, NJ, Erlbaum, 1976

Walters ET, Byrne JH: Associative conditioning of single sensory neurons suggests a cellular mechanism for learning. Science 219:405–408, 1983

Walters ET, Carew TJ, Kandel ER: Classical conditioning in Aplysia californica. Proc Natl Acad Sci USA 76:6675–6679, 1979

Walters ET, Carew TJ, Kandel ER: Associative learning in Aplysia: evidence for conditioned fear in an invertebrate. Science 211:504–506, 1981

Watson JB: Psychology as the behaviorist views it. Psychol Rev 20:158–177, 1913

Watson JB: Behaviorism. New York, WW Norton, 1925

Watson JB, Rayner R: Conditioned emotional reaction. J Exp Psychol 3:1–14, 1920

Watson JD: Molecular Biology of the Gene, 3rd Edition. Menlo Park, CA, WA Benjamin, 1976

Weiss JM: Somatic effects of predictable and unpredictable shock. Psychosom Med 32:397–408, 1970

Weisskopf VF: Bicentennial address: frontiers and limits of physical sciences. American Academy of Arts and Sciences Bulletin 35:4–23, 1981

"NEUROBIOLOGY AND MOLECULAR BIOLOGY"

Eric J. Nestler, M.D., Ph.D.

Reading this incisive and penetrating essay by Eric Kandel for the first time in 20 years offered a fascinating glimpse into the world of neuroscience of the early 1980s and underscored for me the tremendous advances that have been made in the neurosciences over the last two decades. When I first read the article in 1983, I had just completed my Ph.D. research in Paul Greengard's laboratory at Yale University and was headed off for residency training in psychiatry. I thought a lot about setting up my own laboratory and about which experimental methods were most ripe for new approaches to psychiatric neuroscience.

In his essay "Neurobiology and Molecular Biology: The Second Encounter," Kandel weighed in on a key debate at the time: the role of molecular biology in the neurosciences. Many leading investigators in the neurosciences, whose research focused on the detailed anatomical connections in the central nervous system, on the ionic basis of nerve conductance or on nervous system development, did not envision the value of molecular approaches to the nervous system. Kandel had first described a wave of molecular approaches to neuroscience in the 1960s, which largely involved prominent molecular biologists from other disciplines moving to investigations of the nervous system. He astutely noted that this early period was overly optimistic, in that those involved predicted rapid, transforming advances akin to advances provided by molecular biology in other disciplines. Although such transforming advances did not materialize, this period was important in providing fundamentally new models for neuroscience, such as the use of non-mammalian organisms (*C. elegans, Drosophila*) to study nervous system development and function.

The second encounter with molecular biology, the subject of Kandel's 1983 essay, represented a much more systematic application of molecular methods to neuroscience. At the time of the essay, such studies were largely dominated by molecular cloning techniques and the production of monoclonal antibodies. For the first time, proteins that had been discovered and characterized based solely on some functional activity (e.g., ion channel conductance, neurotransmitter receptor binding) were being cloned. This age also witnessed the first identification of families of novel regulatory proteins that drive the formation and differentiation of neural cells during development. Kandel predicted the degree to which this wave of molecular biology would transform neuroscience and that it would not primarily be by conceptual leaps forward but by providing uniquely powerful tools that would enable neuroscientists to probe their systems at an increasingly penetrating molecular level.

Kandel's essay is impressively prescient in its predictions, and I have to admit that unlike Kandel, I did not fully appreciate the magnitude of these contributions back in 1983, while I was in the thick of experiments at the bench. Kandel foretold, for example, the widespread use of mutational analysis of simple organisms and homology screening of molecular libraries to identify new families of genes involved in nervous system function and development. As another example, he emphasized the importance of using molecular tools to characterize changes in gene expression during development and in the adult animal to understand how the nervous system adapts and changes over time.

Indeed, in rereading Kandel's essay, it is very impressive to see just how far the field has come in 20 years. In the early 1980s, only one ion channel (the nicotinic acetylcholine receptor from skeletal muscle) was cloned and its subunit structure delineated. Today, hundreds of ion channels have been cloned, some have even been crystallized, and detailed information is available concerning the molecular mechanisms governing channel gating. Mutations in many of these channels have been found to be the cause of a range of neurological disorders. In the early 1980s, neurotransmitter release was understood at a descriptive level: Ca^{2+} influx during the nerve impulse triggers the translocation of transmitter-filled vesicles to the presynaptic membrane where the transmitter is released via exocytosis. Today, this process has been elucidated with an impressive degree of molecular detail, where Ca^{2+} binding to target proteins triggers cascades of protein-protein interactions that control vesicle trafficking and fusion. In the early 1980s, the notion that the phosphorylation of neural proteins regulates nerve conductance and synaptic transmission was still controversial. Today, protein phosphorylation is known as the dominant molecular mechanism by which all types of neural proteins are regulated. These are just some of the advances in neuroscience achieved over the past two decades that would not have been possible without the extraordinary tools of molecular biology.

Equally striking in Kandel's review is one major area of knowledge where our progress has been less dramatic: understanding precisely how neural circuits produce complex behavior. This goal is of particular importance to Kandel, myself, and our many colleagues in psychiatry as we strive to explain the neural basis of mental disorders. Clearly, some critical progress has been made; for example, through the explosive use of conventional and, more recently, inducible cell-targeted mouse mutants, viral vectors, antisense oligonucleotides, RNAi, and related tools, we have seen extraordinary advances in the ability to relate individual proteins within particular brain structures to complex behavior. Yet the precise circuit mechanisms by which these proteins, through altered functioning of individual nerve cells, give rise to most types of complex behavior remain almost as elusive as they were 20 years ago.

This cuts to the heart of a central theme in Kandel's elegant overview to this current volume. Are we simply waiting for still additional methodological advances to enable us to gain a neural understanding of complex behavior, or is such a reductionist approach inherently limited? I strongly agree with Kandel's notion that neuroscience will one day provide a mechanistic understanding of complex behavior under normal and pathological conditions. In taking stock of where we've come as a field since 1983, I remain as optimistic as ever that we will achieve this goal, and I look forward to reading about our field's progress in this and other remaining challenges two decades from now!

CHAPTER 5

NEUROBIOLOGY AND MOLECULAR BIOLOGY

The Second Encounter

Eric R. Kandel, M.D.

As this symposium illustrates, the recent application of molecular genetics to cellular neurobiology is generating a great deal of excitement. Although this excitement is in many ways unique, for many of us who have been working in neurobiology it is accompanied by a sense of déjà vu. The sense that we have been here before is accurate, since this present contact between neurobiology and molecular biology is in fact the second, not the first, encounter between the two disciplines. To put into perspective the recent impact of molecular genetics on neurobiology, I will divide this summary into two parts. I will begin with some comments about the first encounter—the historical origins of the relationship between molecular biology and neural science viewed from a personal and obviously limited perspective. These origins set the tradition that has culminated in this symposium. Second, I will

This article was originally published in the *Cold Spring Harbor Symposia on Quantitative Biology,* Volume 48, 1983, pp. 891–908.

use the issues raised by this symposium to highlight the major themes emerging in contemporary molecular neurobiology. Although in this summary I restrict citations mostly to the papers of this volume, I use these papers as a starting point for considering other issues, which for brevity I will describe without further citation.

The Return of Molecular Biology

The first encounter between neurobiology and molecular biology dates to the late 1960s. At that time, several distinguished molecular biologists believed that many of the interesting problems in their field were close to being solved; they turned to the brain as their next problem, as their descendants are now doing. During the preceding two decades, molecular biology had enjoyed an enormous increase in technical capabilities and explanatory power. This molecular approach to biological problems had several roots: the classical genetics of T. H. Morgan and his disciples in America; the examination of the structure of ordered biological polymers by X-ray crystallography that was introduced by Astbury and Bragg in England; and finally, the application of the thinking used in modern physics to problems of biology, especially characterized by the speculations of Schrödinger (*What Is Life?*) and the work of Max Delbrück and his associates. All of these intellectual precursors shared an experimental approach that depended on model building and therefore on a willingness to study preparations that best exemplified the phenomena of interest. This led to a search for conveniently simple systems that provided abundant material. Thus, geneticists interested in inheritance in higher organisms first studied *Drosophila* and *Escherichia coli*; crystallographers first analyzed keratin and hemoglobin; and molecular biologists interested in replication of DNA examined bacterial viruses. Although the impetus was to understand complex phenomena, study was governed by optimization of simple experimental systems and by the presumed universality of the phenomena chosen for study.

With this approach, the flow of genetic information from the nucleus to the protein-synthetic machinery of the cell was elegantly outlined between 1950 and 1965. Implicit in Watson and Crick's discovery of the double helical structure of DNA is the insight it provided into the nature of replication. This soon led to the discovery of mRNA, the deciphering of the genetic code, and an understanding of the mechanism of protein synthesis.

By 1965, we were well on the way to understanding the informational biochemistry of gene expression because of the development of the Jacob-Monod model of the operon. In this model, a structural gene that codes for a specific protein is regulated by a promoter element that contains a DNA sequence called the operator. The structural gene is normally blocked from being transcribed by a repressor protein that is bound to the operator sequence of the promoter element. But the gene can be switched on rapidly by a small signal molecule produced by cellular metabolism that binds to and removes the repressor protein. These small molecules ultimately determine the rate of transcription of the structural gene. The insight that gene function is not fixed but can be regulated by the environment through small molecules (such as inducers) provided a coherent intellectual framework for understanding much of bacterial physiology. In addition, this model suggested the first molecular explanation of cellular differentiation during eukaryotic embryogenesis. According to this view (now known to be slightly oversimplified), every cell in the body contains all the genes of the genome. Development, thus, would result from the appropriate switching on and off of particular patterns of genes in different cells.

To many workers, it then seemed that most of biology, including development, could be inferred, in principle if not in detail, from rules already at hand. The rules, the argument went, had been derived from viruses and bacterial cells, but the code was universal, and evolution conservative. Many could not help agreeing with Monod that an elephant is an *E. coli* writ large. As a result, these biologists felt that only one major frontier remained—the brain, and within it, development and the biology of mentation: cognition, perception, thought, and learning.

Although time has shown this view to be overly optimistic, neurobiology benefited from this optimism, for within a few years a number of talented molecular biologists migrated into neurobiology: Francis Crick, J.P. Changeux, Sidney Brenner, Seymour Benzer, Cyrus Levinthal, Gunther Stent, and Marshall Nirenberg, for example. Their enthusiasm immediately brought many younger people into the field (some of whom were at this symposium—Regius Kelly, Louis Reichardt, and Douglas Fambrough) who infused neurobiology with new perspectives and methods.

This first encounter was characterized by the same experimental approaches that had served molecular biology so well: model building, the selection of convenient experimental preparations endowed with abundant material for study, and, most novel for neurobiology, the use of mutational genetics. An outstanding example of a preparation rich in substances of neurobiological interest is the electric organ of *Torpedo* and eel used originally by David Nachmansohn (1959) to study the biochemical components of cholinergic transmission. This starting material has yielded detailed structural information about the nicotinic acetylcholine receptor (AChR), the enzymes responsible for the synthesis and degradation of acetylcholine (ACh), and the cholinergic vesicle. Various other preparations were introduced into neurobiology explicitly because they were useful for mutational analysis, including tumor cell lines, neuroblastoma and PC12 cells, and simple organisms such as *C. elegans* and *Drosophila,* whose short life cycles make them suitable for genetic analysis. Interest also focused on isogenic lines of fish and mice that promised to shed light on how genes determine the specificity of connections in the vertebrate brain.

After the initial excitement, however, these first émigrés encountered difficult footing in the new terrain. In 1965, good systems for carrying out mutational analyses of the nervous system did not exist. As a result, the early pioneers spent much effort and ingenuity developing systems that were new to neurobiology. A full decade and longer passed before the potential and promise of the first encounter were fulfilled—before the emphasis moved from developing systems to answering important questions about the systems. Although the methods of mutational analysis ultimately made an impact on neurobiology, these methods did not prove *immediately* applicable. As a result, the influence of the first migration was gradual rather than dramatic, leading to an evolution rather than a revolution in neurobiology. At times during those years, many workers may have felt that molecular neurobiology would never reach a lively generative phase—that rapid pace of progress that had made the rest of molecular biology so exciting. There was continued movement; the problems were becoming progressively better defined, more interesting, and more accessible; the standard of work within all of neurobiology was rising; but progress was slow.

As this symposium has illustrated, in the last 3 years we have benefited from a second encounter—a return of molecular biology. This renewal of interest has come with the development of a variety of powerful molecular techniques: recombinant DNA, DNA- and protein-sequencing methods, and monoclonal antibodies. Complementing these developments in molecular biology, patch-clamp techniques have allowed electrophysiologists to measure currents through single ion channels.

The second encounter, however, differs from the first in several important respects. Neurobiology now has a stronger tradition in molecular biology. The work of the first generation of émigrés took hold and a variety of well-defined and well-studied systems are available for mutational analysis. The neurobiology of *Drosophila* and *C. elegans* has come of age. Most important, the questions currently answerable on the molecular level have been greatly clarified. In addition, the techniques of recombinant DNA are applicable to a much broader range of preparations than are those of mutational analysis. Moreover, the fact that at least some neurobiologically interesting genes are conserved in evolution raises the possibility that one might be able to benefit routinely from the mutational advantages of *Drosophila* by cloning genes in *Drosophila,* and then by using those clones to screen the genomic libraries of higher animals.

Furthermore, cloning offers the possibility of transforming an *E. coli* into a family of electric organs, for example. Because of this capability, the gene products from even the smallest neurons might be harvested in abundance. If one is interested in a particular molecular component of a cell, cloning techniques could be used to produce the material of interest in amounts sufficient for biochemical analysis. For instance, this approach could be used to characterize the Na$^+$ channel protein, which has been difficult to study because it represents much less than 0.1% of the nerve cell's total protein.

The new technology can also elucidate the changes in gene expression that certainly take place as the nervous system develops and that are likely to underlie long-term forms of synaptic plasticity. Libraries of cDNA can be probed with nucleotide sequences from nerve cells, both at different stages of development and from the mature animal under different protocols of training to assess changes in mRNA synthesis.

It is now also possible to delineate the organization of particular genes. If one assumes that neurobiological processes are mediated by universal molecular mechanisms, the preparations at hand can be used to determine whether there are brain-specific varieties of molecules within a class and, within this class, whether different neurons use different molecular entities. Which components are shared or common, and which are diverse?

Given the fact that the terrain looks inviting for the return of molecular biology, how has neurobiology been affected in the 3 years since the second encounter? This symposium attests that the progress has been encouraging and that we not only have learned much, but have learned it more rapidly than one might have expected. With the development of new techniques and the recruitment of excellent scientists trained in a new set of disciplines, the landscape of certain segments of neurobiology is beginning to change. In addition, and perhaps more important in the long term, a critical shift in attitude has taken place within neurobiology. Neurobiology is beginning to overcome an intellectual barrier that has separated it from the rest of biological science, a barrier that has existed because the language of neurobiology has been based heavily on neuroanatomy and electrophysiology and only modestly on the more universal biological language of biochemistry and molecular biology. Until 3 years ago, most molecular biologists felt that merely being interested in the central question posed by neurobiology—how does the brain operate?—was insufficient for starting work in the field and that even to begin work required an extensive knowledge of neuroanatomy or electrophysiology. This meeting has shown that this need not be so—at least at the outset. I am not here describing, much less advocating, lack of adequate preparation. A thorough understanding of the issues confronting the study of the brain is clearly needed. To work in a particular area of the nervous system, one has to come to grips with its structure and physiology. But one now can *begin* to work on molecular aspects of a problem without being intimidated by the formidable facts of electrophysiology or overwhelmed by the wealth of fine detail in brain anatomy. Since in principle the methodologies of recombinant DNA and monoclonal antibodies can be applied to any system of interest, some gifted newcomers have already made interesting contributions to neurobiology by selecting systems in which the anatomical and physiological detail is limited or straightforward.

Moreover, this is only the tip of the iceberg. As progress accelerates, the barriers that have traditionally separated neurobiology from cell biology will be reduced even more. Two further consequences are likely to result from this change in the landscape of neurobiology. First, talented scientists from other areas of biology will increasingly be attracted to neurobiology because its intrinsically fascinating problems will be posed in ways that lend themselves to molecular approaches. Second, we in neurobiology will begin to appreciate that some of the problems we find fascinating are not unique to the nervous system and might be more profitably studied elsewhere.

On the other hand, this symposium has also illustrated that recombinant DNA and hybridoma methodologies are techniques, not conceptual schemes. There will be life in neurobiology after cloning. The basic questions that confront the study of the brain continue to be: How do nerve cells work? How do their interactions produce thought, feeling, perception, movement, and memory? New techniques are interesting for neurobiology only insofar as they help answer these questions. It is clear that the techniques of molecular genetics will prove to be of great value, but it is also clear that these techniques cannot go it alone; additional approaches will be needed.

Let me now turn to consider some specific issues addressed in the symposium and to use them as a springboard for reviewing some of the major themes in current molecular neurobiology.

Molecular Neurobiology: From Molecules to Behavior

Channel Proteins

Membrane proteins endow nerve cells with signaling capabilities

The distinctive electrical signaling capabilities of nerve cells derive from two families of specialized membrane proteins called channels and pumps that allow ions to cross the membrane. Pumps actively transport ions against an electrochemical gradient and therefore require metabolic energy. Channels allow ions to move rapidly down their electrochemical gradient and do not require metabolic energy.

Channel proteins, in turn, are grouped into two classes: 1) nongated channels that are always open and 2) gated channels that can open and close.

Voltage-gated channels sense the electrical field and are opened by changes in membrane potential. Chemically gated channels open when ligands such as transmitter or hormone molecules are bound to them. Neurons vary in the types of channels they possess. Even different regions of a single neuron can have different types of channels.

The current understanding of signaling in nerve cells originates from the ionic hypothesis formulated in mathematical terms by Hodgkin, Huxley, and Katz in 1952. According to this theory, the resting and action potentials result from unequal distribution of K^+, Na^+, and Cl^- across the membrane. The Na^+ pump maintains the concentration of Na^+ inside the axon approximately 20 times lower than that on the outside. The resting membrane has nongated channels (called leakage channels) permeable to K^+, and the resting potential of nerve cells is therefore close to the equilibrium potential for K^+ (approximately –80 mV). The small deviation from the equilibrium potential for K^+ results from a slight permeability of the leakage channels to Na^+ and Cl^-. An axon membrane is able to generate an action potential because it contains two independent voltage-gated channels, one for Na^+ and the other for K^+. Both are closed at rest and are opened with depolarization. Depolarization gates the Na^+ channel, admitting some Na^+ into the cell, which in turn causes further depolarization; this opens up more Na^+ channels and gives rise to a regenerative process that drives the membrane potential toward the Na^+ equilibrium potential of about +55 mV. Depolarization also opens K^+ channels, but with a delay. K^+ channels allow K^+ to move out of the cell, and this event, together with the inactivation of the Na^+ channel, repolarizes the cell and terminates the action potential.

Over the last several years, the ionic hypothesis has been extended by the finding of additional ion channels in the cell body and in the terminal regions of the nerve cell that are not present in its axon. For example, nerve terminals and cell bodies contain voltage-gated Ca^{++} channels. The opening of these channels is responsible for the influx of the Ca^{++} necessary for the exocytotic release of transmitter by synaptic vesicles. In muscle cells, opening of the Ca^{++} channels is a crucial step in initiating contraction. Moreover, in addition to the K^+ channel described by Hodgkin and Huxley (1952), called the delayed K^+ channel, several other types of gated K^+ channels have been found in both the nerve terminals and cell bodies. These include the early K^+ channel and the Ca^{++}-activated K^+ channel.

Synaptic transmission in its simplest form represents an extension of this set of mechanisms. It uses channels that are gated chemically rather than by voltage. For example, at the nerve-muscle synapse in vertebrates, Fatt and Katz (1951) and Takeuchi and Takeuchi (1966) showed that synaptic transmission involves the gating of a channel that passes small cations—primarily Na^+ and K^+—when ACh binds to the channel.

The best-understood membrane protein is the
ion channel activated by ACh

The initial findings of Fatt and Katz and Takeuchi and Takeuchi opened up the study of the molecular properties of the channel gated by ACh. Here the progress has been remarkable. I still remember discussions in the early 1960s of whether the AChR was a protein or a lipid. When this issue was settled, the question persisted until 1970 as to whether the AChR and acetylcholinesterase (AChE) are the same molecule. On the basis of studies that showed the esterase to be a peripheral rather than an integral membrane protein that does not react with affinity labels or with ligands highly specific for the receptor, we now know that the receptor and the esterase are different proteins.

In addition, studies by Katz and Miledi (1970) and by Anderson and Stevens (1973) using noise analysis, and subsequent patch-clamp studies by Neher and Sakmann (1976), have delineated the elementary currents that flow when a single AChR channel changes from a closed to an open conformation in response to ACh. Each channel opens briefly (on an average for 1 msec) in the presence of ACh and gives rise to an all-or-nothing square pulse of inward current that allows about 20,000 Na^+ ions to move into the cell (Anderson and Stevens 1973; Katz and Miledi 1970; Neher and Sakmann 1976). The resulting transport rate of 10^7 ions/sec is 1,000 times greater than that of carrier-mediated transport mechanisms such as that by valinomycin. These measurements have confirmed the basic idea of Hodgkin, Huxley, and Katz—long thought to be correct—that ions can cross the membrane through transmembrane pores.

We are now also beginning to learn something about the molecular biology of the AChR. The work of Karlin, Lindstrom, Raftery, and others has shown that the receptor protein is an asymmetrical molecule with five subunits divided into four types (2 α, 1 β, 1 γ, 1 δ). Each α subunit binds one ACh molecule (Karlin et al.). This is consistent with the earlier pharmacological finding that two molecules of ACh are necessary to gate the channel. Each of the four types of subunits is encoded by a different mRNA and therefore by a different gene (Anderson and Blobel; Numa et al.; Raftery et al.). Indeed, each of the genes for the four types of subunits has now been cloned (Numa et al.; Patrick et al.) and there is direct evidence that both copies of the α subunit are transcribed from a single gene (Numa et al.). The biochemical difference between the two α subunits results from posttranslational modifications, although the exact nature of the modifications remains unclear (Hall et al.; Karlin et al.; Lindstrom et al.; Merlie et al.; Numa et al.; Raftery et al.).

A comparison of the complete nucleotide sequences of the subunits reveals a substantial homology among them, consistent with the notion that they all arose from a single ancestral protein (Numa et al.; Raftery et al.). An obvious possibility is that the ancestral AChR consisted of a homo-oligomer and that later gene duplication and divergence led to the evolution of the gene family that now encodes for the various subunits of the contemporary nicotinic AChR.

Sequence data and related immunological, biochemical, and structural information on the AChR are also beginning to give us some ideas of how the subunits are oriented in the membrane (Anderson and Blobel; Changeux et al.; Fairclough et al.; Karlin et al.; Numa et al.; Patrick et al.). Each of the four subunits is a transmembrane protein (Anderson and Blobel; Changeux et al.; Raftery et al.). The aminoterminal region of each subunit is thought to lie on the extracellular side of the membrane, and this region of the α subunits is likely to contain the recognition sites for ACh, which are certainly extracellular. Earlier affinity-labeling studies had shown that the ACh-binding sites contain cysteine residues (Karlin et al.), and on the basis of the sequence data, it has been possible to pick out the cysteine residues that are also probably components of these sites (Numa et al.). As we shall see below, the disposition of the carboxyl terminus is still not clear (Fairclough et al.; Numa et al.).

Electron microscopic studies indicate that the five chains are arranged around the central channel (Fairclough et al.; Karlin et al.). Since the sequence homology extends through most of the primary structure of the subunits, each subunit is likely to have a similar structural motif. As a result, each subunit probably makes a similar contribution to the total structure Fairclough et al.; Numa et al.). For example, Numa's data suggest that each subunit has four extended hydrophobic regions. Each of these hydrophobic regions is believed to traverse the membrane once. If that is so, each subunit threads through the membrane four times (Changeux et al.; Hershey et al.; Numa et al.; Patrick et al.). The hydrophobic transmembrane domains are postulated to link hydrophilic domains that extend beyond the surfaces of the membrane into the cytoplasm on one side and the extracellular space on the other. The extracellular domain of each chain is about 25 kD and the cytoplasmic domains are smaller and of variable size.

One possibility that was entertained a few years ago was that the channel (ionophore) and the recognition site for ACh (the receptor) might represent different and separable polypeptide chains. But current structural information (including negative-stain electron microscopy and image reconstruction) suggests that all subunits contribute to and are positioned around the channel like the staves of a barrel. Conductance studies suggest that the channel narrows to a diameter of 6 Å (Hille 1977). Since the channel is only weakly selective—it excludes anions but is permeable to monovalent and divalent cations as well as nonelectrolytes—it is thought to be a water-filled neutral pore without fixed charge. Numa has therefore proposed that the walls of the channel might be made up of the polar side chains of the helices of the inferred transmembrane segments and that these side chains (primarily the hydroxyl oxygens of threonine and serine residues) bestow upon the channel its cation selectivity.

An alternative model has been advanced by Stroud and his colleagues on the basis of a search, using Fourier analysis, for the periodicities that characterize the amphipathic secondary structure (Fairclough et al.). According to Stroud's model, each subunit has not four but five helical transmembrane segments. Four are identical to Numa's, and the fifth helix is believed to be hydrophobic on one face and hydrophilic on the other. This structure suggested to Stroud that the fifth α helix forms the walls of the ion channel. The existence of a fifth transmembrane segment in this model would have an additional consequence: it would cause the carboxyl terminus of the subunits to lie on the cytoplasmic side of the membrane. This also is in contradistinction to Numa's model of four transmembrane segments, which places the carboxyl terminus together with the amino terminus on the external surface. It should be possible to distinguish experimentally between the two models.

Monoclonal antibodies to subunits of the AChR have contributed importantly to all aspects of the study of the receptor: its synthesis, assembly, conformation, and the structure of its subunits (Lindstrom et al.). These studies also have had a key role in elucidating the molecular nature of myasthenia gravis. This disease of neuromuscular function is characterized by muscular weakness that is increased by activity and relieved, sometimes dramatically, by rest. Modern immunological techniques have shown that myasthenia is an autoimmune disease resulting from self-produced antibodies to AChR. These antibodies lead to a higher turnover of AChRs by cross-linking them as well as by facilitating their endocytosis (Lindstrom et al.). As a result, the affected skeletal muscles of patients with myasthenia gravis contain fewer AChRs than do those of normal people. In view of the clinical importance of the AChR, it is fortunate that the receptor is highly conserved through evolution; its gene has been isolated from humans (Numa et al.) as well as from *Drosophila* (Ballivet et al.), where it might be studied effectively.

Although we now know a great deal about the nicotinic AChR, we still know little on the molecular level about how the structure of the channel is expressed in its function. In addition to the problem of ion selectivity, which I will consider below, other key questions must be addressed. First, how is the binding of ACh transduced into opening of the channel? Does the transduction process explain why the total mass of the receptor protein is so large (250 kD) and why the protein is divided into five chains? It is clear from studies of ionophoric antibiotics (such as gramicidin A) and of bacteriorhodopsin (Dunn et al.) that one can build a perfectly good channel with only one small polypeptide chain. Second, how is the receptor assembled? Is it by self-assembly, or are other proteins involved? Third, how is gene expression for the subunits regulated during development and following denervation?

These questions illustrate a point that I will return to repeatedly. Defining nucleotide sequence is an important step toward achieving a molecular understanding of neuronal function, but it is only a beginning. It will be essential to combine information derived from molecular genetic techniques with insights gained from cell biological, biophysical, and structural approaches. In particular, sequence data must be tied to structural biochemistry, on the one hand, and to function, on the other. Indeed, it will not be easy to study the molecular mechanisms by which the AChR channels work (how permeation occurs, for example). This difficulty stems from the fact that, unlike organic molecules, the substrates—the ions that move through the various ACh channels—cannot be altered for specificity studies (although in the case of the Na^+ or K^+ channels much has been learned by using ions of different size, shape, and charge). Thus, the tricks that are possible in the study of enzyme mechanisms—based on the use of substrate analogs—cannot be applied to ion channels. However, photoactivated affinity labels of the channel have been used to identify the subunits that contribute to the channel of the AChR (Changeux et al.; Karlin et al.).

Site-directed mutagenesis, which has been used in the case of bacterio-rhodopsin to alter the gene products at specific molecular loci (Dunn et al.), can assist in the analysis of channels. With this form of molecular genetic analysis, each subunit of the ligand-gated channel might be analyzed in terms of the contribution that a particular peptide sequence makes to the various aspects of permeation. The most direct approach to these problems is likely to come from studies in which site-directed mutagenesis is used to alter, in defined ways, the structure of the genes for the subunits. These altered genes or their mRNAs can then be introduced into nonneuronal cells capable of expressing them—such as oocytes or cell lines (Barnard et al.). If the approach works, it can be used to elucidate the nature of the recognition sites on the channel for the transmitter and the selectivity sites within the channel for the ion, as well as other components crucial to permeation. The availability of cloned individual subunits will also make it possible to use reconstitution systems to examine the mechanisms by which subunits assemble and the functions they perform.

The nicotinic AChR at the vertebrate nerve-muscle synapse is the best-studied AChR. But ACh also interacts with other receptors, which control other ion channels. The predominant AChRs in the vertebrate central nervous system show greater sensitivity to muscarine and atropine than to nicotine and *d*-tubocurarine and are therefore called muscarinic receptors. There are several different muscarinic receptors (Birdsall et al.). One, for example, produces its excitatory action not by opening a cation channel for Na^+ and K^+ but by closing a channel to K^+. What are the structures of these muscarinic receptors? Do they resemble the nicotinic receptor? The availability of cloned nicotinic receptor genes now might make it possible to probe the genomic libraries of animals to see whether there are homologous subunits of the muscarinic receptor.

In addition to the nicotinic and muscarinic receptors of vertebrates, at least three other AChRs are present in invertebrates such as *Aplysia*, and each of these receptors controls different ionic channels (Na^+ and K^+; Cl^- and K^+). It will be fascinating to see whether there is a structural or ontogenic logic to this large family of AChR channels—the nicotinic, the muscarinic, and the several invertebrate receptors. Comparative analysis of their sequence and additional structural information might offer important clues to one of the central problems of channel function, ion selectivity: How do some ACh channels select for K^+ alone while others select for Na^+ and K^+, and still others only for Cl^-.

We still know little about the structure of the Na^+, K^+, and Ca^{++} channels

I have so far reviewed what we know of the structure of the AChR channel and have pointed out some of the gaps remaining in our knowledge that must be filled before we understand this channel thoroughly. When it comes to the structure of the Na^+ channel, the various K^+ channels, and the Ca^{++} channel, unfortunately even less is known, although some information is likely to emerge soon for the Na^+ channel. However, studies of these channels illustrate how much voltage- and patch-clamp experiments have contributed to our understanding of kinetic properties, gating, and channel modulation. For example, we know that the Na^+ channel (unlike the AChR channel) is highly discriminating in its selectivity for ions. It is 10 times more selective for Na^+ than for K^+. The Ca^{++} channel is 10 times more selective still, being 100 times more selective for Ca^{++} than for either Na^+ or K^+. In addition to the selectivity of the Na^+ channel, we know that it exists in three functional states: closed, open, and inactivated. Patch-clamp analysis by Aldrich and Stevens has revealed that the inactivated state is accessible from both the resting and the active states but that open channels move into the inactive state about 100 times more rapidly than do closed or resting channels.

Until recently, we thought that the channels contributing to the action potential were gated only by voltage (as is the case with the Na^+ channel), whereas channels that produce synaptic actions were gated only by a transmitter (as is the case with the ACh channel). We have now learned that this old rule has several exceptions. Single-channel and other biophysical analyses have shown that some channels that contribute to the action potential are also modulated by transmitters. Particularly interesting is the finding that two of these dual-purpose channels are modulated by their transmitter through a cAMP-dependent protein phosphorylation—the Ca^{++} channel in the heart modulated by adrenergic agonists (Reuter et al.; Tsien et al.) and the K^+ channel in sensory neurons of *Aplysia* modulated by serotonin (Camardo et al.). Although these studies provide direct evidence for cAMP-dependent protein phosphorylation in modulating ion channels in excitable membranes, a key problem still remains: What is the substrate or substrates that are actually phosphorylated? Does the kinase phosphorylate the channel protein itself or does it modify a regulatory protein closely associated with the channel?

Although single-channel analysis has contributed much to our understanding of the kinetics of the Na^+, Ca^{++}, and K^+ channels, we still know little about the molecular details. A good beginning is being made in the case of the Na^+ channel, however. This has been possible because of the finding of ligands with high affinity and specificity (TTX, saxitoxin, and batrachotoxin), as well as antibodies for this channel. The Na^+ channels isolated from mammalian synaptosomes, from mammalian muscle, and from eel electroplax and brain, all contain a large polypeptide of 250–300 kD that is glycosylated (Agnew et al.; Catterall et al.; Fritz et al.). In the mammalian brain, this peptide is called the α subunit; it is isolated together with two smaller subunits, β_1 (39 kD) and β_2 (37 kD). β_2 is linked to the α subunit by disulfide bonds. The Na^+ channel in sarcolemma from mammalian muscle has three small components (39 kD, 38 kD, and 47 kD), in addition to a large peptide.

To what degree are the Na^+ channels from these various sources related? Does the existence of the large peptide in each of them indicate that all of these Na^+ channels share a major subunit? To answer this question it will be necessary to determine parts of the amino acid sequence of the large peptide. Just as Raftery's data on the partial amino acid sequence of the AChR opened the way for the cloning of this molecule, so now some sequence data are badly needed to move analysis of the Na^+ channel to the next level.

The K^+ and Ca^{++} channels pose even greater problems for molecular analysis because, unlike the Na^+ channel, there were until recently no comparable ligands or antibodies for these channels. The dihydroxypyridines, however, are a new class of drugs thought to interact specifically with Ca^{++} channels (Gengo et al.; Gould et al.) and thus may aid in their isolation. But attention is focused at the moment on one of the K^+ channels, the early K^+

channel, which is altered in the *Drosophila* mutant called *shaker* (L.Y. Jan et al.; Salkoff). *Shaker* mutants show spontaneous, nonfunctional movements under certain circumstances that are due to prolonged action potentials in nerve and muscle cells. These abnormal action potentials are the result of a single gene mutation that deprives *shaker* of early K$^+$ channels. Were this mutation to exist only in mice, the problem might have to rest for a while, but in *Drosophila*, specific techniques now make it possible to isolate mutant genes. A particularly effective technique is transposon tagging, whereby the gene of interest is both mutated and marked by the insertion of moveable (transposable) genetic elements (transposons). A transposable genetic element in *Drosophila* that is especially useful is the P element because it becomes highly mobile when males from a strain that carries the P element are mated with females from a strain that does not. This type of mating leads to hybrid dysgenesis, a process that greatly increases mutations in the offspring. Because the P element is inserted into new sites in the genome of the offspring, these mutations allow one to screen dysgenic flies for the *shaker* defect phenotypically (L.Y. Jan et al.). *Shaker* mutants isolated in this way indeed contain P elements close to the *shaker* locus (L.Y. Jan et al.). One should now be able to isolate the P element and the surrounding nucleotide sequences, and thereby to isolate segments of the gene for the early K$^+$ channel.

Isolation of any K$^+$ channel might lead to the isolation of other K$^+$ channels if they share sequence homology. A comparative approach here would be of particular interest because there are at least five identified K$^+$ channels on the membranes of nerve cells. It will be fascinating to see to what degree the various kinetically distinct K$^+$ channels are related. Do they share any subunits? Perhaps the K$^+$ channels, like the AChR channels, are made up of several subunits and all K$^+$ channels will be found to share all but one of them, the unique subunit giving each class of K$^+$ channel its particular voltage and time parameters. Characterization of all the K$^+$ channels can also suggest functional interrelationships among them, and site-directed mutagenesis can help to specify the nature and physiological relevance of their differences.

One of the most challenging tasks in channel physiology for the next 5 years is clearly to understand the nicotinic AChR better and to move beyond it to other channels. Some general molecular rules, about channel selectivity, kinetic properties, and voltage and transmitter gating, should underlie the structure and function of membrane channels, and a detailed comparison of the family of AChR channels and of the various K$^+$ channels may well lead us to them. Outlines of some of these rules are already emerging from single-channel analysis. Combining in situ mutagenesis with single-channel analysis, on the one hand, and with modern structural analysis

(X-ray, neutron, and electron diffraction; electron microscopy; and nuclear magnetic resonance), on the other, should prove immensely instructive.

Synaptic Transmission: It Now Seems as If Neurons Use Two Major Classes of Synaptic Transmitters

Neurons have available to them two means for intercellular communication: electrical and chemical. A particular membrane specialization is used for each. Electrical transmission involves a well-characterized membrane protein called connexon, which forms channels within specializations called gap junctions that connect the cytoplasm of the pre- and postsynaptic cells. At gap junctions, electrical current generated by the action potential in one cell flows directly across into the connected cell.

Chemical transmission is more specialized; in addition to distinctive zones of apposition between the two neighboring cells, it involves molecular machinery in the presynaptic neuron for the storage and release of transmitter as well as receptor molecules in the postsynaptic cell. An action potential that invades the presynaptic terminal of a neuron activates voltage-gated Ca^{++} channels. The resulting inflow of Ca^{++} allows synaptic vesicles, each containing several thousand molecules of transmitter, to bind to specialized release sites called active zones. Once released, the transmitter diffuses across the synaptic cleft, where it binds to a receptor and opens (or closes) chemically gated channels, initiating current flow in the postsynaptic cell. Depending on the channels gated by the transmitter, the current flow will produce excitation or inhibition.

Whereas something is now known about the receptors to a variety of transmitters and the ion channels they control (see, for example, Barker et al.; Sakmann et al.), we are only beginning to understand the presynaptic molecular machinery that controls vesicle binding and transmitter release (Goldin et al.; Kelly et al.). A significant conceptual advance has come from the appreciation that vesicle release is intimately tied to vesicle mobilization, the process by which vesicles are loaded onto release sites. In turn, mobilization of vesicles is related to axonal transport and, at its origin in the Golgi apparatus of the cell body, transport of vesicles is related to membrane sorting. Kelly's data on membrane sorting indicate that there are two pathways for externalizing products manufactured in the Golgi apparatus: a constitutive pathway and a regulated pathway. Only the regulated pathway is used by material stored in synaptic vesicles. Synaptic vesicles move from the Golgi apparatus into the axon by interacting with specific transport elements. A given vesicle must select, for example, whether it will be transported to the terminal regions of the axon for release or to the dendrite for insertion into the dendritic membrane. For different vesicles to move to different regions of the cell, two conditions must exist: 1) there must be several transport systems, and 2) secretory vesicles must have a recognition molecule for selecting the appropriate system. After they are carried by axonal transport to the nerve terminal, synaptic vesicles can be loaded into active zones for release. Ca^{++} entry promotes fusion of the vesicle to the membrane, presumably by means of calmodulin or a similar Ca^{++}-binding protein.

Small molecule transmitters

A nice example of how an analysis of the underlying molecular logic of a system can lead to much broader insights comes from studies of the genes encoding the synthetic enzymes for small molecule transmitters (Joh et al.; Mallet et al.; O'Malley et al.). Catecholamines are synthesized through a pathway consisting of four well-characterized and related enzymes. The first step of the pathway, the amino acid tyrosine, is converted to L-dopa by the enzyme tyrosine hydroxylase (TH). Second, L-dopa is decarboxylated to dopamine by aromatic L-amino acid decarboxylase. Third, the enzyme dopamine β-hydroxylase (DBH) converts dopamine to norepinephrine. Finally, the last enzyme in the pathway, phenylethanolamine N-methyltransferase (PNMT), catalyzes the synthesis of epinephrine from norepinephrine. There are two interesting features about the regulation of these enzymes. First, not all nerve cells that release catecholamines express all four enzymes, although the neurons in the adrenal medulla and neurons that release epinephrine do so. But neurons that synthesize norepinephrine do not express the enzyme PNMT, and neurons that release dopamine express neither PNMT nor DBH. The phenotypic expression of catecholamine-synthesizing enzymes in neurons (and in chromaffin cells) can be independently regulated so that one cell type can express one of the four enzymes and not the others. Second, insofar as a neuron expresses one or more of the genes of this pathway, that expression is coordinately regulated. Conditions that alter the synthesis of one enzyme also change the synthesis of the others. For example, neural activity in the noradrenergic neurons of the locus coeruleus causes an increase in the synthesis of norepinephrine, and this is reflected in a coordinately regulated increase in the expression of TH, DBH, and PNMT (Joh et al.; Mallet et al.; O'Malley et al.).

Peptide transmitters

Most of the emphasis in the study of chemical synaptic transmission has been on small transmitter molecules: norepinephrine, serotonin, ACh, and various amino acids or closely related substances (Barker et al.; Cull-Candy; Sakmann et al.). But the number of potential signaling substances is actually much greater, as recently became clear with the discovery that peptides, ranging in length from 2 to 100 amino acids, also serve as chemical transmitters within the nervous system. A given peptide can function in three overlapping ways: 1) as a neurotransmitter by acting over very short distances (300–500 Å) on neighboring nerve cells, 2) as a local hormone by diffusion over somewhat larger distances (1–2 mm), and 3) as a neurohormone by being released into the bloodstream to act on distant targets. These several functions of peptides (and of certain biogenic amine transmitters) indicate that the conventional distinction between hormone and transmitter no longer holds in any rigorous sense. Thus, superimposed upon the anatomical precision of the pattern of neuronal connections is an equally precise but spatially more separated pattern of chemically addressed interactions determined by the peptide transmitters and their receptors on target cells.

The distinction between small molecule transmitters and peptides seems fundamental, and it is likely to be important for understanding brain function. As is the case with small molecule transmitters, peptides are released from the nerve terminals of neurons in a Ca^{++}-dependent manner, and after they are released they act upon specific receptors in the postsynaptic cell. Although the actions of these two classes clearly overlap, peptides and small molecule transmitters tend to differ in their modes of action. Peptides are extremely active and are effective in concentrations as low as 10^{-10} M. On the other hand, conventional small molecule transmitters must be present in concentrations as much as 10^5 times greater before they are effective. In addition, small molecule transmitters are rapidly and effectively removed or degraded. For example, the enzyme AChE hydrolyzes ACh, and there are specific high-affinity uptake mechanisms to remove transmitters from the synaptic cleft. These mechanisms are apparently not available for peptide transmitters: no uptake mechanisms have yet been discovered and no rapidly acting degradative enzymes exist (although some peptidases are specific [see Mason et al., for example]). As a result, peptides tend to have a much more lasting effect on neurons than the small molecule transmitters that often can act rapidly (Y. N. Jan et al.). Even more fundamental is the fact that small molecule transmitters are typically synthesized in the presynaptic terminal by a series of biosynthetic enzymes (such as choline acetyltransferase for ACh) and are therefore immediately available for release. In contrast, the precursors of peptide transmitters are synthesized on ribosomes in the cell body and the peptides must then be transported to the nerve terminal. Finally, peptides often coexist with other (small molecule) transmitters (Dodd et al.; see also Hökfelt et al. 1980).

Particularly interesting is the finding that peptides are characteristically synthesized from a large precursor molecule (a polyprotein) that often contains within itself the sequence for other neuroactive peptides (Buck et al.;

Herbert et al.; Mahon and Scheller; Roberts et al.). The discovery that polyproteins are precursors of peptides was made by Herbert, Roberts, and their colleagues, when they showed that ACTH derives from a much larger precursor, pro-opiomelanocortin, which also contains α-, β-, and γ-MSH; β-lipotropin; and the enkephalins. Depending on the nature of its proteolytic processing, the same polyprotein can yield different sets of peptides in different cells. For example, pro-opiomelanocortin is expressed in both the intermediate and the anterior lobes of the pituitary; in the anterior lobe it is processed to ACTH, but in the intermediate lobe all of the ACTH is further processed to α-MSH. Pro-opiomelanocortin is also produced in the arcuate nucleus of the hypothalamus where it is processed to a mixture of ACTH and α-MSH. In addition to differences in processing the same precursors, cells in the anterior and intermediate lobes of the pituitary express the same gene for pro-opiomelanocortin but respond differently to the same stimulus (glucocorticoid) simply because the cells in the intermediate lobe lack glucocorticoid receptors (Roberts et al.). It therefore becomes important to localize a given peptide to its cells of origin. For this purpose, antibodies to the peptide or labeled probes for its nucleotide sequence can be used in immunocytochemical or in situ hybridization studies (Buck et al.; Mahon and Scheller; Roberts et al.).

The study of polyprotein precursors has revealed an underlying unity in the organization of peptide transmitters (or putative transmitter candidates). Since the discovery of met- and leu-enkephalins by Hughes, Kosterlitz, and their colleagues in 1975, 18 additional peptides have been found that possess opioid-like analgesic activity when injected into the brain. All of the opioid peptides are extensions of the carboxyl terminus of either met- or leu-enkephalin. Moreover, all 18 peptides derive from only three precursors: pro-opiomelanocortin, proenkephalin, and prodynorphin (Herbert et al.; Rossier). The three precursors show remarkable regularities. They are almost identical in length, and the biologically active peptides are almost exclusively confined to the carboxyterminal half of the precursor. The active domain of each precursor is flanked on both sides by pairs of basic amino acid residues, creating potential cleavage sites for trypsin-like enzymes.

These similarities have two implications. First, the existence of the flanking basic residues has led to the discovery within polyproteins of several completely novel and previously unanticipated peptides (Buck et al.; Evans et al.; Herbert et al.; Mahon and Scheller; Sutcliffe et al.). This finding in turn suggests one rational strategy in the search for novel peptides: the precursor molecules of each known peptide can be cloned and then explored for potentially new peptide sequences outlined by flanking basic cleavage sites. (For another strategy, see Sutcliffe et al., this volume.) In a beautiful application of this general approach, a new mammalian peptide was discovered encoded on the same gene with the calcitonin coding sequence. Immunocytochemistry was then used to show that this peptide is present in neurons of the taste pathway (Evans et al.).

The second implication of the similarities among peptide transmitters is that each family may have evolved from a single ancestral gene. For example, because the opioid peptides have common opioid activity and share structurally similar repeated sequences, it seems likely that these peptides arose from a series of duplications. Herbert et al. have suggested that an enkephalin gene repeating unit of 48 bases may be the building block for all three genes of the three opioid precursors. Duplication and rearrangements of this building block may have led to the various opioid genes that exist today.

The realization that peptides are created from polyproteins raises another question: Why are polyproteins used as precursors? In principle, polyproteins provide an opportunity for diversity because they can be processed or expressed differently in different cells. They also offer a mechanism for coordinated expression and release and, in the case of gene families, for the release of different combinations of peptides. The coordinated release of diverse peptides in various combinations, acting throughout the nervous system, may be important in orchestrating different aspects of a behavior. Work by Earl Mayeri and his colleagues (1979; Rothman et al. 1983) on the component peptides of the precursor for the egg-laying hormone (ELH) in *Aplysia* has indicated that these peptides can indeed produce different neuronal actions. Some peptides from the precursor excite some cells, while other peptides inhibit other cells. The possibility is intriguing that different peptides of a polyprotein are involved in mediating complex behavioral sequences. By means of their several constituent peptides, polyproteins could ensure that the various neuronal circuits responsible for the different facets of a stereotyped behavior are activated in a coherent manner. This notion, consistent with findings in *Aplysia* (Buck et al.; Mahon and Scheller), is also supported in mammals, where the complex responses to stress involve the action not simply of ACTH but also of β-endorphin, γ-MSH, and corticotropin-releasing factor. It will therefore be important to see exactly how (on what neurons and on what peripheral targets) these several peptides act in producing the response to stress.

Thus, it is becoming clear that to make sense of the peptides we will have to look not only at genes but also at behavior. To correlate peptides with behavior will in turn require that we look beyond single peptides to the family of peptides produced by a precursor. It will also be profitable to examine peptides that have already been linked to certain behaviors—such as cholecystokinin to satiety, and angiotensin to thirst. Perhaps the precursors to these peptides contain additional peptides whose function can further enlighten our understanding of these instinctive (homeostatic) behaviors.

Although peptides are providing an additional dimension to our understanding of neural action, many of the peptides now identified are in an early stage of analysis. Only in a very few cases has release of a peptide from a neuron been convincingly demonstrated, or has the action of a peptide been analyzed on the cellular level. We are still in the age of peptides looking for functions. Simple behavioral systems in vertebrates and invertebrates should prove extremely useful here (Buck et al.; Y.N. Jan et al.; Mahon and Scheller). Moreover, whereas we now know the structures of many peptides, we know little of the receptors for peptides. Yet understanding these receptors is crucial, since with peptides, as with small molecules, the ultimate action is determined by the molecular nature of the receptor. How do these receptors work? Do they use second messengers? Are they encoded in gene families? Many peptides seem to have diversified from an ancestral gene. What about their receptors? Did they evolve independently of the peptides, or coordinately with them?

I stress the significance of peptides because they constitute an important bridge between molecular neurobiology and the neurobiology of information processing and behavior. In the excitement generated by progress in molecular neurobiology, we should not lose sight of why molecules such as the Na^+ channel and the AChR channel are fascinating. The interest in them lies not only in their properties as intrinsic membrane proteins but also in the fact that these proteins must be understood before we can explain the workings of the brain: how we and simple animals move, behave, and learn. There is some concern that the dramatic progress in molecular biology, which one can confidently predict, will lead to a separation of the cellular aspects of neurobiology from the aspects concerned with information processing. If this were to happen, it could in fact lead to a fragmentation of neuroscience that would be regrettable because much of the beauty of the field consists in its unity and scope. To my mind, the study of peptide precursors illustrates once again that deep scientific insights are often synthetic, not divisive. Knowledge of the nucleotide sequence of the gene that encodes the peptide is likely to illustrate principles of behavior that would be difficult to extract from study of either the gene products or behavior in isolation.

Development: Cell Lineage, Axon Outgrowth, Cell Recognition, and Synapse Formation

The nervous systems of all animals develop as a specialization of the ectoderm of the body surface. In some animals, neurogenesis then proceeds by proliferation and differentiation in situ. In other animals, proliferation and differentiation occur at different sites: neurons first proliferate and then migrate to their definitive location. In nematode worms, the ectodermal cells in the body wall give rise to the neural epithelium. Within this neural epithelium, primitive neuroblasts lose contact with the inner and outer surfaces of the ectoderm, round up, and typically proliferate in situ, giving rise to clones of progeny. Other neurons, common in the nervous system of vertebrates, develop in the ciliated columnar ectoderm of the neural tube or the neural crest, then withdraw from the mitotic cycle and migrate over varying distances to the ultimate destinations of their cell bodies. After they have reached their destination, neurons start to spin out their axons, which then travel sometimes over considerable distances and by complex routes to their appropriate region of the brain. Finally, in that region, the growth cone of the neuron seeks out the appropriate target cell through what appears to be chemical trial and error. The outgrowing axon often bypasses thousands of candidate neurons and, in a selection process that presumably involves rather precise recognition, selects only the correct target (or one of a small population of correct targets). The recognition event in turn initiates a new series of steps involving competition with other outgrowing fibers and eventually leading to the differentiation and stabilization of the synapse.

The papers at the symposium leave little doubt that we are on the brink of an important era in the study of neural development.

First, with the increase in experimental detail about the sequence of events during synapse formation over the last 15 years, we have become more sophisticated about the questions involved. Before dissociated cell culture was available, many scientists thought that the critical insights into development could be achieved by studying what was believed to be one all-or-none step: synapse formation. We now realize that synapse formation is not one step but a family of steps that probably begins with cell lineage and extends to the maturation of the functioning synapse. It seems likely, from what we already know, that each step has its own subroutines, each controlled by particular molecules acting on a cell at a particular time and priming it for the next step (Nirenberg et al.; see also Rubin et al. 1980).

Second, we are about to move from descriptive phenomenology to molecular analysis. Several advances presented at the meeting are particularly promising:

Cell lineage can now be studied with remarkable precision

In the nematode worm *C. elegans*, it has been possible to trace the entire lineage of all the cells of the nervous system in the living animal by using Nomarski differential interference contrast microscopy (Horvitz et al.; Sulston; J.G. White et al., personal communication). These studies have shown that the differentiation of most (but not all) nerve cells in the nematode is largely autonomous and relatively independent of cell-cell interactions. If a given precursor is killed, all cells derived from that precursor will typically be absent in the adult, and all other nerve cells will be present. But no obvious rules relate cell lineage to final cell type. Individual members of a class of neurons (for example, serotonergic or dopaminergic cells) do not come from a common ancestor but from distant independent lineages. However, neurons of a given class are generated as homologous descendants of the separate progenitors, each of which undergoes the same pattern of division and generates the same complement of cell types. In general, neuronal differentiation does not require migration. Each neuron is born close to the final position of its cell body, and the choices for cell-cell interactions are limited. One of the functions of cell lineage seems to be to put the correct cell in the correct place at the right time, so that its outgrowing processes select only the appropriate targets from all available neighborhoods (J. G. White et al., personal communication). Some nerve cells do migrate and have a wider range of choices (Sulston).

A promising approach to the molecules that are important in determining the developmental program of specific cells, including cells of known behavioral function, is the use of mutants of cell lineage (Horvitz et al.). Work on *C. elegans* may also help to clarify aspects of the individuality of neurons, which I will consider later. *C. elegans* is ideally suited to this problem because each of the 273 cells of its nervous system is unique (Sulston; J. G. White et al., personal communication). It will be interesting to see to what degree the defining characteristics of these cells are reflected in distinctive patterns of gene expression.

*Pathfinding seems to involve selective adhesion
of growth cones to specific substrates*

To establish connections with specific targets, neurons send out axonal growth cones that navigate from their site of origin at one cell body to the site of the target. Studies of growth cone navigation in the grasshopper embryo suggest that these cones use a precise set of pathfinding processes to seek their target (Bastiani et al.; Bentley et al.; Raper et al.). As an embryonic limb begins to emerge from the body wall, it is initially devoid of neurons. The first axons to appear in the limb (the pioneer fibers) are the axons of sensory neurons that arise within the limb epithelium. As the growth cones of these pioneer fibers reach the central nervous system, the growth cones of the first motor axons start to emerge and run in the opposite direction. These two sets of axons thereby establish the initial scaffolding for subsequent axon outgrowth. Raper et al. have proposed that the axonal pathways established by the pioneer axons are distinctly labeled on their cell surface. Later growth cones are thought to be differentially determined in their ability to choose which labeled pathways to follow (Bastiani et al.; Raper et al.). Some labeled pathways and axon fascicles can be characterized by specific monoclonal antibodies (Hockfield et al.; Raper et al.).

To analyze the mechanisms that regulate pioneer fiber outgrowth and that allow later outgrowing axons to follow specific pathways will require a variety of molecular and mutational approaches. *Drosophila* mutants have already been used successfully to manipulate axon morphology and consequently synapse formation (Wyman and Thomas). The finding that the *Drosophila* embryo bears strong resemblances to that of the grasshopper is therefore encouraging, for it suggests that one might be able to use the wealth of experimental strategies and mutants available in *Drosophila* to explore the possible existence and the functional roles of specific cell-surface molecules in the initial pioneer fibers as well as in the later outgrowing axons (Raper et al.).

Nerve cell adhesion molecules have been
characterized at the molecular level

The direction of a field can be dramatically altered when the analysis moves from description to an examination of the action of specific molecules. A clear example of this is occurring in the study of cell adhesion, which is important for various types of cell-to-cell contact. In turn, direct cell-to-cell contact is thought to be essential for various steps in development, from the initial separation of neuronal from nonneuronal precursors, to the subsequent migration of neurons, and to the later stages of differentiation that include axonal pathfinding and synapse formation. Despite the importance of cell-to-cell contact throughout development, very little was known about the molecular events underlying it until recently. A common notion has been that neurons (and nonneuronal cells) display on their surface or secrete into the extracellular matrix macromolecules important for cell-cell interactions.

How do these signals work? There is a range of possibilities. At one extreme, Letourneau (1975) has shown that neurite outgrowth will occur merely if the growth cone is provided with an effective surface for adhesion. At the other extreme, highly specific mechanisms involving recognition as well as adhesion could exist. For example, molecules on the surface of one cell could be recognized by specific receptors on the surface of the outgrowing growth cones of other cells. Alternatively, the signal might be secreted by one cell and internalized by the outgrowing cell, so that it would act from within the second cell to influence the direction of neurite outgrowth.

Over the last decade, Edelman and his colleagues, in particular Rutishauser, have used developing chick and mouse nerve cells to isolate and characterize the first clearly identified nerve cell adhesion molecules (N-

CAMs). N-CAMs are high-molecular-weight glycoproteins (180–250 kD) that contain large amounts of the charged sugar, sialic acid. Different regions of the brain are thought to contain different forms of N-CAM, distinguished by different sialic acid residues. In addition, during development, N-CAM of high sialic acid content (embryonic N-CAM) is converted to a form much lower in sialic acid (adult N-CAM). This conversion occurs at different times in different parts of the developing nervous system. On the basis of these findings, Edelman has suggested that structurally distinct N-CAMs could play a role in both early and late development. Early in development, N-CAM might be important for segregating cell types. Once cells are committed to a given neuronal lineage, N-CAM might function in adhesion as well as in the recognition required for appropriate cell types to interact in the various regions of the brain.

There is as yet no direct evidence for the participation of N-CAM in specific recognition, but its importance for adhesion is supported by two experiments. First, N-CAM is present in the neural crest when sensory neurons initially appear, but it is transiently lost during migration while fibronectin in the pathway undergoes a concomitant increase. After the sensory neurons reach their target in the dorsal root ganglion, N-CAM reappears. Second, transformation of nerve cells by Rous sarcoma virus alters both cell-to-cell adhesion and expression of N-CAM.

We will now need to distinguish between adhesiveness (which may be a quite general property of stickiness that closely related cells could show) and cell recognition (which may require a higher degree of specificity). Consequently, we need to know the degree of specificity introduced into N-CAM by glycosylation. Alterations in the number of sialic acid residues seem to result in altered adhesiveness, but it has not been shown that such changes in adhesiveness will change the nature of the recognition events. Conceivably, N-CAM could serve to stabilize cell-to-cell contact after recognition has been achieved by a different set of molecules. Alternatively, N-CAM might be important for recognition as well.

How many molecular species are likely to be used for cell adhesion and recognition functions? The discovery of a variety of other factors involved in cell-to-cell adhesion and recognition (Goridis et al.; Lander et al.; Matthew and Patterson; Schachner et al.; Stallcup et al.) strongly indicates that at least several molecular species are involved, and there are probably more. However, if the finding proves general that adhesion molecules can exist in a large number of different forms (because of posttranslational modifications such as glycosylation), the informational potential of a few recognition molecules would be greatly enhanced and the task of analyzing developmental processes simplified.

The extracellular matrix contains proteoglycan important for neurite outgrowth

Molecules that function in cell adhesion and recognition are found not only in neurons but also in the nonneuronal cells of the extracellular matrix to which outgrowing neurons attach as they migrate and differentiate. Whereas it was once thought that the extracellular matrix merely filled the space between cells, it is now clear that the matrix serves as a substrate for outgrowing processes, secreting molecules that provide important clues for development. For example, when cerebellar granule cells or neural crest cells are cultured, their ability to adhere to and migrate on fibronectin (a protein secreted by the fibroblasts) correlates well with their migratory pattern in the animal. Several matrix factors have now been found that stimulate outgrowth from particular classes of neurons. One of these, a heparan-sulfate proteoglycan, is secreted by cultured corneal endothelial cells and promotes rapid outgrowth of neurites from sympathetic or sensory neurons when they are attached to a substrate. The molecule acts quite specifically: it stimulates only neurons that send axons to the periphery; nerve cells whose axons are restricted to the CNS do not respond to it (Lander et al.). A monoclonal antibody has now been developed that blocks this type of neurite outgrowth, and this antibody also binds to a heparan-sulfate proteoglycan (Matthew and Patterson). The appearance of immunoreactivity is correlated with axonal outgrowth.

Another example of the contributions of the extracellular matrix to outgrowth can be found in the basal lamina present between the pre- and postsynaptic elements at the nerve-muscle synapse. The basal lamina contains several polypeptides (the most potent of which is 80 kD) that direct the reformation of the presynaptic specialization and the final position of outgrowing axons during regeneration. The polypeptides also direct the infoldings and aggregation of AChRs in the plasma membrane of regenerating myofibers (Nitkin et al.; Sanes and Chiu). It will be important to relate these findings to the initial outgrowth of motor neurons in early development.

The cloning of nerve growth factor may prove important clinically

Although several neuron-specific regulatory signals have now been delineated, the first to be identified continues to be the best understood. Nerve growth factor (NGF) was discovered in 1951 by Rita Levi-Montalcini and Viktor Hamburger. It is essential for the survival of sympathetic neurons and certain sensory neurons and stimulates the outgrowth of their processes. Antibodies to NGF produce an immunosympathectomy—a selective destruction of sympathetic neurons. In a sense, the discovery of NGF and the first isolation of chemical transmitter substances marked the beginning of the molecular exploration of the nervous system—predating by a decade the analysis of the AChR. In the ensuing 30 years, research on NGF has provided a model of the level of understanding that we need to attain about the action of other factors governing the development and function of the nervous system.

Recombinant techniques and sensitive immunoassays have focused renewed attention on NGF (Darling et al.; Thoenen et al.; Ullrich et al.). The physiological site of origin of NGF, long uncertain, has now been shown to be in tissues innervated by sympathetic neurons. The density of innervation is correlated with the levels of NGF. The endogenous NGF is transported retrogradely and is accumulated in sympathetic ganglia. cDNA probes of nucleotide sequences coding for NGF should now be useful for exploring the role of NGF in the CNS and its roles, if any, in familial dystonia and other neurological diseases of humans.

The Generation of Macromolecular Complexity in the Brain

Neurobiologists, like other intellectuals, often divide themselves on ideological grounds into two groups: reductionists and holists. Reductionists (called cellular connectionists in neurobiology) tend to think that the brain is best studied on the cellular level because 1) the cell is the fundamental signaling unit of the nervous system, and 2) the cells of the brain are not identical (as are the parenchymal cells of the liver): one nerve cell often differs quite remarkably from the next. Holists think the brain should be studied only as a whole because the whole is much more than the sum of its parts. New principles emerge by looking at the brain as a whole that cannot be inferred by looking at its constituent cells.

Over the last two decades, much of what is interesting in neurobiology, and all that I have so far reviewed in the symposium, has been done within a conceptual framework of cellular connectionism. But, surprisingly, biochemists first coming into the field often resort to a more global approach, not philosophically—because biochemists are reductionists and clearly want to understand how molecules and cells work—but for expedience, because to work effectively biochemists need lots of starting material, which can most readily be obtained by grinding up whole brain or regions of the brain. However, in most cases, as they begin to appreciate that the brain is not the liver but a remarkably heterogeneous organ made up of many cell types and subtypes, most biochemists typically move either toward improving the purity of their system or toward finding a better one.

This symposium has shown that under some (I would think rare) circumstances, valuable information can come from studies of the brain as a whole tissue. First, these studies have shown that the brain expresses more genes than any other tissue of the body (Hahn et al.; Sutcliffe et al.). The kidney or liver expresses between 10,000 and 20,000 distinct mRNA sequences; the brain is thought to express at least four times as many.

In addition to expressing a greater number of genes, there is some evidence that the cells of the brain make extensive use of an unusual class of mRNA that is encountered only infrequently in other cells. Whereas in other tissues the mRNA that is translated into protein almost invariably is polyadenylated (poly[A]$^+$), Hahn and his colleagues and Chikaraishi et al. have found that the brain contains a large number of mRNAs lacking this poly(A) tail (poly[A]$^-$ mRNA). Of further interest is the finding that poly(A)$^-$ messages are not abundant at birth but become prominent only during postnatal development, suggesting that they have a special role in late stages of development. Thus, despite its preliminary nature, the finding of abundant poly(A)$^-$ mRNA in the brain and the possibility that it codes for proteins other than those coded by poly(A)$^+$ messages are intriguing and potentially important. But to demonstrate in a compelling manner the functional significance of this unusual form of mRNA, it will be essential to show that these poly(A)$^-$ mRNAs encode different proteins than do poly(A)$^+$ mRNAs.

I hasten to add an obvious cautionary comment: Not everything that is brain-specific need be important. The mammalian genome is thought to contain about 100,000 genes. Perhaps 30% of these genes may be specific to brain. One would like to think that all of them will prove to be equally interesting, but I doubt it. Some brain-specific genes will certainly prove much more important than others. Moreover, we may never be able to understand the brain completely, in all of its detail. We will want therefore to focus on some key problems and to explore them thoroughly. A corollary to this argument is that although the brain is the organ of mentation, not every brain-specific mRNA need code (or, indeed, is likely to code) for a protein involved in a *higher* mental function. On the other hand, many mental functions that we consider fascinating probably utilize the same proteins that other cells of the body use for different purposes.

The possibility that an unusual type of mRNA processing may be common in the brain raises the question of whether other specialized processes are exploited by the brain more extensively than elsewhere. Is there, for example, genomic rearrangement in the brain? In immune systems, several functional gene domains are recombined to generate a diverse set of antibody genes. Rearrangement of DNA in neurons—were it found to be nonrandom and combinatorial—could also be important for expanding the informational potential of the brain's genome.

The brain expresses more genes than other organs, but, like other organs, it can also achieve macromolecular complexity in posttranscriptional ways. Good evidence now exists for at least three classes of additional mechanisms: 1) alternative processing of mRNA, 2) alternative processing of protein precursors, and 3) covalent modifications of mature proteins (phosphorylation, methylation, glycosylation, etc.).

Evans and his colleagues have found that a single calcitonin gene is capable of generating two mRNAs and two gene products by differential processing of the mRNA. In the cells of the thyroid gland, the calcitonin gene transcribes an mRNA in which a poly(A) site is selected that is part of the calcitonin exon. In brain cells, another polyadenylation site is recognized, and this site is adjacent to the exon from a peptide (the CGR peptide) so that in the brain only this peptide gene and not the calcitonin gene is transcribed.

We have already considered alternative precursor processing as a way to produce different families of opioid peptides in different cells (Herbert et al.). The same peptide can also exist in native and covalently modified form. A common modification is phosphorylation. At least 10% of all brain-specific proteins can be phosphorylated on serine or threonine residues. Some of these (such as protein 1 or synapsin I) are common to all vertebrate nerve cells (DeGennaro et al.). This protein is composed of two peptides of 86 kD and 80 kD and is associated with synaptic vesicles (DeGennaro et al.). It may be a protein involved in vesicle mobilization or retrieval and is especially interesting because its state of phosphorylation is modulated by several chemical transmitters and by impulse activity. Other phosphoproteins are common to broad classes of cells (for example, cells receiving dopaminergic input), and some are specific to a single cell type, such as the Purkinje cell of the cerebellum (Lewis et al.). In contrast to the ubiquity of phosphorylation, methylation and adenylation are used much less frequently.

Neuronal diversity and neuronal recognition

The complexity of the nervous system emerges from many developmental steps that encompass determination, differentiation, formation of connections, cell death, synapse elimination, and fine-tuning of the remaining synapses. These steps reflect lineage, early developmental history, later competitive interactions, and other epigenetic modifications. The resulting heterogeneity of cellular properties in the brain is often referred to as neuronal diversity. Although descriptively correct, the phrase is in some ways unfortunate because it cannot help but bring to mind that much better understood instance of biological diversity, antibody diversity. The term *antibody diversity* refers to the fact the B lymphocytes make millions of different antibodies that bind to different antigenic determinants. Antibody diversity involves a change in a single product, immunoglobulin. The genetic mechanism that generates antibody diversity operates on a common family of proteins with a common function. In contrast, neurons are not diverse in only one way but have a family of diversities. Let me illustrate this by indicating the different classes of apparently independent functions that contribute to differences between neurons.

First, neurons differ in their signaling capabilities. Some neurons generate action potentials; others do not. Neurons that generate action potentials may be silent, or they may be spontaneously active even in the absence of input (like the pacemaker cells of the heart). Cells that are spontaneously active can fire either regularly or in bursts. All these (and other, more subtle) differences can be traced to the family of specific ion channels present in the membrane of a given neuron and, as we have seen, neurons differ in their ion channels.

Second, neurons differ in the chemical transmitters that they synthesize. Some neurons synthesize one of a large family of small molecule transmitters; other neurons synthesize one and often several peptide transmitters; still others synthesize combinations of these transmitters. Expression of any particular transmitter, whether it is a small molecule or a peptide, commits the cell to a whole pattern of differentiation. This pattern includes not only the biosynthetic enzymes for the transmitter but also characteristic membrane systems consisting of vesicles and uptake mechanisms, both within the vesicles and in the external membrane.

Third, neurons differ from one another in the connections they make with their target cells.

Fourth, neurons differ in the connections they receive and therefore in the receptors they have to small molecule transmitters, peptides, and hormones. Not only do cells respond differently to different transmitters, but they can respond differently to the same transmitter. For example, some receptors to a given transmitter molecule are linked to adenylate cyclase; others are not (Schramm et al.).

Fifth, neurons differ in structure, in the size of the cell body (which can vary from 5 μm to 100 μm in vertebrates and from 5 μm to 1,000 μm in invertebrates), in the presence or absence of axons, and in the number and shape of dendrites (Lasek et al.; Matus et al.). The shape of the neuron is an external manifestation of the molecular structure of its cytoskeleton (Baitinger et al.; Ginzburg et al.; Lasek et al.; Matus et al.; Weber et al.), and this too differs between cells.

Finally, neurons are thought to have distinctive recognition molecules that distinguish neurons in the various regions of the brain from each other during development and that also distinguish the position of each cell or cell grouping within each region. Surface recognition molecules are thought to be important for allowing cells to interconnect with their appropriate targets.

It is in the context of the last distinction that the concept of neuronal diversity is currently used most often. This usage derives from the fact that some developmental neurobiologists were struck by the similarity between the recognition problem faced by antibodies and that faced by nerve cells during development. Studies ranging from those of Cajal at the beginning of the century to the recent work of Sperry have indicated that neurons are connected to each other in a precise way. This precision suggested to Sperry and to other (but by no means all) developmental biologists that each neuron, or at least small groups of postsynaptic cells, may have a molecular individuality that allows recognition by appropriate outgrowing presynaptic neurons.

Even if we assume that this type of diversity in recognition molecules actually exists (and there is as yet no hard evidence to support that assumption), most probes that explore differences between neurons cannot distinguish between diversity based on cell recognition and diversity due to any one of the other factors, genetic and epigenetic, that affect the generation and modification of specific proteins that cause neurons to differ from one another. Thus, insofar as differences between neurons are explored, it will be important for the time being to specify the dimension of diversity to which those differences belong. In particular, recognitional diversity should, for the present, be distinguished from other forms of diversity, and it should be sought primarily in the developing brain. Now, it is conceivable that the other forms of diversity are secondary to recognition. Perhaps nerve cells develop specific surface molecules after birth for the purpose of recognition, and perhaps other differences derive from this early marking. But these possibilities must all be demonstrated.

How do we define a cell type in the brain?

The extent of neuronal diversity raises one of the deepest questions in neurobiology: What constitutes a cell type in the nervous system? Unlike muscle or liver, not only does the vertebrate brain consist of a very large number of nerve cells but even the most cursory examination of histological sections with a Golgi stain suggests that the brain contains many different types of cells. The current estimate is that there are about 1,000–10,000 different cell types, but there may be many more. For example, nerve cells of the same differentiated set do not connect with the same targets. The millions of Purkinje cells of the cerebellum connect not to one but to several quite different deep cerebellar nuclei. Should these subpopulations be considered different types of cells? Is there any cellular feature that correlates with output connections? Are the cells also distinguished by their inputs? Their location in the cerebellum? Is any of these features correlated with other cellular properties?

Here a family of molecular questions needs to be probed: How does one distinguish a particular cell type from another on the molecular level? How many different proteins are necessary to generate a cell type? At what level of regulation (transcriptional or posttranscriptional) is this difference established? How directly is cell type related to cell lineage in vertebrates? How directly is cell type related to particular recognition molecules? How many cell types are there *actually* in the vertebrate nervous system?

This last question is nicely illustrated with an example from the vertebrate retina. Anatomical studies using Golgi stains (until recently the major method for distinguishing cell type) have long encouraged the belief that there is only one type of amacrine cell in the retina (although Cajal early on pointed out that amacrine cells could be distinguished from one another by subtle differences in the pattern of their dendritic branching). By adding immunocytochemical and other criteria to those of Cajal, we are now able to recognize about 25 distinct classes of amacrine cells (Brecha et al. 1983). Similarly, there are several types of receptor, bipolar, and horizontal cells, and some 20 classes of ganglion cells. All told, the retina, once thought to be a simple structure consisting of only five cell types, is now known to contain about 75 cell types. The differences are based in each case on at least two independent criteria!

By analyzing subtle but fundamental differences between cells, we might be able to arrive at a fingerprint of a cell—a multifactorial definition of what constitutes a cell type. A particularly good strategy would be to seek genes and gene products that regulate the differentiation of neuronal cell types. This could perhaps best be done by examining *otherwise similarly appearing cells* such as Purkinje cells, or the granule cells of the cerebellum with two questions in mind: 1) Can reliable differences be recognized on the molecular level? 2) What functional consequences do these differences have? Monoclonal antibodies, which have primarily been used to explore distinctions between very different cells, could now be used to detect differences between otherwise similarly appearing cells. Moreover, plus/minus screening of a cDNA library made from particular regions in the brain (such as the cerebellum) against labeled cDNA made from mRNA derived from subgroups of Purkinje cells could lead to the discovery of differences in gene expression among Purkinje cells. With this approach, one might also uncover regulatory proteins that determine differences between cell types.

By analogy to other cells of the body, it would appear very likely that different cell types express different sets of genes and that a wide variety of cell types might be specified by a few genes coding for regulatory proteins, which in turn are expressed in various combinations. To give but one dramatic example, alterations in regulatory genes are thought to result in homeotic mutants of *Drosophila*, where wholesale transformations in body parts—the formation of a complete and complex alternative structure—occur following a single gene mutation. Thus, a single gene mutation converts a part of the body that normally makes an antenna into one that makes a leg by switching precursor antenna cells into making leg proteins. The proteins encoded by these mutant genes are each thought to control a large number of other genes. A number of simple schemes have been outlined whereby many different cell types can be specified by combinations of a few regulatory proteins (see Gierer 1974). The *notched* locus has homeotic properties and is known to grossly alter neuronal development. One possible outcome of cloning the *notch* locus (Kidd et al. 1983) might be an insight into a class of genes with a particularly important role in controlling neuronal differentiation.

As is evident from this discussion of cell types, the questions posed about neurons are not unique to molecular studies of brain development but are being asked in all areas of developmental biology. Molecular techniques are extremely powerful for analyzing the structure of individual molecules, but we are only beginning to develop techniques that examine how these molecules interact during development to produce a functioning cell. For a deeper understanding of the mechanisms that give rise to the different cell types of the vertebrate brain, we will need to be able to define and subtly manipulate these molecular systems.

Molecular heterogeneity can be detected with monoclonal antibodies

Since neurons differ from one another in a variety of ways, molecular heterogeneity should be readily demonstrable with hybridoma technology, which can generate specific antibodies using a complex antigen. This has indeed proved to be the case. Screens of the nervous system with libraries of monoclonal antibodies raised to the nervous system show a remarkable antigenic heterogeneity (Barnstable et al.; Hockfield et al.; McKay et al.; Zipser et al.; S. Benzer, personal communication). Some antigens are specific for neurons, some specific for glial cells, some for axons, some for cell bodies; some are unique to a specific cell type, others are shared by several cell types; the cell types may in turn prove to be related. These antibodies have therefore provided superb markers for certain cell types (see, for example, Raff et al.), and they should make possible the isolation and ultimate purification of many neuron-specific proteins. This step is badly needed. Our understanding of neurons on the molecular level is seriously limited by the fact that we know very few of their important proteins. The discovery of many new antigenic determinants, and the handle they provide to uncovering new proteins, is only the first and probably the easiest step. It will be more difficult to relate these proteins to function. One strategy for accomplishing this goal is to cast a wide net. We need to know the biochemical properties of these antigens and their distribution in the cell. The new molecular techniques make it easier to gather this information, but we also must have assay systems with which we can explore directly whether a particular gene or gene product alters a neuronal property of interest.

Interest in the biochemical functions of antigens related to neuronal differentiation should not blind us to other potentially interesting outcomes of the growing availability of monoclonal antibodies. For example, monoclonal antibodies raised against the cat spinal cord identify subsets of neurons organized in columns in the primate visual cortex (Hockfield et al.). Markers of this kind might allow us to analyze the early development of cortical columns, which current techniques have been unable to address because they require that connections between the eye and the cortex already exist. Similarly, the availability of several antibodies that distinguish different neuronal types in the *Drosophila* eye (S. Benzer, personal communication) may allow us to probe retinal development.

Perception, Behavior, and Learning

An ultimate aim of neuroscience is to provide an intellectually satisfying set of explanations, in molecular terms, of normal mentation, perception, motor coordination, feeling, thought, and memory. In addition, neuroscience would ultimately like to account for the disorders of function produced by neurological and psychiatric diseases.

I have so far outlined how the papers of this symposium and the recent progress in molecular neurobiology have contributed to our understanding of the signaling of nerve cells and their patterns of development and interconnection. The patterns of interconnection established by genetic and developmental processes, in turn, determine the capability for motor coordination and perception. Thus, the molecular information that we are beginning to obtain on how neurons interconnect will provide details essential for explaining behavior.

Take the simplest case of an elementary reflex—a sensory stimulus producing a motor response. We need to understand how information flows from the transduction of the sensory stimulus to the initiation of the movement. As I have indicated, we are gaining a fairly good appreciation of the molecular mechanisms of central synaptic transmission and neuromuscular transmission. By contrast, we have not, until recently, had comparable insights into sensory transduction. But substantial progress has now been made in understanding transduction by visual and chemical stimuli.

Advances in elucidating the molecular basis of visual transduction

(Dunn et al.; Stryer) have come from two sources: from studies of bacterio-rhodopsin and of rhodopsin in the vertebrate retina. Bacteriorhodopsin is contained in the purple membrane, a specialized patch in the membrane of *Halobacteria*. A protein of 248 amino acids, it contains a light-absorbing prosthetic group (or chromophore) called retinal, which is identical to that found in the rod photoreceptor cells of the vertebrate eye. The complete amino acid sequence of bacteriorhodopsin is now known from both protein and gene sequencing (Dunn et al.). Three-dimensional reconstruction of the protein by electron microscopy and low-angle electron diffraction analysis suggests that the protein transverses the membrane seven times, forming an α helix each time. The exact length of each helix is not known, but it is thought to be about 30 amino acids (Dunn et al.). A single photon of light excites the chromophore, causing it to change in conformation; in so doing, the chromophore transfers two H^+ ions out of the cell. The key question now being addressed is: What path does the ion take through the membrane in moving from the inside to the outside of the cell? To approach this question, Khorana and his colleagues are using site-directed mutagenesis (replacing the natural nucleotide sequence with specifically synthesized oligonucle-otides) to produce specific amino acid replacements.

In rod photoreceptor cells of vertebrates, the rhodopsin is not located in the external membrane but in the membrane of the disk, an intracellular or-ganelle. Nonetheless, photoisomerization of the 11-*cis* retinal chromophore of rhodopsin to the all-*trans* form gives rise to a potential change (a hyper-polarization) in the external (plasma) membrane of the photoreceptor neu-ron that is essential for signaling. How is this accomplished? Presumably, a chemical messenger—a transmitter or an ion—must carry information from the disk membrane to the external membrane. Evidence points strongly to the participation of two messengers: Ca^{++} and cGMP. In the dark, cGMP de-polarizes the membrane by keeping Na^+ channels open. Light activates a phosphodiesterase, which hydrolyzes cGMP. This is thought to have a role in closing the Na^+ channels and hyperpolarizing the cell (although here more quantitative data are still needed). A single photo-excited rhodopsin molecule activates several hundred molecules of phosphodiesterase. The photoactivation of the phosphodiesterase is mediated by transducin, a pe-ripheral membrane protein whose activation requires the exchange of GDP for GTP (Stryer). In its dependence on GTP and its sensitivity to inhibition by cholera toxin, transducin resembles the G protein of the adenylate cyclase system. This similarity suggests that the transducin and the adenylate cy-clase systems of other tissues belong to the same family of signal-coupling proteins and that the activation of cGMP phosphodiesterase by light resem-bles the activation of adenylate cyclase by hormones or transmitters.

A remarkably good understanding is now also being achieved into chemotransduction from studies of bacteria (Adler; Koshland et al.). Bacteria (*E. coli, Salmonella*) have different chemoreceptors for different attractant and repellent sugars. A few of these receptors are methyl-accepting (chemotaxis) proteins whose degree of covalent modification is proportional to stimulus intensity. They generate an excitatory signal—the nature of which is still not known—which determines frequency of tumbling: the changes in the direction of rotation of the flagella that move the bacterium. In response to a positive gradient of attractant, the tumbling is suppressed; the flagella rotate counterclockwise for long periods, moving the bacterium in a straight path. For an escape response to a repellent, the flagella rotate clockwise, causing the bacterium to tumble. The response of the bacterium can adapt over time, even though the attractant or repellent is still present. This adaptation results from a change in the methylation of the methyl-accepting chemotaxis proteins.

Thus, as in the adenylate cyclase and transducin systems, chemoreception in bacteria involves more than sensing and recognition of the ligand by the receptor. In each case, the receptors are part of a complex of molecules that initiate a cascade of events both in series and in parallel. In the case of the aspartate receptor (Koshland et al.), the three key functions—recognition, signal transduction, and adaptation—can be separated from each other by the techniques of in situ mutagenesis.

Recent studies have indicated that in the multicellular nervous systems of invertebrates and vertebrates there is, imposed upon the network of nerve cells and interconnections that control a behavior, a set of regulatory processes that can alter the excitable properties of nerve cells and modify the strength of their connections. These regulatory processes are activated by experience, such as learning, and result in the modification of behavior.

Learning refers to the modification of behavior by the acquisition of new information about the world; *memory* refers to the retention of the information. A given learning process can produce both long- and short-term memory. We are beginning to see in invertebrates how simple neural circuits give rise to elementary forms of behavior and how these behaviors can be modified (Aceves-Piña et al.; Kandel et al.; Schwartz et al.). Insights have come from genetic studies in *Drosophila* and from cell-biological studies in *Aplysia* and other opisthobranch mollusks into simple forms of learning and the short-term memory for each. In the three forms that have been studied, habituation, sensitization, and classical conditioning, the learning has been pinpointed to specific neurons and has been shown to involve changes in both cellular properties and synaptic strength. In the instances of short-term memory so far analyzed, the changes in synaptic strength lead to a change in the amount of transmitter released. Altered transmitter release in turn is caused by a modulation of ion channels in the presynaptic terminal. In both

Drosophila and *Aplysia*, sensitization and classical conditioning seem to involve aspects of the same molecular machinery. Short-term memory has been shown to be independent of new protein synthesis and to involve covalent modification of preexisting protein by means of cAMP-dependent protein phosphorylation (Aceves-Piña et al.; Camardo et al.; Kandel et al.; Schwartz et al.). In classical conditioning, this cascade is amplified, whereas in sensitization it is not. It is noteworthy that covalent modification of preexisting proteins also produces behavioral adaptation (this time by methylation) in bacteria (Adler; Koshland et al.).

Although we are beginning to understand aspects of the molecular changes underlying short-term memory, we know little about long-term memory. An important clue has been provided by Craig Bailey and Mary Chen (1983), who have found that long-term memory in *Aplysia* is associated with structural changes in the synapses. It is therefore possible that new protein synthesis is required to produce these changes (Schwartz et al.). With recombinant DNA techniques, one should be able to explore the question, Does learning produce long-term alterations in behavior by regulating gene expression?

Perspectives

As this last question and the many earlier questions that I have posed illustrate, we will be confronting in the nervous system some of the most difficult and profound problems in biology. The early émigrés from molecular biology were overly optimistic in 1965 in thinking that all but the biology of the brain could be inferred from the principles at hand. But they were correct in thinking that the nervous system is one of the last frontiers of biology and that insights into its cellular and molecular mechanisms are likely to be particularly penetrating and unifying. For in studying the molecular biology of the brain, we are taking another important step in a philosophical progression to which experimental biology has become almost inexorably committed since Darwin. In Darwin's time, it was difficult to accept that the human form was not uniquely created but evolved from lower animals. More recently, there has been difficulty with the narcissistically even more disturbing notion that the mental processes of humans have also evolved from those of animal ancestors and that mentation is not ethereal but can be explained in terms of nerve cells and their interconnections. The next challenge, which this symposium and modern neurobiology have opened up for us, is the possibility—indeed, the likelihood—that many molecules important for the higher nervous functions of humans may be conserved in evolution and found in the brains of much simpler animals, and, moreover, that some of these molecules may not even be unique to the cells of the brain but may be used generally by cells throughout the body. The merger of molecular biology and neurobiology that the two encounters have accomplished is therefore more than a merger of methods and concepts. Ultimately, molecular neurobiology, the joining of the disciplines, represents the emerging conviction that a coherent and biologically unified description of mentation is possible.

Acknowledgments

I have benefited from the comments on earlier drafts of this summary by James H. Schwartz, Sally Muir, Arthur Karlin, and Richard Axel.

References

Anderson CR, Stevens CF: Voltage clamp analysis of acetylcholine produced end-plate current fluctuations at frog neuromuscular junction. J Physiol 235:655–691, 1973

Bailey CH, Chen M: Morphological basis of long-term habituation and sensitization in Aplysia. Science 220:91–93, 1983

Brecha N, Eldred W, Kuljis RO, et al: Identification and localization of biologically active peptides in the vertebrate retina, in Progress in Retinal Research. Edited by Osborne NN, Chader GJ. New York, Plenum, 1983

Fatt P, Katz B: An analysis of the end-plate potential recorded with an intra-cellular electrode. J Physiol 115:320–370, 1951

Gierer A: Molecular models and combinatorial principles in cell differentiation and morphogenesis. Cold Spring Harb Symp Quant Biol 38:951–961, 1974

Hille B: Ionic basis of resting and action potentials, in Handbook of Physiology; the Nervous System, Part 1, Vol 1. Edited by Kandel ER. Bethesda, MD, The American Physiological Society, 1977, p 261

Hodgkin AL, Huxley AF: A quantitative description of membrane current and its application to conduction and excitation in nerve. J Physiol 117:500–544, 1952

Hodgkin AL, Huxley AF, Katz B: Measurement of current-voltage relations in the membrane of the giant axon of Loligo. J Physiol 116:424–448, 1952

Hökfelt T, Johansson O, Ljungdahl Å, et al: Peptidergic neurones. Nature 284:515–521, 1980

Hughes J, Smith TW, Kosterlitz HW, et al: Identification of two related pentapeptides from the brain with potent opiate agonist activity. Nature 258:577–580, 1975

Katz B, Miledi R: Membrane noise produced by acetylcholine. Nature 226:962–963, 1970

Kidd S, Lockett TJ, Young MW: The notch locus of Drosophila melanogaster. Cell 34:421–433, 1983

Letourneau PC: Cell-to-substratum adhesion and guidance of axonal elongation. Dev Biol 44:92–101, 1975

Mayeri E, Brownell P, Branton WD: Multiple, prolonged actions of neuroendocrine bag cells on neurons in Aplysia, I. effects on bursting pacemaker neurons. J Neurophysiol 42:1165–1184, 1979

Nachmansohn D: Chemical and Molecular Basis of Nerve Activity. New York, Academic Press, 1959

Neher E, Sakmann B: Single-channel currents recorded from membrane of denervated frog muscle fibres. Nature 260:799–802, 1976

Rothman BS, Mayeri E, Brown RO, et al: Primary structure and neuronal effects of α-bag cell peptide, a second candidate neurotransmitter encoded by a single gene in bag cell neurons of Aplysia. Proc Natl Acad Sci USA 80:5753–5757, 1983

Rubin LL, Schuetze SM, Weill CL, et al: Regulation of acetylcholinesterase appearance at neuromuscular junctions in vitro. Nature 283:264–267, 1980

Takeuchi A, Takeuchi N: On the permeability of the presynaptic terminal of the crayfish neuromuscular junction during synaptic inhibition and the action of γ-amino-butyric acid. J Physiol 183:433, 1966

"NEURAL SCIENCE"

Steven E. Hyman, M.D.

The work of Eric Kandel stands as an inspiration to psychiatry because it connects the experiential and biological levels of analysis with each other (Kandel 1998). In so doing, this work suggests a serious forward path for an eventual understanding of the mechanisms by which psychiatric treatments—especially psychotherapies—might act. That there might be such a connection seems uncontroversial today, but at the time when Kandel began his psychiatric training, links between psyche and brain could only be imagined, and were occasionally denied. Indeed, throughout the mid-twentieth century, many important figures in psychiatry treated neuroscience as almost irrelevant to understanding either illness or treatment. Partly as a result, the typical career path for a person interested both in serious academic psychiatry and in fundamental neuroscience was to give up one or the other. As evidenced by the papers collected here, Kandel never abandoned psychiatry. Although he devoted his career to the bench, not the ward or consulting room, he reached out to psychiatry at regular intervals to remind its practitioners of the important connections that could be established (Kandel 1998).

While openly confessing Cartesians (who would declare mind and brain to be completely different substances requiring special mechanisms to interact) were rare in late-twentieth century psychiatry, all too many psychiatrists behaved day to day as if Descartes had been right in his dualism. While by no means a universal view, many psychiatrists in the middle and even the end of the twentieth century divided disorders into those that were "biological" and others that resulted from experiences during development. For "biological" disorders, medication would be the treatment, whereas for those based on life experience, the answer would lie in psychotherapy. To some degree, this distinction remains enshrined in the *Diagnostic and Statistical Manual of Mental Disorders*, Text Revision (American Psychiatric Association 2000), in its categorical separation of personality disorders (thought to be experiential in origin) from other psychiatric disorders on its own diagnostic axis. While such a diagnostic structure would not be agreed to today de novo, it exists as a fossil record of the thinking of the 1970s. The group of colleagues who we might describe as "crypto-Cartesians" might have agreed that a brain is required either to administer psychotherapy or to benefit from it, but viewed the brain as a rather general substrate about which detailed understandings might at best serve as a distraction from clinical matters at hand (very much as Kandel describes the training environment at the Massachusetts Mental Health Center in the introduction to this volume).

The implication for psychiatry in Kandel's work and that of others who have worked on brain plasticity is that life experience and indeed all types of learning, including psychotherapy, influence thinking, emotion, and behavior by modifying synaptic connections in particular brain circuits. Moreover, as many scientists have shown, these circuits are shaped over a lifetime by multiple complexly interacting factors including genes, illness, injury, experience, context, and chance.

Clearly, we have a long way to go before we can claim understanding of the precise cellular mechanisms and neural circuits involved in psychopathology and its treatment, but substantial progress has been made in understanding the fundamental mechanisms by which memories are inscribed in neural circuits, as the following essay shows. This type of progress in basic neuroscience combined with the rise of cognitive neuroscience, brain imaging, progress in genetics (albeit slow), and, above all, open-minded pragmatism about treatment modalities in a younger generation of psychiatrists, has led to the steady, if not yet complete, emergence of a post-Cartesian psychiatry. In some sense, psychiatry as a field is now ready to grapple with the work of Kandel and other scientists who have elucidated the mechanisms by which the brain is altered by experience in health and in disease.

Besides the undercutting of dualist approaches to mind and brain that is at the core of Kandel's experimental work, there is an additional take-home message for psychiatry in the following essay, "Neural Science: A Century of Progress and the Mysteries That Remain," in which the authors take on no less ambitious a task than summarizing the highlights of neuroscience from its very beginnings to the present with some predictions as to its most fruitful future directions. Beginning with the first page of the essay, the authors distinguish two approaches to neuroscience: a top-down, or holistic, approach to problems versus a bottom-up, or reductionist, approach to problems. The essay makes it compellingly clear not only that both approaches are needed but that they must interact if progress is to be made in understanding cognition, emotion, the control of behavior, and the underpinnings of psychiatric illness. That should not be a very controversial point. It must be added, however, that progress comes only when the right approach is taken to the problem at hand. The kind of reductionism to which the essay refers is a scientific approach that is appropriate at a certain stage of problem solving; it is not a philosophical goal or a worldview. In other words, the experimental reductionism of Kandel does not represent the goal of explaining all of human behavior in terms of more and more fundamental components, such as individual cells, genes, molecules, atoms, or quarks. Rather, the point is to break down problems into tractable components, with the ultimate goal of understanding how all of the components come together—in full recognition of the fact that identifying and characterizing the individual parts does not explain higher-level phenomena. (Here we have to credit Descartes, who recommended this approach to science.) As the following essay illustrates, perhaps most clearly in its extensive discussion of the visual system, it is not possible to make progress without effective reductionist approaches, but ultimately, purely reductionist explanations will not answer our most fundamental questions.

Psychiatry has too often treated holism and reductionism as if they must be opposed to each other instead of being necessarily complementary approaches to be wielded wisely as a particular problem dictates. Taking a reductionist approach to understanding a psychiatric illness through genetics or neuropathology is not a denial of the importance of the whole person or the psychosocial context in which he or she functions but an effective route toward understanding. Kandel's career illustrates the success that comes from a disciplined approach to science. Had he taken a prematurely holistic approach to learning and memory, the results would likely have been superficial and ultimately unsatisfying. Knowing Eric as I do, I am quite certain that what he was and is most interested in are the highest integrated aspects of thought and emotion and how memory contributes to them. However, he disciplined himself to ask the most penetrating questions that were still tractable. Kandel was courageous enough to select as a model organism for the initial stage of his career *Aplysia californica,* a creature neither well known nor attractive—and presumably not even tasty (others interested in the neurobiology of behavior chose to work on the lobster). He chose *Aplysia* for the best of reductionist reasons: the organism was complex enough to exhibit simple forms of learning, but its nervous system was simple enough to be thoroughly analyzed. This organism provided a platform from which to gain a mechanistic understanding of memory, especially simple forms such as sensitization. Through years of painstaking investigation, Kandel and his colleagues were able to provide information that proved relevant to higher organisms, and indeed, through their more recent efforts on a mammalian model, the mouse, they have been able to apply what was initially learned from *Aplysia*.

It should be noted that even in disciplines that from the point of view of a psychiatrist might seem inherently fully reductionist, such as cell biology, the dialectic between reductionism and holism is playing itself out today. It turns out that the important protein building blocks of cells do not work in isolation nor can their function within even an individual cell be understood one molecule at a time. What has become clear is that the molecular components of cells function within complexly interacting networks that exhibit compensation, redundancy, and adaptation. We cannot understand the brain—or individual cells—without knowing the building blocks and their properties, but we cannot understand cells, organs, the brain, or behavior by just knowing their component parts.

References

American Psychiatric Association: Diagnostic and Statistical Manual of Mental Disorders, Fourth Edition. Washington, DC, American Psychiatric Association, 1994

Kandel ER: A new intellectual framework for psychiatry. Am J Psychiatry 155:457–469, 1998

CHAPTER 6

NEURAL SCIENCE

A Century of Progress and the Mysteries That Remain

Thomas D. Albright, Ph.D.

Thomas M. Jessell, Ph.D.

Eric R. Kandel, M.D.

Michael I. Posner, Ph.D.

Introduction

The goal of neural science is to understand the biological mechanisms that account for mental activity. Neural science seeks to understand how the neural circuits that are assembled during development permit individuals to perceive the world around them, how they recall that perception from memory, and, once recalled, how they can act on the memory of that perception. Neural science also seeks to understand the biological underpinnings of our emotional life, how emotions color our thinking, and how the regulation of emotion, thought, and action goes awry in diseases such as depression, mania, schizophrenia, and Alzheimer's disease. These are enormously complex

This article was originally published in *Cell*, Volume 100, and *Neuron*, Volume 25, 2000, pp. S1–S55.

problems, more complex than any we have confronted previously in other areas of biology.

Historically, neural scientists have taken one of two approaches to these complex problems: reductionist or holistic. Reductionist, or bottom-up, approaches attempt to analyze the nervous system in terms of its elementary components, by examining one molecule, one cell, or one circuit at a time. These approaches have converged on the signaling properties of nerve cells and used the nerve cell as a vantage point for examining how neurons communicate with one another, and for determining how their patterns of interconnections are assembled during development and how they are modified by experience. Holistic, or top-down, approaches focus on mental functions in alert, behaving human beings and in intact experimentally accessible animals and attempt to relate these behaviors to the higher-order features of large systems of neurons. Both approaches have limitations, but both have had important successes.

The holistic approach had its first success in the middle of the nineteenth century with the analysis of the behavioral consequences following selective lesions of the brain. Using this approach, clinical neurologists, led by the pioneering efforts of Paul Pierre Broca, discovered that different regions of the cerebral cortex of the human brain are not functionally equivalent (Ryalls and Lecours 1996; Schiller 1992). Lesions to different brain regions produce defects in distinctively different aspects of cognitive function. Some lesions interfere with comprehension of language, others with the expression of language; still other lesions interfere with the perception of visual motion or of shape, with the storage of long-term memories, or with voluntary action. In the largest sense, these studies revealed that all mental processes, no matter how complex, derive from the brain and that the key to understanding any given mental process resides in understanding how coordinated signaling in interconnected brain regions gives rise to behavior. Thus, one consequence of this top-down analysis has been initial demystification of aspects of mental function: of language perception, action, learning, and memory (Kandel et al. 2000).

A second consequence of the top-down approach came at the beginning of the twentieth century with the work of the Gestalt psychologists, the forerunners of cognitive psychologists. They made us realize that percepts, such as those that arise from viewing a visual scene, cannot simply be dissected into a set of independent sensory elements such as size, color, brightness, movement, and shape. Rather, the Gestaltists found that the whole of perception is more than the sum of its parts examined in *isolation*. How one perceives an aspect of an image, its shape or color, for example, is in part determined by the context in which that image is perceived. Thus, the Gestaltists made us appreciate that to understand perception we needed not only to understand the physical properties of the elements that are perceived, but more importantly, to understand how the brain reconstructs the external world in order to create a coherent and consistent internal representation of that world.

With the advent of brain imaging, the holistic methods available to the nineteenth-century clinical neurologist, based mostly on the detailed study of neurological patients with defined brain lesions, were enhanced dramatically by the ability to examine cognitive functions in intact, behaving normal human subjects (Posner and Raichle 1994). By combining modern cognitive psychology with high-resolution brain imaging, we are now entering an era when it may be possible to address directly the higher-order functions of the brain in normal subjects and to study in detail the nature of internal representations.

The success of the reductionist approach became fully evident only in the twentieth century with the analysis of the signaling systems of the brain. Through this approach, we have learned the molecular mechanisms through which individual nerve cells generate their characteristic long-range signals as all-or-none action potentials and how nerve cells communicate through specific connections by means of synaptic transmission. From these cellular studies, we have learned of the remarkable conservation of both the long-range and the synaptic signaling properties of neurons in various parts of the vertebrate brain—indeed, in the nervous systems of all animals. What distinguishes one brain region from another and the brain of one species from the next is not so much the signaling molecules of their constituent nerve cells but the number of nerve cells and the way they are interconnected. We have also learned from studies of single cells how sensory stimuli are sorted out and transformed at various relays and how these relays contribute to perception. Much as predicted by the Gestalt psychologists, these cellular studies have shown us that the brain does not simply replicate the reality of the outside world but begins at the very first stages of sensory transduction to abstract and restructure external reality.

In this review, we outline the accomplishments and limitations of these two approaches in attempts to delineate the problems that still confront neural science. We first consider the major scientific insights that have helped delineate signaling in nerve cells and that have placed that signaling in the broader context of modern cell and molecular biology. We then go on to consider how nerve cells acquire their identity, how they send axons to specific targets, and how they form precise patterns of connectivity. We also examine the extension of reductionist approaches to the visual system in an attempt to understand how the neural circuitry of visual processing can account for elementary aspects of visual perception. Finally, we turn from reductionist to holistic approaches to mental function. In the process, we confront some of the enormous problems in the biology of mental functioning that remain elusive, problems in the biology of mental functioning that have remained completely mysterious. How does signaling activity in different regions of the visual system permit us to perceive discrete objects in the visual world? How do we recognize a face? How do we become aware of that perception? How do we reconstruct that face at will, in our imagination, at a later time and in the absence of ongoing visual input? What are the biological underpinnings of our acts of will?

As the discussions below attempt to make clear, the issue is no longer whether further progress can be made in understanding cognition in the twenty-first century. We clearly will be able to do so. Rather, the issue is whether we can succeed in developing new strategies for combining reductionist and holistic approaches in order to provide a meaningful bridge between molecular mechanisms and mental processes: a true molecular biology of cognition. If this approach is successful in the twenty-first century, we may have a new, unified, and intellectually satisfying view of mental processes.

The Signaling Capabilities of Neurons

The Neuron Doctrine

Modern neural science, as we now know it, began at the turn of the century when Santiago Ramón y Cajal provided the critical evidence for the *neuron doctrine,* the idea that neurons serve as the functional signaling units of the nervous system and that neurons connect to one another in precise ways (Ramón y Cajal 1894, 1906/1967, 1911/1955). Ramón y Cajal's neuron doctrine represented a major shift in emphasis to a cellular view of the brain. Most nineteenth-century anatomists—Joseph von Gerlach, Otto Deiters, and Camillo Golgi, among them—were perplexed by the complex shape of neurons and by the seemingly endless extensions and interdigitations of their axons and dendrites (Shepherd 1991). As a result, these anatomists believed that the elements of the nervous system *did not* conform to the *cell theory* of Schleiden and Schwann, the theory that the cell was the functional unit of all eukaryotic tissues.

The confusion that prevailed among nineteenth-century anatomists took two forms. First, most were unclear as to whether the axon and the many dendrites of a neuron were in fact extensions that originated from a *single* cell. For a long time they failed to appreciate that the cell body of the neuron, which housed the nucleus, almost invariably gave rise to two types of extensions: to *dendrites* that serve as input elements for neurons and that receive information from other cells, and to an *axon* serves as the output element of the neuron and conveys information to other cells, often over long distances. Appreciation of the full extent of the neuron and its processes came ultimately with the histological studies of Ramón y Cajal and from the studies of Ross Harrison, who observed directly the outgrowth of axons and dendrites from neurons grown in isolation in tissue culture.

A second confusion arose because anatomists could not visualize and resolve the cell membrane and therefore they were uncertain whether neurons were delimited by membranes throughout their extent. As a result, many believed that the cytoplasm of two apposite cells was continuous at their points of contact and formed a syncytium or reticular net. Indeed, the neurofibrils of one cell were thought to extend into the cytoplasm of the neighboring cell, serving as a path for current flow from one cell to another. This confusion was solved intuitively and indirectly by Ramón y Cajal in the 1890s and definitively in the 1950s with the application of electron microscopy to the brain by Sanford Palay and George Palade.

Ramón y Cajal was able to address these two questions using two methodological strategies. First, he turned to studying the brain in newborn animals, where the density of neurons is low and the expansion of the dendritic tree is still modest. In addition, he used a specialized silver staining method developed by Camillo Golgi that labels only an occasional neuron, but labels these neurons in their entirety, thus permitting the visualization of their cell body, their entire dendritic tree, and their axon. With these methodological improvements, Ramón y Cajal observed that neurons, in fact, are discrete cells, bounded by membranes, and inferred that nerve cells communicate with one another only at specialized points of appositions, contacts that Charles Sherrington (1897) was later to call *synapses*.

As Ramón y Cajal continued to examine neurons in different parts of the brain, he showed an uncanny ability to infer from static images remarkable functional insights into the dynamic properties of neurons. One of his most profound insights, gained in this way, was the *principle of dynamic polarization*. According to this principle, electrical signaling within neurons is unidirectional: the signals propagate from the receiving pole of the neuron—the dendrites and the cell body—to the axon, and then along the axon to the output pole of the neuron—the presynaptic axon terminal.

The principle of dynamic polarization proved enormously influential because it provided the first functionally coherent view of the various compartments of neurons. In addition, by identifying the directionality of information flow in the nervous system, dynamic polarization provided a logic and set of rules for mapping the individual components of pathways in the brain that constitute a coherent neural circuit (Figure 6–1). Thus, in contrast to the chaotic view of the brain that emerged from the work of Golgi, Gerlach, and Deiters, who conceived of the brain as a *diffuse nerve net* in

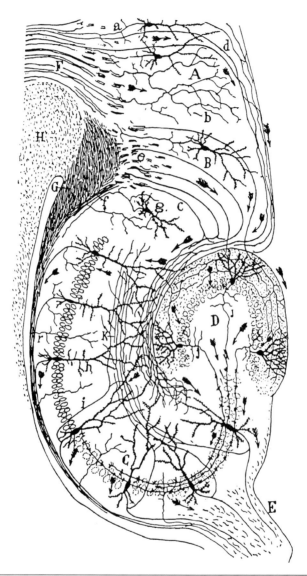

FIGURE 6–1. Ramón y Cajal's illustration of neural circuitry of the hippocampus.

A drawing by Ramón y Cajal based on sections of the rodent hippocampus, processed with a Golgi and Weigert stain. The drawing depicts the flow of information from the entorhinal cortex to the dentate granule cells (by means of the perforant pathway) and from the granule cells to the CA3 region (by means of the mossy fiber pathway) and from there to the CA1 region of the hippocampus (by means of the Schaffer collateral pathway).

Source. Based on Ramón y Cajal 1911/1955.

which every imaginable type of interaction appeared possible, Ramón y Cajal focused his experimental analysis on the brain's most important function: the processing of information.

Sherrington (1906) incorporated Ramón y Cajal's notions of the neuron doctrine, of dynamic polarization, and of the synapse into his book *The Integrative Action of the Nervous System*. This monograph extended thinking about the function of nerve cells to the level of behavior. Sherrington pointed out that the key function of the nervous system was integration; the nervous system was uniquely capable of weighing the consequences of different types of information and then deciding on an appropriate course of action based upon that evaluation. Sherrington illustrated the integrative capability of the nervous system in three ways. First, he pointed out that reflex actions serve as prototypic examples of behavioral integration; they represent coordinated, purposeful behavior in response to a specific input. For example in the flexion withdrawal and cross-extension reflex, a stimulated limb will flex and withdraw rapidly in response to a painful stimulus while, as part of a postural adjustment, the opposite limb will extend (Sherrington 1910). Second, since each spinal reflex—no matter how complex—used the motor neuron in the spinal cord for its output, Sherrington developed (1906) the idea that the motor neuron was the *final common pathway* for the integrative actions of the nervous system. Finally, Sherrington discovered (1932)—what Ramón y Cajal could not infer—that not all synaptic actions were excitatory; some could be inhibitory. Since motor neurons receive a convergence of both excitatory and inhibitory synaptic input, Sherrington argued that motor neurons represent an example—the prototypical example—of a cellular substrate for the integrative action of the brain. Each motor neuron must weigh the relative influence of two types of inputs, inhibitory and excitatory, before deciding whether or not to activate a final common pathway leading to behavior. Each neuron therefore recapitulates, in elementary form, the integrative action of the brain.

In the 1950s and 1960s, Sherrington's last and most influential student, John C. Eccles (1953), used intracellular recordings from neurons to reveal the ionic mechanisms through which motor neurons generate the inhibitory and excitatory actions that permit them to serve as the final common pathway for neural integration. In addition, Eccles, Karl Frank, and Michael Fuortes found that motor neurons had a specialized region, the initial segment of the axon, which served as a crucial integrative or decision-making component of the neuron (Eccles 1964; Fuortes et al. 1957). This component summed the total excitatory and inhibitory input and discharged an action potential if, and only if, excitation of the motor neuron exceeded inhibition by a certain critical minimum.

The findings of Sherrington and Eccles implied that each neuron solves

A

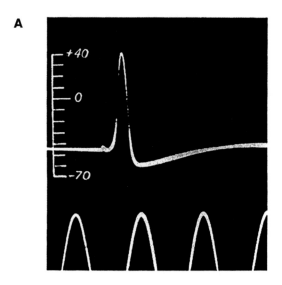

B

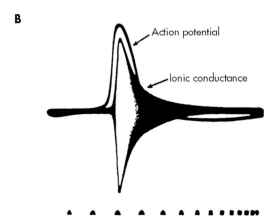

C

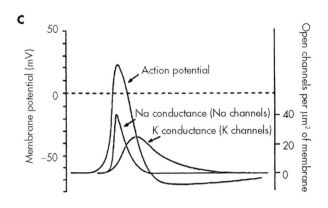

FIGURE 6-2. The action potential *(opposite page)*.

(A) This historic recording of a membrane resting potential and an action potential was obtained by Alan Hodgkin and Andrew Huxley with a capillary pipette placed across the membrane of the squid giant axon in a bathing solution of seawater. Time markers (500 Hz) on the horizontal axis are separated by 2 msec. The vertical scale indicates the potential of the internal electrode in millivolts; the seawater outside is taken as zero potential.

(B) A net increase in ionic conductance in the membrane of the axon accompanies the action potential. This historic recording from an experiment conducted in 1938 by Kenneth Cole and Howard Curtis shows the oscilloscope record of an action potential superimposed on a simultaneous record of the ionic conductance.

(C) The sequential opening of voltage-gated Na^+ and K^+ channels generates the action potential. One of Hodgkin and Huxley's great achievements was to separate the total conductance change during an action potential, first detected by Cole and Curtis (Figure 6–2B), into separate components that could be attributed to the opening of Na^+ and K^+ channels. The shape of the action potential and the underlying conductance changes can be calculated from the properties of the voltage-gated Na^+ and K^+ channels.

Source. (A) From Hodgkin AL, Huxley AF: "Action Potentials Recorded From Inside a Nerve Fiber." *Nature* 144:710–711, 1939. (B) Modified from Kandel et al. 2000. (C) From Kandel ER, Schwartz JH, Jessell T: *Principles of Neural Science,* 4th Edition. New York, McGraw-Hill, 2000.

the competition between excitation and inhibition by using, at its initial segment, a *winner takes all* strategy. As a result, an elementary aspect of the integrative action of the brain could now be studied at the level of individual cells by determining how the summation of excitation and inhibition leads to an integrated, all-or-none output at the initial segment. Indeed, it soon became evident that studies of the motor neuron had predictive value for all neurons in the brain. Thus, the initial task in understanding the integrative action of the brain could be reduced to understanding signal integration at the level of individual nerve cells.

The ability to extend the analysis of neuronal signaling to other regions of the brain was, in fact, already being advanced by two of Sherrington's contemporaries, Edgar Adrian and John Langley. Adrian (1957) developed methods of *single unit analysis* within the central nervous system, making it possible to study signaling in any part of the nervous system at the level of single cells. In the course of this work, Adrian found that virtually all neurons use a conserved mechanism for signaling *within* the cell: the *action potential.* In all cases, the action potential proved to be a large, all-or-none, regenerative electrical event that propagated without fail from the initial segment of the axon to the presynaptic terminal. Thus, Adrian showed that what made one cell a sensory cell carrying information of vision and another cell a motor cell carrying information about movement was not the nature of the action potential that each cell generated. What determined function was the neural circuit to which that cell belonged.

Sherrington's other contemporary, John Langley (1906), provided some of the initial evidence (later extended by Otto Loewi, Henry Dale, and Wilhelm Feldberg) that, at most synapses, signaling *between* neurons—*synaptic transmission*—was chemical in nature. Thus, the work of Ramón y Cajal, Sherrington, Adrian, and Langley set the stage for the delineation, in the second half of the twentieth century, of the mechanisms of neuronal signaling—first in biophysical (ionic), and then in molecular terms.

Long-range signaling within neurons: the action potential

In 1937, Alan Hodgkin found that an action potential generates a local flow of current that is sufficient to depolarize the adjacent region of the axonal membrane, in turn triggering an action potential. Through this spatially interactive process along the surface of the membrane, the action potential is propagated without failure along the axon to the nerve terminal (Figure 6–2A). In 1939, Kenneth Cole and Howard Curtis further found that when an all-or-none action potential is generated, the membrane of the axon undergoes a change in ionic conductance, suggesting that the action potential reflects the flow of ionic current (Figure 6–2B).

Hodgkin, Andrew Huxley, and Bernhard Katz extended these observations by examining which specific currents flow during the action potential. In a landmark series of papers in the early 1950s, they provided a quantitative account of the ionic currents in the squid giant axon (Hodgkin et al. 1952). This view, later called the *ionic hypothesis,* explained the resting membrane potential in terms of voltage-insensitive (nongated or leakage) channels permeable primarily to K^+ and the generation and propagation of the action potential in terms of two discrete, voltage-gated conductance pathways, one selective for Na^+ and the other selective for K^+ (Figure 6–2C).

The ionic hypothesis of Hodgkin, Huxley, and Katz remains one of the deepest insights in neural science. It accomplished for the cell biology of neurons what the structure of DNA did for the rest of biology. It unified the cellular study of the nervous system in general, and in fact, the study of ion channels in general. One of the strengths of the ionic hypothesis was its generality and predictive power. It provided a common framework for all electrically excitable membranes and thereby provided the first link between neurobiology and other fields of cell biology. Whereas action potential signaling is a relatively specific mechanism distinctive to nerve and muscle cells, the permeability of the cell membrane to small ions is a general feature shared by all cells. Moreover, the ionic hypothesis of the 1950s was so precise in its predictions that it paved the way for the molecular biological explosion that was to come in the 1980s.

Despite its profound importance, however, the analysis of Hodgkin, Huxley, and Katz left something unspecified. In particular, it left unspecified the molecular nature of the pore through the lipid membrane bilayer and the mechanisms of ionic selectivity and gating. These aspects were first addressed by Bertil Hille and Clay Armstrong. In the late 1960s, Hille devised procedures for measuring Na^+ and K^+ currents in isolation (for a review, see Hille et al. 1999). Using pharmacological agents that selectively block one but not the other ionic conductance pathway, Hille was able to infer that the Na^+ and K^+ conductance pathways of Hodgkin and Huxley corresponded to independent ion channel proteins. In the 1970s, Hille used different organic and inorganic ions of specified size to provide the first estimates of the size and shape of the pore of the Na^+ and the K^+ channels. These experiments led to the defining structural characteristic of each channel—the *selectivity filter*—the narrowest region of the pore, and outlined a set of physical-chemical mechanisms that could explain how Na^+ channels are able to exclude K^+ and conversely, how K^+ channels exclude Na^+.

In parallel, Armstrong addressed the issue of gating in response to a change in membrane voltage. How does an Na^+ channel open rapidly in response to voltage change? How, once opened, is it closed? Following initial experiments of Knox Chandler on excitation contraction coupling in muscle, Armstrong measured minute "gating" currents that accompanied the movement, within the transmembrane field, of the voltage sensor postulated to exist by Hodgkin and Huxley. This achievement led to structural predictions about the number of elementary charges associated with the voltage sensor. In addition, Armstrong discovered that mild intracellular proteolysis selectively suppresses Na^+ channel inactivation without affecting voltage-dependent activation, thereby establishing that activation and inactivation involve separate (albeit, as later shown, kinetically linked) molecular processes. Inactivation reflects the blocking action of a globular protein domain, a "ball," tethered by a flexible peptide chain to the intracellular side of the channel. Its entry into the mouth of the channel depends on the prior activation (opening) of the channel. This disarmingly simple "mechanical" model was dramatically confirmed by Richard Aldrich in the early 1990s. Aldrich showed that a cytoplasmic aminoterminal peptide "ball" tethered by a flexible chain does indeed form part of the K^+ channel and underlies its inactivation, much as Armstrong predicted.

Until the 1970s, measurement of current flow was carried out with the voltage-clamp technique developed by Cole, Hodgkin, and Huxley, a technique that detected the flow of current that followed the opening of thou-

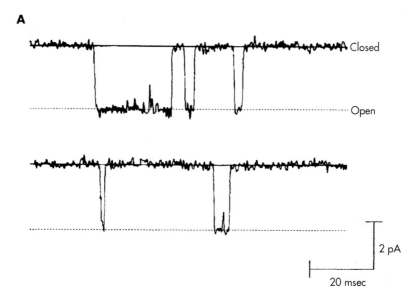

A

Closed

Open

2 pA

20 msec

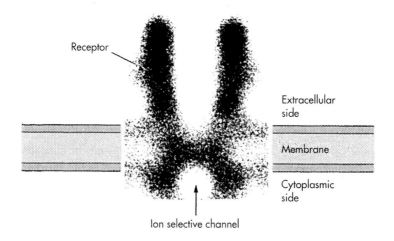

B

Receptor

Extracellular side

Membrane

Cytoplasmic side

Ion selective channel

FIGURE 6–3.   The conductance of single ion channels and a preliminary view of channel structure *(opposite page).*

(A) Recording of current flow in single ion channels. Patch-clamp record of the current flowing through a single ion channel as the channel switches between its closed and open states.

(B) Reconstructed electron microscope view of the ACh receptor-channel complex in the fish *Torpedo californica.* The image was obtained by computer processing of negatively stained images of ACh receptors. The resolution is 1.7 nm, fine enough to visualize overall structure but too coarse to resolve individual atoms. The overall diameter of the receptor and its channel is about 8.5 nm. The pore is wide at the external and internal surfaces of the membrane but narrows considerably within the lipid bilayer. The channel extends some distance into the extracellular space.

Source.   (A) Courtesy of B. Sakmann. (B) Adapted from studies by Toyoshima and Unwin; from Kandel ER, Schwartz JH, Jessell T: *Principles of Neural Science,* 4th Edition. New York, McGraw-Hill, 2000.

sands of channels. The development of patch-clamp methods by Erwin Neher and Bert Sakmann revolutionized neurobiology by permitting the characterization of the elemental currents that flow when a single ion channel—a single membrane protein—undergoes a transition from a closed to an open conformation (Neher and Sakmann 1976) (Figure 6–4A). This technical advance had two additional major consequences. First, patch clamping could be applied to cells as small as 2–5 µm in diameter, whereas voltage clamping could only be carried out routinely on cells 50 µm or larger. Now, it became possible to study biophysical properties of the neurons of the mammalian brain and to study as well a large variety of nonneuronal cells. With these advances came the realization that virtually all cells harbor in their surface membrane (and even in their internal membranes) Ca^{2+} and K^+ channels similar to those found in nerve cells. Second, the introduction of patch clamping also set the stage for the analysis of channels at the molecular level, and not only voltage-gated channels of the sort we have so far considered but also of ligand-gated channels, to which we now turn.

Short-range signaling between neurons: synaptic transmission

The first interesting evidence for the generality of the ionic hypothesis of Hodgkin, Huxley, and Katz was the realization in 1951 by Katz and Paul Fatt that, in its simplest form, chemical synaptic transmission represents an extension of the ionic hypothesis (Fatt and Katz 1951, 1952). Fatt and Katz found that the synaptic receptor for chemical transmitters was an ion channel. But rather than being gated by voltage as were the Na^+ and K^+ channels, the synaptic receptor was gated chemically, by a ligand, as Langley, Dale,

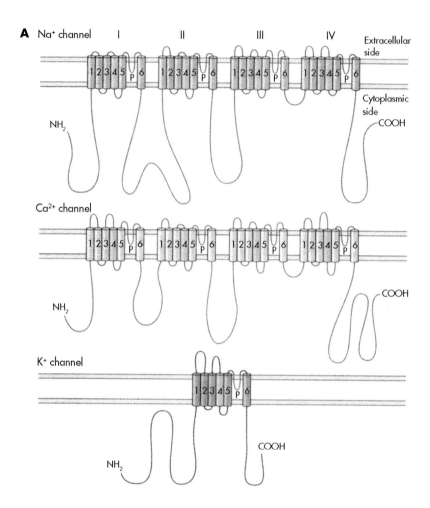

A   Na⁺ channel

B   ACh, GABA, and glycine receptor subunit

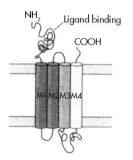

Glutamate receptor subunit

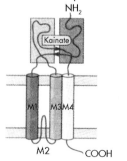

FIGURE 6–4. The membrane topology of voltage- and ligand-gated ion channels *(opposite page)*.

(A) The basic topology of the α subunit of the voltage-gated Na$^+$ channel, and the corresponding segments of the voltage-gated Ca^{2+} and K$^+$ channels. The α subunit of the Na$^+$ and Ca^{2+} channels consists of a single polypeptide chain with four repetitions of six membrane-spanning α helical regions. The S4 region, the fourth membrane-spanning α helical region, is thought to be the voltage sensor. A stretch of amino acids, the P region between the fifth and sixth α helices, dips into the membrane in the form of two strands. A fourfold repetition of the P region is believed to line the pore. The shaker type K$^+$ channel, by contrast, has only a single copy of the six α helices and the P region. Four such subunits are assembled to form a complete channel.

(B) The membrane topology of channels gated by the neurotransmitters ACh, GABA glycine, and kainate (a class of glutamate receptor ligand).

Source. (A) Adapted from Catterall 1988 and Stevens 1991. (B) From Kandel ER, Schwartz JH, Jessell T: *Principles of Neural Science,* 4th Edition. New York, McGraw-Hill, 2000.

Feldberg, and Loewi had earlier argued. Fatt and Katz and Takeuchi and Takeuchi showed that the binding of acetylcholine (ACh), the transmitter released by the motor nerve terminal, to its receptors leads to the opening of a new type of ion channel, one that is permeable to *both* Na$^+$ and K$^+$ (Takeuchi and Takeuchi 1960) (Figure 6–3). At inhibitory synapses, transmitters, typically γ-aminobutyric acid (GABA) or glycine, open channels permeable to Cl$^-$ or K$^+$ (Boistel and Fatt 1958; Eccles 1964).

In the period 1930 to 1950, there was intense controversy within the neural science community about whether transmission between neurons in the central nervous system occurred by electrical or chemical means. In the early 1950s Eccles, one of the key proponents of electrical transmission, used intracellular recordings from motor neurons and discovered that synaptic excitation and inhibition in the spinal cord was mediated by chemical synaptic transmission. He further found that the principles of chemical transmission derived by Fatt and Katz from studies of peripheral synapses could be readily extended to synapses in the nervous system (Brock et al. 1952; Eccles 1953, 1964). Thus, during the 1960s and 1970s the nature of the postsynaptic response at a number of readily accessible chemical synapses was analyzed, including those mediated by ACh, glutamate, GABA, and glycine (see, for example, Watkins and Evans 1981). In each case, the transmitter was found to bind to a receptor protein that directly regulated the opening of an ion channel. Even prior to the advent in the 1980s of molecular cloning, which we shall consider below, it had become clear, from the biochemical studies of Jean-Pierre Changeux and of Arthur Karlin, that in ligand-gated channels the transmitter binding site and the ionic channel constitute different domains within a *single* multimeric protein (for reviews see Changeux et al. 1992; Cowan and Kandel 2000; Karlin and Akabas 1995).

As with voltage-gated channels, the single channel measurements of Neher and Sakmann (1976) brought new insights into ligand-gated channels. For example, in the presence of ligand, the ACh channel at the vertebrate neuromuscular junction opens briefly (on average for 1–10 msec) and gives rise to a square pulse of inward current, roughly equivalent to 20,000 Na$^+$ ions per channel per msec. The extraordinary rate of ion translocation revealed by these single channel measurements confirmed directly the idea of the ionic hypothesis—that ions involved in signaling cross the membrane by passive electrochemical movement through aqueous transmembrane channels rather than through transport by membrane carriers (Figure 6–3A).

Following the demonstration of the chemical nature of transmission at central as well as peripheral synapses, neurobiologists began to suspect that communication at all synapses was mediated by chemical signals. In 1957, however, Edwin Furshpan and David Potter (1957) made the discovery that transmission at the giant fiber synapse in crayfish was electrical. Subsequently, Michael Bennett (1972) and others showed that electrical transmission was widespread and operated at a variety of vertebrate and invertebrate synapses. Thus, neurobiologists now accept the existence of two major modes of synaptic transmission: *electrical,* which depends on current through gap junctions that bridge the cytoplasm of pre- and postsynaptic cells; and *chemical,* in which pre- and postsynaptic cells have no direct continuity and are separated by a discrete extracellular space, the synaptic cleft (Bennett 2000).

The Proteins Involved in Generating Action Potentials and Synaptic Potentials Share Features in Common

In the 1980s, Shosaku Numa, Lily Yeh Jan, Yuh Nung Jan, William Catterall, Steven Heineman, Peter Seeburg, Heinrich Betz, and others cloned and expressed functional voltage-gated Na$^+$, Ca^{2+}, and K$^+$ channels, as well as the ligand-gated receptor channels for ACh, GABA, glycine, and glutamate (Armstrong and Hille 1998; Green et al. 1998; Numa 1989). Prior biophysical studies already had taught us much about channels, and as a consequence molecular cloning was in a position rapidly to provide powerful new insights into the membrane topology and subunit composition of both voltage-gated and ligand-gated signaling channel proteins (Armstrong and Hille 1998; Colquhoun and Sakmann 1998). Molecular cloning revealed that all ligand-gated channels have a common overall design and that this design shares features with voltage-gated channels.

Based on sequence identity, ligand-gated channels can be divided into two superfamilies: 1) receptors for glutamate (of the NMDA [N-methyl-D-aspartic acid] and non-NMDA classes) and 2) receptors for other small molecule transmitters: nicotinic ACh, 5-hydroxytryptamine, GABA, glycine, and ATP (Green et al. 1998) (Figure 6–6). Of these, the most detailed information is again available on the nicotinic ACh receptors of skeletal muscle (Figure 6–3B). This receptor is made up of four distinct subunits, α, β, γ, and δ, with the α subunit represented twice in a five-subunit channel ($\alpha_2\beta\gamma\delta$). Three-dimensional images reveal a channel made up of the five subunits surrounding the water-filled channel pore (Figures 6–3B and 6–4). Much as predicted by Hille, the channel appears to be divided into three regions: a relatively large entrance region on the external surface; a narrow transmembrane pore, only a few atomic diameters wide, which selects for ions on the basis of their size and charge; and a large exit region on the internal plasma membrane surface.

The first of the voltage-sensitive channels to be cloned, the brain Na^+ channel, was found to consist of one large (α) and two smaller (β) subunits. The α subunit is widely distributed and is the major pore-forming subunit essential for transmembrane Na^+ flux, whereas the smaller subunits are regulatory and are expressed only by subsets of cells (where they participate in channel assembly and inactivation). The α subunit consists of a single peptide of about 2,000 amino acids with four internally repeated domains of similar structures. Each domain contains six putative membrane-spanning segments, S1 to S6, which are thought to be α helical, and a reentrant P loop. The P loop connects the S5 and S6 segments and forms the outer mouth and selectivity filter of the channel.

The voltage-gated Ca^{2+} channels are similar to the Na^+ channel in their overall design. However, each of the cloned K^+ channels encodes only a single domain, of about 600 amino acids, containing the six putative transmembrane regions and the P loop. As might be predicted from this structure, four of these subunits are required to form a functional channel (either as homo- or as heterotetramers).

The wealth of sequence information that emerged from molecular cloning illustrated the remarkable conservation of channel molecules, and in turn demanded information on the structure of these channels. One of the recent successes of ion channel biology has been the first steps in the elucidation of ion channel structure. The first ion channel structure to be revealed was that of a K^+ channel (called KcsA) from the bacterium, *Streptomyces lividans*. The amino acid sequence of KcsA shows it to be most similar to the inward rectifier type of K^+ channel that contributes to the regulation of the resting membrane potential. The amino acid sequence of these channels predicts only two transmembrane domains connected by a P loop, in

A₁

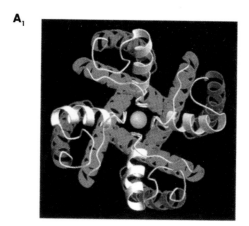

A₂

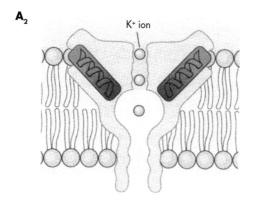

K⁺ ion

B

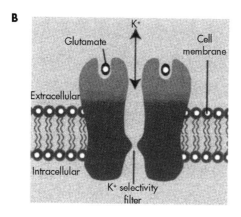

K⁺

Glutamate

Cell membrane

Extracellular

Intracellular

K⁺ selectivity filter

FIGURE 6–5. The crystal structure of a bacterial inward-rectifying K⁺ channel and a glutamate receptor *(opposite page)*.

(A₁) A view of the bacterial K⁺ channel in cross section in the plane of the membrane. The four subunits are shown, with each subunit depicted in a different color. The membrane-spanning helices are arranged as an inverted teepee.

(A₂) A side view of the channel illustrating three K⁺ ions within the channel. The pore helices contribute a negative dipole that helps stabilize the K⁺ ion in the water-filled inner chamber. The two outer K⁺ ions are loosely bound to the selectivity filter formed by the P region.

(B) Schematic depiction of a bacterial ligand-gated glutamate receptor channel with a K⁺ channel pore. The extracellular regions of the channel show sequence similarity to the ligand-binding domains of glutamate receptors (red in the figure here). The pore region resembles an inverted potassium channel pore (blue).

Source. (A₂) From Doyle DA, Morais Cabral J, Pfuetzner RA, et al: "The Structure of the Potassium Channel: Molecular Basis of K⁺ Conduction and Selectivity." *Science* 280:69–77, 1998. (B) Image courtesy of E. Gouaux; see Chen et al. 1999.

contrast to the more familiar voltage-gated K⁺ channels, which have six transmembrane domains. When reconstituted in lipid bilayers, KcsA forms a tetramer. The 3.2 Å resolution crystal structure reported by Roderick MacKinnon and his colleagues revealed that the tetramer has two transmembrane-spanning α helices connected by the P region (Doyle et al. 1998) (Figures 6–5A and 6–5B).

In retrospect it was remarkable how accurately this structure had been anticipated by the earlier biophysical studies of Hille and Armstrong. Hille and Armstrong had, for example, correctly predicted the selectivity filter to be a narrow region near the outer face of the membrane lined by polar residues. One surprise, however, is that the channel pore is not lined by hydrophilic amino acid side chains but by the carbonyl backbone of conserved amino acids, containing glycine-tyrosine-glycine residues that are characteristic of nearly all K⁺-selective channels. The narrow channel in the selectivity filter rapidly broadens in hourglass fashion to form a "lake" roughly halfway through the membrane, in which 60–100 water molecules diffuse the charges of K⁺ ions residing in this cavity. Four short α helices in the P loops have their helix dipole negative electrostatic fields focused on the cavity, further stabilizing the K⁺ ion poised at the selectivity filter. Finally, a long water-filled hydrophobic channel tunnels to the cytoplasm.

MacKinnon's compelling images even visualized two K⁺ ions within the selectivity filter. Thus, a total of three K⁺ ions are positioned at distinct sites within the pore, each separated from the other by about 8 Å. This view of a single pore capable of accommodating three K⁺ ions was precisely as predicted by Hodgkin some 50 years earlier. MacKinnon's structure thus provided explanations for K$^+$ channel selectivity and conduction. What we lack, however, is an insight into the mechanisms of voltage-dependent gating.

The membrane subunits of many voltage-dependent potassium channels associate with additional proteins known as the β subunits (Isom et al. 1994). One function of β subunits is to modify the gating of K$^+$ channels. MacKinnon and his colleagues have now gone on to provide the structure of the β subunit of a voltage-dependent K$^+$ channel from eukaryotic cells (Gulbis et al. 1999). Like the integral membrane components of the potassium channel, the β subunits have a fourfold symmetrical structure. Surprisingly, each subunit appears similar to an oxidoreductase enzyme, complete with a nicotinamide cofactor active site. Several structural features of the enzyme active site, including its location with respect to the fourfold axis, imply that it may interact directly or indirectly with the K$^+$ channel's voltage sensor. Thus, the oxidative chemistry of the cell may be intrinsically linked to changes in membrane potential by the interaction of the α and β subunits of the voltage-dependent K$^+$ channels.

The expression of ligand-gated receptors also is not limited to multicellular organisms. For example, it has become evident recently that even prokaryotes have functional ligand-gated glutamate receptors. Eric Gouaux and his colleagues (Chen et al. 1999) have cloned and expressed a glutamate-gated channel from the cyanobacterium *Synechocystis* PCC 6803, and in so doing have provided a further surprise: the receptor has a transmembrane structure similar to that of KcsA and forms a K$^+$-selective pore. Thus, this receptor is related both to the inward rectifier K$^+$ channels and to eukaryotic glutamate receptors (Figure 6–5B). The extracellular region bears sequence homology to the ligand-binding domains of glutamate receptors, whereas the pore region bears resemblance to an inverted K$^+$ channel. This finding has led Gouaux and his colleagues to propose a prokaryotic glutamate receptor as the precursor of eukaryotic receptors. In addition, this receptor provides a missing link between K$^+$ channels and glutamate receptors, and indicates that both ligand- and voltage-gated ion channels have a similar architecture, suggesting that they both derive from a common bacterial ancestor.

Synaptic Receptors Coupled to G Proteins Produce Slow Synaptic Signals

In the 1970s, evidence began to emerge from Paul Greengard and others that the neurotransmitters that activate ligand-gated (*ionotropic*) channels to produce rapid synaptic potentials lasting only milliseconds—glutamate, ACh, GABA, serotonin—also interact with a second, even larger class of receptors (termed *metabotropic* receptors) that produce slow synaptic responses that persist for seconds or minutes (for a review, see Nestler and Greengard 1984). Thus, a single presynaptic neuron releasing a single transmitter can produce a variety of actions on different target cells by activating distinct ionotropic or metabotropic receptors.

Molecular cloning revealed that these slow synaptic responses are transduced by members of a superfamily of receptors with seven transmembrane-spanning domains, which do not couple to ion channels directly but do so indirectly by means of their coupling to G proteins. G proteins couple this class of receptors to effector enzymes that give rise to second messengers such as cAMP, cGMP, diacylglycerol, and metabolites of arachidonic acid. G proteins and second messengers can activate some channels directly. More commonly, these messengers activate further downstream signaling molecules, often a protein kinase that regulates channel function by phosphorylating the channel protein or an associated regulatory protein (for review see Nestler and Greengard 1984). The family of G protein–coupled seven transmembrane-spanning receptors is remarkably large, and its members serve not only as receptors for small molecule and peptide transmitters, but also as the sensory receptors for vision and olfaction.

The study of slow synaptic potentials mediated by second messengers has added several new features to our understanding of chemical transmission. Four of these are particularly important. First, second messenger systems regulate channel function by acting on cytoplasmic domains of channels. This type of channel regulation can be achieved in three different ways: 1) through the phosphorylation of the channel protein by a second messenger-activated protein kinase, 2) through the interaction between the channel protein and a G protein activated by the ligated receptor, or 3) by the direct binding to the channel protein of a cyclic nucleotide, as is the case with the ion channels of photoreceptor and olfactory receptor cells gated by cAMP or cGMP. Second, by acting through second messengers, transmitters can modify proteins other than the channels, thereby activating a coordinated molecular response within the postsynaptic cell. Third, second messengers can translocate to the nucleus and modify transcriptional regulatory protein, in this way controlling gene expression. Thus, second messengers can covalently modify preexisting proteins as well as regulate the synthesis of new proteins. This latter class of synaptic action can lead to long-lasting structural changes at synapses. Finally, we are beginning to appreciate functional differences in slow synaptic actions. Whereas fast synaptic actions are critical for routine behavior, slow synaptic actions are often *modulatory* and act upon neural circuits to regulate the intensity, form, and duration of a given behavior (Kandel et al. 2000).

Chemical Transmitter Is Released From the
Presynaptic Terminal in Multimolecular Packets

In addition to providing initial insights into the structure and function of the
ligand-gated postsynaptic receptors responsible for postsynaptic transmission, Katz and Fatt also provided the groundwork for a molecular analysis
of transmitter release from the presynaptic terminals with the discovery of
its *quantal* nature (reviewed in B. Katz 1969). Katz, with Fatt and Jose del
Castillo, discovered that chemical transmitters, such as ACh, are released
not as single molecules but as multimolecular packets called *quanta*. At the
neuromuscular junction each quantum comprises about 5,000 molecules of
transmitter (del Castillo and Katz 1954; Fatt and Katz 1952). Each quantum
of ACh (and of other small molecule transmitters such as glutamate or
GABA) is packaged in a single small organelle, the *synaptic vesicle*, and is released by exocytosis at specialized release sites within the presynaptic terminal called the *active zones*. In response to a presynaptic action potential, each
active zone generally releases 0 or 1 quantum, in a probabilistic manner
(Figure 6–6). Synapses that release large quantities of transmitter to evoke a
large postsynaptic response, such as the synapse between nerve and muscle,
contain several hundred active zones (Heuser 1977) (Figures 6–8A and 6–
8B). In the central nervous system, however, many presynaptic terminals
contain only a single active zone.

Fatt and Katz (1952) discovered that synapses release quanta spontaneously, even in the absence of activity, giving rise to *spontaneous miniature synaptic potentials*. For a single active zone, the rate of spontaneous release is
quite low, around 10^{-2} per second. In response to a presynaptic action potential, the rate of release is dramatically, but transiently, elevated to around
1,000 per second. Within a few milliseconds, the quantal release rate then
decays back to its low resting level. We know from the work of Katz and
Ricardo Miledi as well as from the studies of Rodolfo Llinas that intracellular
Ca^{2+} is the key signal that triggers the increase in release. When the action
potential invades the terminal, it opens voltage-gated Ca^{2+} channels that are
enriched near the active zone. The resultant influx of Ca^{2+} produces localized accumulations of Ca^{2+} (to >100 µM) in microdomains of the presynaptic terminal near the active zone release site. The local increase in Ca^{2+}
concentration greatly enhances the probability of vesicle fusion and transmitter release. Many presynaptic terminals also have ionotropic and metabotropic receptors for transmitters, and these, in turn, modulate Ca^{2+} influx
during an action potential and thus modify transmitter release.

Kinetic analyses suggest that the exocytotic release of neurotransmitter
from synaptic vesicles involves a cycle composed of at least four distinct
steps: 1) the transport (or mobilization) of synaptic vesicles from a reserve

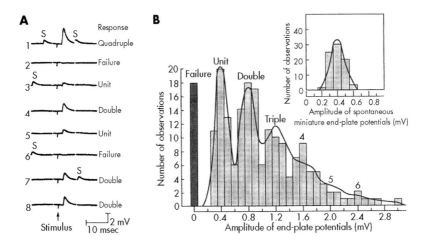

FIGURE 6-6. The quantal nature of neurotransmitter release.

Neurotransmitters are released in fixed unitary increments, or quanta. Each quantum of transmitter produces a postsynaptic potential of fixed amplitude. The amplitude of the postsynaptic potential depends on the quantal unit amplitude multiplied by the number of quanta of transmitter.

(A) Intracellular recordings show the change in potential when eight consecutive stimuli of the same size are applied to a motor nerve. To reduce transmitter output and to keep the end-plate potentials small, the tissue was bathed in a Ca^{2+}-deficient (and Mg^{2+}-rich) solution. The responses to the stimulus vary. Two impulses produce complete failures, two produce unit potentials, and the others produce responses that are approximately two to four times the amplitude of the unit potential. The spontaneous miniature end-plate potentials (S) are similar in size to the quantal unit potential.

(B) The quantal nature of neurotransmitter release. After recording many end-plate potentials, the number of responses at each amplitude was counted and plotted. The distribution of responses falls into a number of peaks. The first peak, at 0 mV, represents release failures. The first peak at 0.4 mV represents the unit potential, the smallest elicited response. This unit response is the same amplitude as the spontaneous miniature potentials (inset). The other peaks in the histogram occur at amplitudes that are integral multiples of amplitude of the unit potential. The solid line shows a theoretical Gaussian distribution fitted to the data of the histogram. Each peak is slightly spread out, reflecting the fact that the amount of transmitter in each quantum varies randomly about the peak. The distribution of amplitudes of the spontaneous miniature potentials, shown in the inset, also fits a Gaussian curve (solid line).

Source.   (A) Adapted from Liley 1956. (B) Adapted from Boyd and Martin 1956.

pool (tethered to the cytoskeleton) to a releasable pool at the active zone; 2) the docking of vesicles to their release sites at the active zone; 3) the fusion of the synaptic vesicle membrane with the plasma membrane during exocytosis, in response to a local increase in intracellular Ca^{2+}; and 4) the retrieval and recycling of vesicle membrane following exocytosis.

A major advance in the analysis of transmitter release was provided by the biochemical purification and molecular cloning of the proteins that participate in different aspects of the vesicle release cycle (Figure 6–7). Paul Greengard's work on the synapsins and their role in short-term synaptic plasticity, the work of Thomas Südhof and Richard Scheller on vesicle-associated proteins, and the work of Pietro De Camilli on membrane retrieval have each contributed seminally to our current view of the dynamics of synaptic vesicle mobilization, docking, and release (for reviews, see Bock and Scheller 1999; Fernandez-Chacon and Südhof 1999). Although we now know most of the molecular participants, at present we still do not have a precise understanding of the molecular events that control any of the four kinetic stages of release. In some instances, however, we have a beginning.

By reconstituting the vesicle cycling system in a test tube, James Rothman and his colleagues have succeeded in identifying proteins that are essential for vesicle budding, targeting, recognition, and fusion (Nickel et al. 1999; Parlati et al. 1999; Söllner et al. 1993). Based on these studies, Rothman and colleagues have proposed an influential model, according to which vesicle fusion requires specialized *donor proteins* (*vesicle snares* or *v-snares*) intrinsic to the vesicle membrane that are recognized by and bind to specific *receptor proteins* in the target membrane (*target snares* or *t-snares*).

Rothman, Scheller, and their colleagues have found that two proteins located in the nerve terminal plasma membrane—syntaxin and SNAP-25—appear to have the properties of plasma membrane t-snares, whereas synaptobrevins/VAMP (vesicle-associated membrane protein), located on the membrane of the synaptic vesicles, have the properties of the donor proteins, or v-snares. The importance of the three snare proteins—VAMP, syntaxin, and SNAP-25—in synaptic transmission was immediately underscored by the findings that these three proteins are the targets of various clostridial neurotoxins, metalloproteases that irreversibly inhibit synaptic transmission. Subsequent reconstitution studies by Rothman and his colleagues showed that fusion could occur with liposomes containing v- and t-snares (Weber et al. 1998). Finally, structural studies by Reinhard Jahn and his colleagues based on quick-freeze/deep-etch electron microscopy and X-ray crystallography demonstrated that VAMP forms a helical coiled-coil structure with syntaxin and SNAP-25 that is thought to promote vesicle fusion by bringing the vesicle and plasma membrane into close apposition (Hanson et al. 1997; Sutton et al. 1998). From these studies it would appear that vesicle

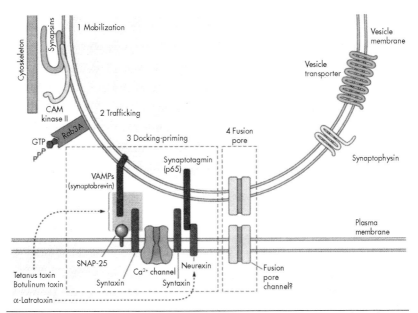

FIGURE 6-7. Some vesicle terminal membrane–associated proteins.

This diagram depicts characterized synaptic vesicle proteins and some of their postulated receptors and functions. Separate compartments are assumed for 1) storage (where vesicles are tethered to the cytoskeleton), 2) the trafficking and targeting of vesicles to active zones, 3) the docking of vesicles at active zones and their priming for release, and 4) release. Some of these proteins represent the targets for neurotoxins that act by modifying transmitter release. VAMP (synaptobrevin), SNAP-25, and syntaxin are the targets for tetanus and botulinum toxins, two zinc-dependent metalloproteases, and are cleaved by these enzymes. α-Latrotoxin, a spider toxin that generates massive vesicle depletion and transmitter release, binds to the neurexins. 1) Synapsins are vesicle-associated proteins that are thought to mediate interactions between the synaptic vesicle and the cytoskeletal elements of the nerve terminal. 2) Rab GTPases appear to be involved in vesicle trafficking within the cell and also in the targeting of vesicles within the nerve terminal. 3) Vesicle docking, fusion, and release appear to involve distinct interactions between vesicle proteins and proteins of the nerve terminal plasma membrane: VAMP (synaptobrevin) and synaptotagmin (p65) are located on the vesicle membrane, and syntaxins and neurexins on the nerve terminal membrane. Arrows indicate potential interactions suggested on the basis of in vitro studies. 4) The identity of the vesicle and plasma membrane proteins that comprise the fusion pore remains unclear. Synaptophysin, an integral membrane protein in synaptic vesicles, is phosphorylated by tyrosine kinases and may regulate release. Vesicle transporters are involved in the concentration of neurotransmitter within the synaptic vesicle.

Source. From Kandel ER, Schwartz JH, Jessell T: *Principles of Neural Science,* 4th Edition. New York, McGraw-Hill, 2000.

fusion uses a helical coiled-coil mechanism analogous to that used for viral fusion proteins (Bock and Scheller 1999; Nickel et al. 1999; Parlati et al. 1999; Söllner et al. 1993). Indeed, VAMP resembles a viral fusion peptide.

One of the most important insights to emerge from research on synaptic vesicle–associated proteins is that sets of molecules similar to those involved in mediating evoked transmitter release are also important for constitutive release. Indeed, homologs of the v- and *t-snares* participate in many aspects of membrane trafficking and constitutive vesicle fusion, including the trafficking of vesicles from the endoplasmic reticulum to the Golgi. Thus, the properties of v- and *t-snares* do not by themselves explain the specific tight Ca^{2+}-dependent regulation of vesicle fusion characteristic of evoked transmitter release from nerve terminals. Südhof has presented evidence that this calcium-dependent step in synaptic vesicle fusion is mediated by the synaptic vesicle proteins, the *synaptotagmins* (or p65). The synaptotagmins contain two domains (C2 domains) homologous to the Ca^{2+} and phospholipid-binding regulatory region of protein kinase C. This property suggested to Südhof that the synaptotagmins might insert into the presynaptic phospholipid bilayer in response to Ca^{2+} influx, thus serving as the Ca^{2+} *sensor* for exocytosis. Indeed, as shown by Charles Stevens, mice lacking the synaptotagmin-1 gene lack the fast synchronized Ca^{2+}-dependent phase of synaptic transmitter, although spontaneous release (which does not depend on Ca^{2+} influx) occurs normally (Fernandez-Chacon and Südhof 1999).

Neurotransmitter is taken up by membrane transporters

Acetylcholine was the first transmitter substance to be identified. In the course of studying its function, it soon became apparent that the action of ACh was terminated by the enzyme acetylcholinesterase. This enzyme is located in the basal membrane in close apposition to the ACh receptor and regulates the amount of ACh available for interaction with the receptor and the duration of its action. Thus, drugs that inhibit the acetylcholinesterase potentiate and prolong the synaptic effects of ACh.

Based upon this set of findings in the cholinergic system, most neurobiologists in the 1950s assumed that all neurotransmitter systems would similarly be inactivated by enzymatic degradation. Thus, when norepinephrine was discovered to be an autonomic transmitter, it was expected that there would be enzymes with a dedicated degradative function. But in 1959, Julius Axelrod and his colleagues found that actions of norepinephrine were terminated not by enzymatic degradation but by a pump-like mechanism that transports norepinephrine back into the presynaptic nerve terminal (Hertting and Axelrod 1961; Iversen 1967). Similar uptake mechanisms were soon found for serotonin and for other amine and amino acid neurotransmitters. The mechanism of enzymatic degradation that inactivates ACh, in fact, turned out to be an exception rather than a rule. Reuptake pumps now have been shown to represent the standard way in which the nervous system inactivates the common amino acid and amine neurotransmitters after they have been released from the synapse. Many therapeutically important drugs, among them antidepressants, are powerful inhibitors of the uptake of norepinephrine and serotonin. Indeed, effective antidepressants such as Prozac are selective inhibitors of the uptake of serotonin.

Peptide transmitters

In addition to small molecules, it is now clear from the work of Thomas Hokfelt and his colleagues that neurons also release small peptides as transmitters. The number of peptides that act in this way exceeds several dozen and raised the question, How do their actions relate to classical neurotransmitters? Originally it was thought that the peptide-containing neurons represented a separate class of cells: neuroendocrine cells. However, Hokfelt and his colleagues showed that peptides and classical small molecule transmitters such as ACh, norepinephrine, and serotonin coexist in individual neurons. Insight into the functional significance of co-transmission has emerged over the last two decades. In the salivary gland, for example, parasympathetic cholinergic neurons contain VIP-like peptides. In contrast, sympathetic norepinephrine neurons contain neuropeptide Y (NPY). In both cases these peptides act to augment the action of the classical transmitter. Thus, VIP induces a phase of vasodilatation and enhances the secretory effects of ACh, while NPY causes phasic vasoconstriction, like norepinephrine (Hokfelt 1991). Gene targeting studies in mice are now beginning to reveal many additional functions for neuropeptide transmitters within the central nervous system.

The Plastic Properties of Synapses

Ramón y Cajal first introduced the principle of *connection specificity:* the idea that a given neuron will not connect randomly to another but that during development a given neuron will form specific connections only with some neurons and not with others. The precision of connections that characterizes the nervous system posed several deep questions: How are the intricate neural circuits that are embedded within the mature nervous system assembled during development? How does one reconcile the properties of a specifically and precisely wired brain with the known capability of animals and humans to acquire new knowledge in the form of learning? And how is knowledge, once learned, retained in the form of memory?

One solution to this problem was proposed by Ramón y Cajal in his 1894 *Croonian Lecture,* in which he suggested that "mental exercise facilitates a greater development of the protoplasmic apparatus and of the nervous collaterals in the part of the brain in use. In this way, preexisting connections between groups of cells could be reinforced by multiplication of the terminal branches of protoplasmic appendix and nervous collaterals. But the preexisting connections could also be reinforced by the formation of new collaterals and protoplasmic expansions."

An alternative solution for memory storage was formulated in 1922 by the physiologist Alexander Forbes. Forbes suggested that memory was sustained not by *plastic changes* in synaptic strength of the sort suggested by Ramón y Cajal but by dynamic reverberating activity within a closed, interconnected loop of self-reexciting neurons. This idea was elaborated by Ramón y Cajal's student Rafael Lorente de Nó (1938), who found examples in his own analyses of neural circuitry and in those of Ramón y Cajal that neurons were often interconnected in the form of closed chains, circular pathways that could sustain reverberatory information.

This view of synaptic plasticity also was seriously challenged by B. Deslisle Burns in his influential book of 1958, *The Mammalian Cerebral Cortex.* Adopting a dynamic view, Burns wrote critically of plasticity mechanisms:

> The mechanisms of synaptic facilitation which have been offered as candidates for an explanation of memory…have proven disappointing. Before any of them can be accepted as the cellular changes accompanying conditioned reflex formation, one would have to extend considerably the scale of time on which they have been observed to operate. The persistent failure of synaptic facilitation to explain memory makes one wonder whether neurophysiologists have not been looking for the wrong kind of mechanisms. (Burns 1958, pp. 96–97)

The distinction between these two ideas—of *dynamic* as opposed to *plastic* changes for memory storage—was first tested experimentally in invertebrates, where studies of nondeclarative memory storage in the marine snail *Aplysia* showed that memory is stored as a plastic change in synaptic strength, not as self-reexciting loops of neurons. These studies found that simple forms of learning—habituation, sensitization, and classical conditioning—lead to functional and structural changes in synaptic strength of specific sensory pathways that can persist for days and that these synaptic changes parallel the time course of the memory process (Castellucci et al. 1970; Kandel and Spencer 1968). These findings reinforced the early ideas of Ramón y Cajal, which have now become one of the major themes of the molecular study of memory storage: Even though the anatomical connections between neurons develop according to a definite plan, the strength and effectiveness are not entirely predetermined and can be altered by experience (Squire and Kandel 1999).

Modern cognitive psychological studies of memory have revealed that memory storage is not unitary but involves at least two major forms: declarative (or explicit) memory and nondeclarative (or implicit) memory. Declarative memory is what is commonly thought of as memory. It is the conscious recall of knowledge about facts and events: about people, places, and objects. This memory requires the medial temporal lobe and a structure that lies deep to it: the hippocampus. Nondeclarative memory such as habituation, sensitization, classical and operant conditioning, and various habits reflect the nonconscious recall of motor and perceptual skills and strategies (Squire and Zola-Morgan 1991). In invertebrates these memories are often stored in specific sensory and motor pathways. In vertebrates these memories are stored, in addition, in three major subcortical structures: the amygdala, the cerebellum, and the basal ganglia (B. Milner et al. 1998).

Behavioral studies of both simple nondeclarative and more complex declarative memories had earlier shown that for each of these forms of memory there are at least two temporally distinct phases: a short-term memory lasting minutes and a long-term memory lasting days or longer (B. Milner 1965; B. Milner et al. 1998). These two phases differ not only in their time course, but also in their molecular mechanism: long-term but not short-term memory requires the synthesis of new proteins. Molecular studies in *Aplysia* and in mice have revealed that these distinct stages in behavioral memory are reflected in distinct molecular phases of synaptic plasticity (Abel et al. 1997; Bourtchouladze et al. 1994; Montarolo et al. 1986). In *Aplysia,* these stages have been particularly well studied in the context of sensitization, a form of learning in which an animal strengthens its reflex responses to previously neutral stimuli, following the presentation of an aversive stimulus (Byrne and Kandel 1996; Carew et al. 1983; Squire and Kandel 1999). The short- and long-term behavioral memory for sensitization is mirrored by the short- and long-term strengthening of the synaptic connections between the sensory neuron and the motor neuron that mediate this reflex. In this set of connections, serotonin, a neurotransmitter released in vivo by interneurons activated by sensitizing stimuli leads to a short-term synaptic enhancement, lasting minutes, which results from a covalent modification of preexisting proteins mediated by the cAMP-dependent protein kinase A (PKA) and by protein kinase C (PKC). By contrast, facilitation lasting several days results from the translocation of PKA and mitogen-activated protein kinase (MAPK) to the nucleus of the sensory neurons, where these kinases activate CREB-1 and derepress CREB-2, leading to the induction of a set of immediate response genes and ultimately resulting in the growth of new synaptic connections (Bartsch et al. 1995, 1998).

A similar cascade of gene induction is recruited for nondeclarative memory storage in *Drosophila* (Dubnau and Tully 1998; Yin and Tully 1996; Yin et al. 1995) and for spatial and object recognition memory, forms of declarative (explicit) memory storage that can be studied in mice (Abel et al. 1997; Bourtchouladze et al. 1994; Impey et al. 1996, 1998, 1999; Silva et al. 1998), indicating that this set of mechanisms may prove to be quite general. In both *Aplysia* and mice, experimental manipulations that reduce the level of the repressor CREB-2 or enhance the level of the activator CREB-1 act to enhance synaptic facilitation and amplify memory storage (Bartsch et al. 1995; Yin et al. 1995). Thus, this set of mechanisms may prove to be quite general and to apply to instances of both declarative and nondeclarative memory in both vertebrates and invertebrates.

The requirement for transcription provided a provisional molecular explanation for the behavioral observation that long-term memory requires the synthesis of new proteins. This requirement, however, posed a cell-biological problem: how can the activation of genes in the nucleus lead to long-lasting changes in the connectivity of those synapses that are active and not in inactive synapses? Recent studies have shown that this synapse-specific, spatially restricted plasticity requires both the activity of the activator CREB-1 in the nucleus as well as local protein synthesis in those processes of the sensory cell exposed to serotonin (Casadio et al. 1999; Martin et al. 1998).

This synapse-specific facilitation can be captured by another synapse of the neuron. Once synapse-specific long-term facilitation has been initiated, stimuli which per se induce only transient facilitation are able to recruit long-term facilitation and the growth of new connections when applied to a second branch (Casadio et al. 1999; Martin et al. 1998). A similar capture of long-term synaptic plasticity has been found in the hippocampus by Frey and Morris (1997). As we have seen, the hippocampus, a region essential for declarative memory, is involved in the storage of memory for objects and space (B. Milner et al. 1998). In 1973, Tim Bliss and Terje Lømo made the remarkable discovery that major synaptic pathways in the hippocampus, including the Schaffer collateral pathway, undergo a long-term form of synaptic plasticity (long-term potentiation, or LTP) in response to a burst of high-frequency stimulation (Figure 6–8). Subsequent studies by Graham Collingridge, Roger Nicoll, and others found that LTP in the Schaffer collateral pathway depends on activation of an NMDA receptor to glutamate in the postsynaptic cell (the pyramidal cell of the CA1 region), resulting in an influx of Ca^{2+} and an activation of the Ca^{2+} calmodulin-dependent protein kinase IIα (CaMKIIα) (for a review, see Collingridge and Bliss 1995).

The correlation between LTP in the Schaffer collateral pathway and spatial memory is not perfect (see, for example, Zamanillo et al. 1999 for an im-

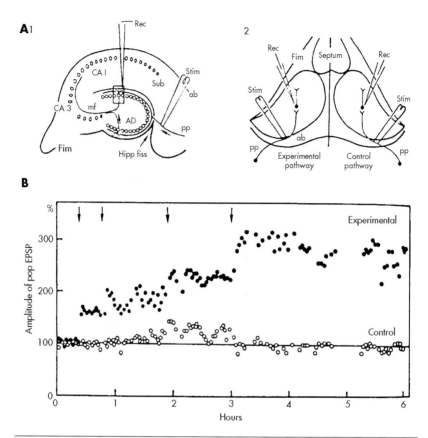

FIGURE 6–8.   The phenomenon of long-term potentiation.

Long-lasting posttetanic potentiation of the hippocampus.

(A) 1) A diagrammatic view of a parasagittal section of the hippocampus showing a stimulating electrode placed beneath the angular bundle (ab) to activate perforant pathway fibers (pp) and a recording microelectrode in the molecular layer of the dentate area (AD). Hipp fiss=hippocampal fissure; Stim=stimulatory electrode; Rec=recording electrode; Fim=fimbria. 2) Arrangement of electrodes for stimulation of the experimental pathway and the control pathway (in the contralateral hippocampus).

(B) Amplitude of the population of excitatory postsynaptic potential (EPSP) for the experimental pathway (filled dots) and ipsilateral control pathway (open dots) as a function of time and of conditioning impulse trains (15/s for 10 s) indicated by arrows. Each value is a computed average of 30 responses. Values are plotted as a percentage of the mean preconditioning value of the population (pop) EPSP.

Source.   From Bliss TVP, Lømo T: "Long-Lasting Potentiation of Synaptic Transmission in the Dentate Area of the Anaesthetized Rabbit Following Stimulation of the Perforant Path." *The Journal of Physiology* 232:331–356, 1973.

portant dissociation). Nevertheless, a variety of experiments have found that interfering with LTP in this pathway (by means of gene knockouts of the NMDA receptor or by the expression of dominant-negative transgenes) commonly interferes both with the representation of space by the neurons of the hippocampus (place cells) and with memory for space in the intact animal (Mayford and Kandel 1999; Tsien et al. 1996) (Figure 6–9). Moreover, enhancing LTP in the Schaffer collateral pathway enhances memory storage for a variety of declarative tasks (Han and Stevens 1999; Tang et al. 1999).

Despite these initial attempts to link LTP to behavioral memory storage, we still lack a satisfactory knowledge about most key facets of hippocampal synaptic plasticity in relationship to memory storage. For example, the facilitation used experimentally to induce LTP involves frequencies of firing that are unlikely to be used normally. The form of LTP used in most experiments therefore is best viewed as a marker for a general capability for synaptic plasticity. How the animal actually uses this capability is not yet known. In addition, although there is agreement that LTP is induced postsynaptically (by the activation of the NMDA receptor and consequent Ca^{2+} influx), there is no consensus on whether the mechanisms of expression are postsynaptic or presynaptic. The persistence of this lack of consensus suggests, as one possibility, that the mechanism for expression of LTP is complex and involves a coordinated pre- and postsynaptic mechanism. Finally, the hippocampus is only one component of a larger medial temporal cortical system. How the components of this system interact and how they relate to neocortical sites of storage is entirely unknown.

A Future for the Study of Neuronal Signaling

Molecular structure, molecular machines, and the integration of signaling pathways

During the last four decades, we have gained great insight from the reductionist approach to neuronal signaling and synaptic plasticity. The molecular characterization of voltage- and ligand-channels and of the many G protein–coupled receptors that we have gained has dramatically advanced the initial insights of Hodgkin, Huxley, and Katz and has revealed a structural unity among the various molecules involved in neural signaling. Elucidation of the primary sequence of these proteins also immediately revealed a commonality in the signaling functions of proteins in neurons and those of other cells. For example, many of the proteins involved in synaptic vesicle exocytosis are used for vesicle transport and for secretion in other cells including yeast. Conversely, bacteriorhodopsin, a bacterial membrane protein, has proven to be the structural prototype for understanding G protein–coupled seven transmembrane–spanning receptors such as those that are activated

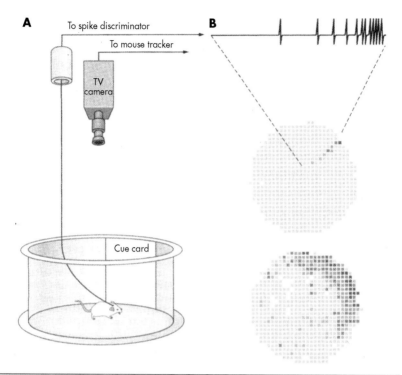

FIGURE 6-9. The detection of place field cells in the mammalian hippocampus.

(A) A recording chamber used to record the firing patterns of place cells. The head of a mouse inside the chamber is attached to a recording cable that is attached to a device able to resolve the timing of action potentials ("spikes") from one or more CA1 pyramidal (place) cells. As the mouse explores the chamber, the location of a light attached to its head is recorded by an overhead TV camera. Its output goes to a tracking device that detects the position of the mouse. The occurrence of spikes as a function of position is extracted and used to form two-dimensional firing-rate patterns that can be analyzed quantitatively or visualized as color-coded firing-rate maps.

(B) The firing patterns from a recording session of a single CA1 hippocampal pyramidal place cell. Darker colors (violet or red) indicate high rates of firing and lighter color (yellow) indicates a low firing rate. Before the recording session the animal was moved and then reintroduced into the circular enclosure. During the recording session, the mouse explores all areas of the enclosure equally. However, each place cell fires only when the mouse is in a specific location. Each time the mouse is returned to the chamber, place cells fire when the animal occupies the same locations that fired those cells previously. The firing pattern for a given cell from a wild-type mouse is stable.

Source. Courtesy of R. Muller.

by light, odorants, and chemical transmitters. Receptors of this class come into play during certain forms of learning and memory, and may even be important in primates for aspects of arousal and attention.

Although we are now only beginning to enter the era of the structural biology of voltage and of ligand-gated channels, we already appreciate that the existing molecular understanding of receptors and of ion channels is remarkably good. In retrospect, however, the obstacles confronted in the study of channels and receptors were comparatively straightforward. The essential properties of receptors and channels are contained within a single molecular entity, and these functions had been well characterized by earlier biophysical and protein chemical studies. Thus, the initial information about primary protein sequence was immediately informative in generating models of transmembrane protein topography and in defining domains that represent the voltage sensor, the ligand-binding domain, the pore, and the inactivation gate. Subsequent site-directed mutagenesis permitted rapid tests of these early predictions, tests that proved surprisingly informative because the structure of channels and receptors predicted the existence of distinct modular domains.

But we now know that many of these receptors, such as the NMDA and AMPA receptors for glutamate, do not function alone but possess specialized cytoplasmic protein domains that serve as platforms for assembling protein machines important for signaling. Thus, in shifting the focus of analysis from the ion channel to cytoplasmic signaling, we are entering a more complex arena of protein-protein interaction and in the interaction between different intracellular signaling pathways where function depends less on the properties of single molecules and intramolecular rearrangement and more on the coordination of a series of molecular events.

Fortunately, in the search for some of the components of these multimolecular machines, such as the presynaptic proteins important for the targeting and docking of vesicles at release sites and the assembly of the molecular machinery for fusion and exocytosis, the study of synaptic transmission will be aided by parallel studies in other areas of cell biology, such as membrane trafficking and viral and cellular fusion events in nonneuronal systems. Thus, despite the new realities and complexities that confront the study of cytoplasmic signaling and transmitter release, it seems safe to predict that these problems will be solved in the near future and that the romantic phase of neuronal signaling, synaptic transmission, and synaptic plasticity will reach closure in the first decades of the twenty-first century.

The great challenge for a reductionist approach in the subsequent decades of the twenty-first century will be of two sorts: first in its application to disease states, and second in its ability to contribute to the analysis of brain systems important for cognition.

Molecular biology of disease

During the last two decades, we have made remarkable progress in analyzing genes important for neurological disorders, especially monogenic diseases. That this progress has been so dramatic encourages one to believe that within the next decade the corpus of neurology may be transformed (for a review, see Cowan et al. 1999). By contrast, progress in understanding the complex polygenic diseases that characterize psychiatry has been noticeably slower.

The analysis of monogenic diseases dates to the beginning of the twentieth century, but it accelerated markedly in 1989 when Louis Kunkel and his associates first succeeded in cloning the gene for Duchenne's muscular dystrophy and found that the protein that it encodes, *dystrophin,* is homologous to α-actinin and spectrin, two cytoskeletal proteins found on the inner surface of the plasma membrane of muscle (Hoffman and Kunkel 1989; Hoffman et al. 1987). Kunkel and his associates were able to show that in severe forms of Duchenne's dystrophy the dystrophic protein (dystrophin) is lacking completely, whereas in a milder form, Becker dystrophy, functional protein is present but in much reduced amounts. Kevin Campbell and his colleagues extended this work importantly by showing that dystrophin is only one component of a larger complex of glycoproteins (the *dystroglycoprotein complex*) that links the cytoskeleton of the sarcoplasm to the extracellular matrix (Straub and Campbell 1997).

A second major step in the analysis of monogenic diseases was taken in 1993 when James Gusella, Nancy Wexler, and their colleagues in the Huntington's Disease Collaborative Research Group isolated the gene responsible for Huntington's disease. In so doing, they discovered that the gene contains an extended series of CAG repeats, thereby placing it together with a number of other important neurological diseases in a new class of disorders: the *trinucleotide repeat diseases.* These repeats were first encountered in the gene responsible for the fragile X form of mental retardation (Kremer et al. 1991; Verkerk et al. 1991). Subsequently, other hereditary disorders of the nervous system were found to have similar repeats. Together, the trinucleotide repeat disorders now constitute the largest group of dominantly transmitted neurological diseases (for reviews see Paulson and Fischbeck 1996; Reddy and Housman 1997; Ross 1997). Based on the nature of their repeats, the trinucleotide repeat disorders can be divided into two groups: type I and type II (Paulson and Fischbeck 1996).

In *type I disorders,* which include Huntington's disease, the number of CAG repeats usually does not exceed 90. The repeats lie within the coding region of the gene, are translated as polyglutamine runs, and seem to cause disease by a gain-of-function mechanism. The observation that the glutamine repeats form β sheets consisting of six to eight residues per strand suggested to Max Perutz and his colleagues (1994) that the repeats could act as a polar zipper that binds and traps other copies of either the same protein or other proteins. This trapping might not only prevent the protein from functioning normally but also could form large aggregates that may be toxic to the cells. In the case of Huntington's disease, Perutz postulated that the accumulation of huntingtin in neurons might lead to the formation of toxic protein aggregates, similar to those observed in Alzheimer's disease or certain prion disorders. Recent studies have indeed shown the existence of such nuclear aggregates, although whether such aggregates reflect the cause or the consequence of the disease remains an unsolved issue.

Type II repeat disorders, which include fragile X, have repeats found in either the 5′ or 3′ untranslated regulatory regions of the gene that result in the mRNA and protein not being expressed. In fragile X, for example, the FMR1 protein is not expressed. The wild-type protein contains RNA-binding motifs (Warren and Ashley 1995) and in one severely retarded patient the mutation was not in the regulatory region but in the coding region. Here a single point mutation in one of the RNA-binding domains is sufficient to cause the disease. These disorders are manifest by attenuated or absent expression of the gene, and the disorder is not progressive but remains fixed from early development onward.

Even in the case of these monogenic neurodegenerative disorders, however, the problem of defining the molecular basis of the disease does not stop with the identification of mutant genes. For several familial forms of neurological diseases, notably Parkinson's disease and amyotrophic lateral sclerosis (ALS or Lou Gehrig's disease), the identification of mutant protein isoforms has not yet resulted in a clearer understanding of the cellular basis of the disease. For example, our appreciation of the fact that gain-of-function mutations in the superoxide dismutase 1 (SOD1) protein underlie certain familial forms of ALS has not revealed the nature of the alteration in the function of this protein. Similarly, the identification of mutated forms of synuclein and Parkin proteins responsible for certain familial cases of Parkinson's disease has left unresolved the issue of how altered forms of these proteins lead to the degeneration of mesencephalic dopaminergic neurons. In addition, for these two disorders and for Huntington's disease, the miscreant proteins are widely expressed by virtually all neurons in the central nervous system, yet in each disease quite distinct classes of neurons undergo degeneration. The advent of more refined methods for translating the information revealed through genomic sequencing to biochemical information about the function of specific proteins in individual classes of neurons, the so-called proteomic approaches, appears to offer considerable promise in resolving these critical issues.

Of all the monogenic diseases, perhaps the most spectacular progress has been made in elucidating the defects that underlie the hereditary myotonias, periodic paralysis, and certain forms of epilepsy. These defects have now been shown to reside in one or another voltage- or ligand-gated ion channels of muscle. These disorders therefore are now referred to as the *channelopathies*—disorders of ion channel function (for review, see Brown 1993; Cowan et al. 1999; Ptácek 1997, 1998). As can be inferred from our earlier discussions, the remarkable progress in understanding these diseases can be attributed directly to the extensive knowledge about ion channel function that was already available.

For example, hyperkalemic periodic paralysis and paramyotonia congenita, two channelopathies due to ion channel disorders that result from mutations in the α subunit of the Na^+ channel, are caused by a number of slightly different dominant mutations that make the Na^+ channel hyperactive by altering the inactivation mechanisms either by changing the voltage dependency of Na^+ activation or by slowing the coupling of activation and inaction (for reviews, see Brown 1993; Ptácek et al. 1997). As was already evident from earlier physiological studies, rapid and complete inactivation of the Na^+ channel is essential for normal physiological functioning of nerve and muscle cells (Catterall 2000). These mutations do not occur randomly but in three specific regions of the channel: the inactivation gate, the inactivation gate receptor, and the voltage sensor regions that have been shown to be functionally important by the earlier biophysical and molecular studies.

In contrast to these particular monogenic diseases, the identification of the genetic basis of other degenerative neurological disorders has been slower. Nevertheless, in some complex diseases such as Alzheimer's disease, appreciable progress has been made recently. This disease begins with a striking loss of memory and is characterized by a substantial loss of neurons in the cerebral cortex, the hippocampus, the amygdala, and the nucleus basalis (the major source of cholinergic input to the cortex). On the cellular level, the disease is distinguished by two lesions: 1) there is an extracellular deposition of neuritic plaques; these are composed largely of β-amyloid ($A\beta$), a 42/43–amino acid peptide; and 2) there is an intracellular deposition of neurofibrillary tangles; these are formed by bundles of paired helical filaments made up of the microtubule-associated protein tau. Three genes associated with familial Alzheimer's disease have been identified: 1) the gene encoding the β-amyloid precursor protein (APP), 2) presenilin 1, and 3) presenilin 2.

The molecular genetic study of Alzheimer's disease has also provided us with the first insight into a gene that modifies the severity of a degenerative disease. The various alleles of the apo E gene serve as a bridge between monogenic disorders and the complexity we are likely to encounter in polygenic disorders. As first shown by Alan Roses and his colleagues, one allele of apolipoprotein E (apo E-4) is a significant risk factor for late-onset Alzheimer's disease, acting as a dose-dependent modifier of the age of onset (Stritt-matter and Roses 1996).

The findings with apo E-4 stand as a beacon of hope for the prospect of understanding the much more difficult areas of psychiatric disorders. Here the general pace of progress has been disappointing for two reasons. First, the diseases that characterize psychiatry, diseases such as schizophrenia, de-pression, bipolar disorder, and anxiety states, tend to be complex, polygenic disorders. Second, even prior to the advent of molecular genetics, neurology had already succeeded in localizing the major neurological disorders to var-ious regions of the brain. By contrast, we know frustratingly little about the anatomical substrata of most psychiatric diseases. A reliable neuropathology of mental disorders is therefore severely needed.

Systems problems in the study of memory and other cognitive states

As these arguments about anatomical substrata of psychiatric illnesses make clear, neural science in the long run faces problems of understanding aspects of biology of normal function and of disease, the complexity of which tran-scends the individual cell and involves the computational power inherent in large systems of cells unique to the brain.

For example, in the case of memory, we have here only considered the cell and molecular mechanisms of memory storage, mechanisms that appear to be shared, at least in part, by both declarative and nondeclarative memory. But, at the moment, we know very little about the much more complex sys-tems problems of memory: how different regions of the hippocampus and the medial temporal lobe—the subiculum, the entorhinal, parahippocam-pal, and perirhinal cortices—participate in the storage of nondeclarative memory and how information within any one of these regions is transferred for ultimate consolidation in the neocortex. We also know nothing about the nature of recall of declarative memory, a recall that requires conscious effort. As these arguments and those of the next sections will make clear, the sys-tems problems of the brain will require more than the bottom-up approach of molecular and developmental biology; they will also require the top-down approaches of cognitive psychology, neurology, and psychiatry. Finally, it will require a set of syntheses that bridge between the two.

The Assembly of Neuronal Circuits

The primary goal of studies in developmental neurobiology has been to clar-ify the cellular and molecular mechanisms that endow neurons with the ability to form precise and selective connections with their synaptic part-ners—a selectivity that underlies the appropriate function of these circuits in the mature brain. Attempts to explain how neuronal circuits are assembled have focused on four sequential developmental steps. Loosely defined, these are: the specification of distinct neuronal cell types; the directed outgrowth of developing axons; the selection of appropriate synaptic partners; and finally, the refinement of connections through the elimination of certain neurons, axons, and synapses. In recent years, the study of these processes has seen enormous progress (Cowan et al. 1997), and to some extent, each step has emerged as an experimental discipline in its own right.

In this section of the review, we begin by describing some of the major advances that have occurred in our understanding of the events that direct the development of neuronal connections, focusing primarily on the cellular and molecular discoveries of the past two decades. Despite remarkable progress, however, a formidable gap still separates studies of neuronal circuitry at the developmental and functional levels. Indeed, in the context of this review it is reasonable to question whether efforts to unravel mechanisms that control the development of neuronal connections have told us much about the functions of the mature brain. And similarly, it is worth considering whether developmental studies offer any prospect of providing such insight in the foreseeable future. In discussing the progress of studies on the development of the nervous system, we will attempt to indicate why such a gap exists and to describe how recent technical advances in the ability to manipulate gene expression in developing neurons may provide new experimental strategies for studying the function of intricate circuits embedded in the mature brain. In this way it should be possible to forge closer links between studies of development and systems-oriented approaches to the study of neural circuitry and function.

The Emergence of Current Views of the Formation of Neuronal Connections

Current perspectives on the nature of the complex steps required for the formation of neuronal circuits have their basis in many different experimental disciplines (Cowan 1998). We begin by discussing, separately, some of the conceptual advances in understanding how the diversity of neuronal cell types is generated, how the survival of neurons is controlled, and how different classes of neurons establish selective pathways and connections.

Inductive Signaling, Gene Expression, and the Control of Neuronal Identity

The generation of neuronal diversity represents an extreme example of the more general problem of how the fates of embryonic cells are specified. Extreme in the sense that the diversity of neuronal cell types, estimated to be in the range of many hundreds (Stevens 1998), far exceeds that for other tissues and organs. Nevertheless, as with other cell types, neural cell fate is now known to be specified through the interplay of two major classes of factors. The first class constitutes cell surface or secreted signaling molecules that, typically, are provided by localized embryonic cell groups that function as organizing centers. These secreted signals influence the pathway of differentiation of neighboring cells by activating the expression of cell-intrinsic determinants. In turn, these determinants direct the expression of downstream effector genes, which define the later functional properties of neurons, in essence their identity. Tracing the pathways that link the action of secreted factors to the expression and function of cell-intrinsic determinants thus lies at the core of attempts to discover how neuronal diversity is established.

The first contribution that had a profound and long-lasting influence on future studies of neural cell fate specification was the organizer grafting experiment of Hans Spemann and Hilde Mangold, performed in the early 1920s (Spemann and Mangold 1924). Spemann and Mangold showed that naive ectodermal cells could be directed to generate neural cells in response to signals secreted by cells in a specialized region of the gastrula-stage embryo, termed the organizer region. Transplanted organizer cells were shown to maintain their normal mesodermal fates but were able to produce a dramatic change in the fate of neighboring host cells, inducing the formation of a second body axis that included a well-developed and duplicated nervous system.

Spemann and Mangold's findings prompted an intense, protracted, and initially unsuccessful search for the identity of relevant neural inducing factors. The principles of inductive signaling revealed by the organizer experiment were, however, extended to many other tissues, in part through the studies of Clifford Grobstein, Norman Wessells, and their colleagues in the 1950s and 1960s (see Wessells 1977). These studies introduced the use of in vitro assays to pinpoint sources of inductive signals, but again failed to reveal the molecular nature of such signals.

Only within the past decade or so has any significant progress been made in defining the identity of such inductive factors. One of the first breakthroughs in assigning a molecular identity to a vertebrate embryonic inductive activity came in the late 1980s through the study of the differentiation of the mesoderm. An in vitro assay of mesodermal induction developed by Peter Nieuwkoop (see Jones and Smith 1999; Nieuwkoop 1997) was used by Jim Smith, Jonathan Cooke, and their colleagues to screen candidate factors and to purify conditioned tissue culture media with inductive activity. This search led eventually to the identification of members of the fibroblast growth factor and transforming growth factor β (TGF-β) families as mesoderm-inducing signals (Smith 1989).

Over the past decade, many assays of similar basic design have been used to identify candidate inductive factors that direct the formation of neural tissue and specify the identity of distinct neural cell types. The prevailing view of the mechanism of neural induction currently centers on the ability of several factors secreted from the organizer region to inhibit a signaling pathway mediated by members of the TGF-β family of peptide growth factors (see Harland and Gerhart 1997). The function of TGF-β proteins, when not constrained by organizer-derived signals, appears to be to promote epidermal fates at the expense of neural differentiation. The constraint on TGF-β–related protein signaling appears to be achieved in part by proteins produced by the organizer, such as noggin and chordin, that bind to and inhibit the function of secreted TGF-β–like proteins. Other candidate neural inducers may act instead by repressing the expression of TGF-β–like genes. However, even now, the identity of physiologically relevant neural inducing factors and the time at which neural differentiation is initiated remain matters of debate.

Some of the molecules involved in the specification of neuronal subtype identity, notably members of the TGF-β, fibroblast growth factor, and Hedgehog gene families, have also been identified (Lumsden and Krumlauf 1996; Tanabe and Jessell 1996). These proteins have parallel functions in the specification of cell fate in many nonneural tissues. Thus, the mechanisms used to induce and pattern neuronal cell types appear to have been co-opted from those employed at earlier developmental stages to control the differentiation of other cells and tissues. Some of these inductive signals appear to be able to specify multiple distinct cell types through actions at different concentration thresholds—the concept of gradient morphogen signaling (Gurdon et al. 1998; Wolpert 1969). In the nervous system, for example, signaling by Sonic hedgehog at different concentration thresholds appears sufficient to induce several distinct classes of neurons at specific positions along the dorsoventral axis of the neural tube (Briscoe and Ericson 1999).

The realization that many different neuronal cell types can be generated in response to the actions of a single inductive factor has placed added emphasis on the idea that the specification of cell identity depends on distinct profiles of gene expression in target cells. Such specificity in gene expression may be achieved in part through differences in the initial signal transduction pathways activated by a given inductive signal. But the major contribution to specificity appears to be the selective expression of different target genes in cell types with diverse developmental histories and thus different responses to the same inductive factor.

The major class of proteins that possess cell-intrinsic functions in the determination of neuronal fate are transcription factors: proteins with the capacity to interact directly or indirectly with DNA and thus to regulate the expression of downstream effector genes. The emergence of the central role of transcription factors as determinants of neuronal identity has its origins in studies of cell patterning in nonneural tissues and in particular in the genetic analysis of pattern formation in the fruit fly *Drosophila*. The pioneering studies of Edward Lewis (1985) on the genetic control of the *Drosophila* body plan led to the identification of genes of the *HOM-C* complex, members of which control tissue pattern in individual domains of the overall body plan. Lewis further showed that the linear chromosomal arrangement of *HOM-C* genes correlates with the domains of expression and function of these genes during *Drosophila* development. Subsequently, Christine Nüsslein-Vollhard and Eric Wieschaus (1980) performed a systematic series of screens for embryonic patterning defects and identified an impressive array of genes that control sequential steps in the construction of the early embryonic body plan. The genes defined by these simple but informative screens could be ordered into hierarchical groups, with members of each gene group controlling embryonic pattern at a progressively finer level of resolution (see St. Johnston and Nüsslein-Volhard 1992).

Advances in recombinant DNA methodology permitted the cloning and structural characterization of the *HOM-C* genes and of the genes controlling the embryonic body plan. The genes of the *HOM-C* complex were found to encode transcription factors that share a 60-amino acid DNA-binding cassette, termed the homeodomain (McGinnis et al. 1984; Scott and Weiner 1984). Many of the genes that control the embryonic body plan of *Drosophila* were also found to encode homeodomain transcription factors and others encoded members of other classes of DNA-binding proteins. The product of many additional genetic screens for determinants of neuronal cell fate in *Drosophila* and *C. elegans* led notably to the identification of basic helix-loop-helix proteins as key determinants of neurogenesis (Chan and Jan 1999). In the process, these screens reinforced the idea that cell-specific patterns of transcription factor expression provide a primary mechanism for generating neuronal diversity during animal development.

The cloning of *Drosophila* and *C. elegans* developmental control genes was soon followed by the identification of structural counterparts of these genes in vertebrate organisms, in the process revealing a remarkable and somewhat unanticipated degree of evolutionary conservation in developmental regulatory programs. The identification of over 30 different families of vertebrate transcriptional factors, each typically comprising tens of individual family members (see Bang and Goulding 1996), has provided a critical molecular insight into the extent of neural cell diversity during vertebrate development. Prominent among these are the homeodomain protein counterparts of many *Drosophila* genes. Vertebrate homeodomain proteins have now been implicated in the control of regional neural pattern, neural identity, axon pathfinding, and the refinement of exuberant axonal projections. The individual or combinatorial profiles of expression of transcription factors may soon permit the distinction of hundreds of embryonic neuronal subsets.

Genetic studies in mice and zebra fish have demonstrated that a high proportion of these genes have critical functions in establishing the identity of the neural cells within which they are expressed. In many cases, the classes of embryonic neurons defined on the basis of differential transcription factor expressions have also been shown to be relevant to the later patterns of connectivity of these neurons. Because of these advances, the problem of defining the mechanisms of cell fate specification in the developing nervous system can now largely be reduced to the issue of tracing the pathway that links an early inductive signal to the profile of transcription factor expression in a specific class of postmitotic neuron—a still daunting, but no longer unthinkable, task.

Control of Neuronal Survival

The tradition of experimental embryology that led to the identification of inductive signaling pathways has also had a profound impact on studies of a specialized, if unwelcome, fate of developing cells: their death.

Many cells in the nervous system and indeed throughout the entire embryo are normally eliminated by a process of cell death. The recognition of this remarkable feature of development has its origins in embryological studies of the influence of target cells on the control of the neuronal number. In the 1930s and 1940s, Samuel Detwiler, Viktor Hamburger, and others showed that the number of sensory neurons in the dorsal root ganglion of amphibian embryos was increased by transplantation of an additional limb bud and decreased by removing the limb target (Detwiler 1936). The target-dependent regulation of neuronal number was initially thought to result from a change in the proliferation and differentiation of neuronal progenitors. A then-radical alternative view, proposed by Rita Levi-Montalcini and Viktor Hamburger in the 1940s, suggested that the change in neuronal number reflected instead an influence of the target on the survival of neurons (Hamburger and Levi-Montalcini 1949). For example, about half of the motor neurons generated in the chick spinal cord are destined to die during embryonic development. The number that die can be increased by removing the target and reduced by adding an additional limb (Hamburger 1975). The phenomenon of neuronal overproduction and its compensation through cell death is now known to occur in almost all neuronal populations within the central and peripheral nervous systems (Oppenheim 1981).

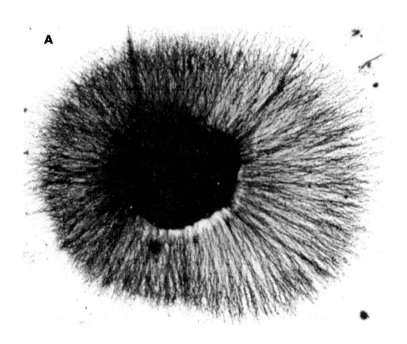

B Neurotrophin receptor interactions

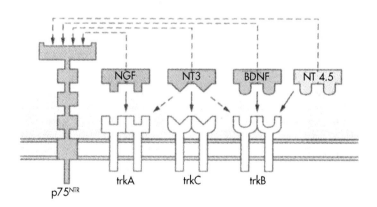

FIGURE 6-10. Growth factors and their receptors *(opposite page)*.

(A) The trophic actions of nerve growth factor on dorsal root ganglion neurons. Photomicrographs of a dorsal root ganglion of a 7-day chick embryo that had been cultured in medium supplemented with nerve growth factor for 24 hours. Silver impregnation. The extensive outgrowth of neurites is not observed in the absence of nerve growth factor.

(B) The actions of neurotrophins depend on interactions with trk tyrosine kinase receptors. Neurotrophins interact with tyrosine kinase receptors of the trk class. The diagram illustrates the interactions of members of the neurotrophin family with distinct trk proteins. Strong interactions are depicted with solid arrows; weaker interactions with broken arrows. In addition, all neurotrophins bind to a low-affinity neurotrophin receptor p75^{NTR}.

Abbreviations: NGF=nerve growth factor; NT=neurotrophin; BDNF=brain-derived neurotrophic factor.

Source. (A) From studies of R. Levi-Montalcini; courtesy of the American Association for the Advancement of Science. (B) From Kandel ER, Schwartz JH, Jessell T: *Principles of Neural Science,* 4th Edition. New York, McGraw-Hill, 2000.

The findings of Levi-Montalcini and Hamburger led to the formulation of the *neurotrophic factor hypothesis:* the idea that the survival of neurons depends on essential nutrient or trophic factors that are supplied in limiting amounts by cells in the environment of the developing neuron, often its target cells (see Oppenheim 1981). This hypothesis prompted Levi-Montalcini and Stanley Cohen to undertake the purification of a neurotrophic activity—an ambitious quest, but one that led eventually to the identification of nerve growth factor (NGF), the first peptide growth factor and a protein whose existence dramatically supported the neurotrophic factor hypothesis (Hamburger 1993; Levi-Montalcini 1966) (Figure 6–10A). The isolation of NGF was a milestone in the study of growth factors and, in turn, motivated searches for additional neurotrophic factors. The efforts of Hans Thoenen, Yves Barde, and others revealed that NGF is but the vanguard member of a large array of secreted factors that possess the ability to promote the survival of neurons (Reichardt and Fariñas 1997).

The best-studied class of neurotrophic factors, which includes NGF itself, are the neurotrophins. Work by Mariano Barbacid, Luis Parada, Eric Shooter, and others subsequently showed that neurotrophin signaling is mediated by the interaction of these ligands with a class of membrane-spanning tyrosine kinase receptors, the trk proteins (see Reichardt and Fariñas 1997) (Figure 6–10B). Nerve growth factor interacts selectively with trkA, and other neurotrophins interact with trkB and trkC. Other classes of proteins that promote neuronal survival include members of the TGF-β family, the interleukin 6–related cytokines, fibroblast growth factors, and hedgehogs (Pettmann and Henderson 1998). Thus, classes of secreted proteins that have inductive activities at early stages of development can also act later to control neuronal survival. Neurotrophic factors were initially considered to promote the survival of neural cells through their ability to stimulate cell metabolism. Quite the contrary. Such factors are now appreciated to act predominantly by suppressing a latent cell suicide program. When unrestrained by neurotrophic factor signaling, this suicide pathway kills cells by *apoptosis*, a process characterized by cell shrinkage, the condensation of chromatin, and eventually cell disintegration (Jacobson et al. 1997; Pettmann and Henderson 1998).

A key insight into the biochemical machinery driving this endogenous cell death program emerged from genetic studies of cell death in C. *elegans* by Robert Horvitz and his colleagues (Hengartner and Horvitz 1994; Metzstein et al. 1998). Over a dozen cell death (*ced*) genes have now been ordered in a pathway that controls cell death in C. *elegans*. Of these genes two, *ced-3* and *ced-4*, have pivotal roles. The function of both genes is required for the death of all cells that are normally fated to die by apoptosis. A third key gene, *ced-9*, antagonizes the activities of *ced-3* and *ced-4*, thus protecting cells from death. Remarkably, this death pathway is highly conserved in vertebrate cells. The ced-3 gene encodes a protein closely related to members of the vertebrate family of caspases, cysteine proteases that function as cell death effectors by degrading target proteins essential for cell viability. The ced-4 gene encodes a protein structurally related to another vertebrate apoptosis-promoting factor, termed Apaf-1. The ced-9 gene encodes a protein that is structurally and functionally related to the Bcl-2–like proteins, some of which also act to protect vertebrate cells from apoptotic death. Apaf-like proteins appear to promote the processing and activation of caspases, whereas some Bcl-2–like proteins interact with Apaf-1/*ced-4* and in so doing, inhibit the processing and activation of caspases.

These findings have revealed a core biochemical pathway that regulates the survival of cells and is thought to serve as the intracellular target of neurotrophic factors. The practical significance of this core cell death pathway has not escaped attention. Pharmacological strategies to inhibit caspase activation are now widely sought after in attempts to prevent the apoptotic neuronal death that accompanies many neurodegenerative disorders.

Axonal Projections and the Formation of Selective Connections

Attempts to unravel how selective neuronal connections are formed in the developing brain have a somewhat different provenance. The electrophysiological studies of John Langley (1897), Charles Sherrington (1906), and others at the turn of the twentieth century, as discussed earlier, had revealed the exquisite selectivity with which mature neuronal circuits function and in the process provided an early hint that their formation may also be a selective process. In parallel, histological studies of the developing brain, applied most decisively by Ramón y Cajal (1911/1955) but also by many others, provided dramatic illustration of embryonic neurons captured in the process of extending dendrites and axons, apparently in a highly stereotyped manner. These pioneering anatomical descriptions provided circumstantial but persuasive evidence that the assembly of neuronal connections is orchestrated in a highly selective manner. By the middle of the twentieth century, many elegant in vivo observations in simple vertebrate organisms had further shown that developing axons extend in a highly reproducible fashion (see Speidel 1933). But even these findings did not result in general acceptance of the idea that the specificity evident in mature functional connections had its basis in selective axonal growth and in selective synapse formation.

An alternative view, advanced most forcefully by Paul Weiss (1941) in the 1930s and 1940s, and termed the *resonance hypothesis*, argued instead that axonal growth and synapse formation were largely random events, with little inherent predetermination. Advocates of the resonance view proposed instead that the specificity of mature circuits emerges largely through the elimination of functionally inappropriate connections, and only at a later developmental stage. This extreme view, however, became gradually less tenable in the light of experiments by Roger Sperry, notably on the formation of topographic projections in the retinotectal system of lower vertebrates. Sperry's studies revealed a high degree of precision in the topographic order of retinal axon projections to the tectum during normal development and further established that this topographic specificity is maintained after experimental rotation of the target tectal tissue—a condition in which the maintenance of an anatomically appropriate connection results in a behaviorally defective neuronal circuit (Sperry 1943; see Hunt and Cowan 1990) (Figure 6–11). Over the subsequent two decades, the consolidation of these early findings led Sperry (1963), in the 1960s, to formulate the chemoaffinity hypothesis, a general statement to the effect that the most plausible explanation for the selectivity apparent in the formation of developing connections is a precise system of matching of chemical labels between pre- and postsynaptic neuronal partners.

Sperry's studies also emphasized the utility of combining embryological manipulation and neuroanatomical tracing methods to probe the specificity of neuronal connectivity. This tradition was extended in the 1970s by Lynn Landmesser and her colleagues to demonstrate the specificity of motor axon projections in vertebrate embryos (Lance-Jones and Landmesser 1981) and

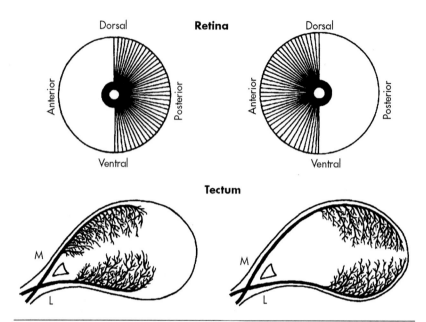

FIGURE 6-11. Sperry's demonstration of topographically specific retinotectal projections.

Anatomical evidence for retinal axon regeneration to original sites of termination in the optic tectum. Sperry's studies showed the pattern of regenerated fibers in the goldfish optic tract and tectum after removal of the anterior (left) or posterior (right) half-retina. The optic nerve was cut at the time of retinal extirpation. The course and termination of the regenerated axons was observed several weeks later, visualized by silver staining. Regenerating axons terminate in appropriate regions despite the availability of additional tectal tissue. *M* and *L* indicate medial and lateral optic tract bundles.

Source.   Adapted from Attardi and Sperry 1963 as illustrated in Purves D, Lichtman JW: *Principles of Neural Development.* Sunderland, MA, Sinauer, 1985.

by Corey Goodman, Michael Bate, and their colleagues in analyses of the stereotyped nature of axonal pathfinding in insect embryos (Bate 1976; Thomas et al. 1984). Thus by the late 1970s, the cellular evidence for a high degree of predetermination and selectivity in axonal growth and synapse formation was substantial, although still not universally accepted (see Easter et al. 1985).

In the 1980s and 1990s, attempts to clarify further the cellular mechanisms of axonal growth and guidance focused on reducing the apparent complexity inherent in the development of axonal projections to a few basic modes of environmental signaling and growth cone response (Goodman and Shatz 1993). As a first approximation, the multitude of cues thought to exist in the environment of a growing axon was proposed to act in one of two ways: 1) at long range, through the secretion of diffusible factors, or 2) at short range, through cell surface-tethered or extracellular matrix-associated factors. In addition, such long- and short-range cues were argued to act either as attractants or local factors permissive for axonal growth or, in a complementary manner, as repellents or factors that inhibit axon extension. What remained unclear after this phase of conceptional reductionism and simplification was the molecular basis of selective axon growth.

The Molecular Era of Axon Growth and Guidance

Today, there is no longer a paucity of molecules with convincing credentials as regulators of axonal growth and guidance (see Tessier-Lavigne and Goodman 1996). This molecular cornucopia is the product of two main experimental approaches: in vertebrate tissues, the biochemical purification of proteins that promote cell adhesion and axonal growth; and in *Drosophila* and *C. elegans*, the application of genetic screens to identify and characterize mutations that perturb axonal projection patterns. Over the past decade, these two complementary approaches have often supplied convergent information and have resulted in the compilation of a rich catalog of molecules with conserved functions in the control of axonal growth in insects, worms, and vertebrates.

An early advance in the molecular characterization of proteins that control axonal growth came with the biochemical dissection of two major adhesive forces that bind neural cells, one calcium independent and the other calcium dependent (Brackenbury et al. 1981). The design of assays to identify neural adhesion molecules based on antibody-mediated perturbation of cell adhesion by Gerald Edelman, Urs Rutishauser, and their colleagues led to the purification of NCAM, a major calcium-independent homophilic cell adhesion molecule (Rutishauser et al. 1982). The widespread expression of NCAM initially argued against a role for this protein in specific aspects of neuronal recognition. The discovery that NCAM is expressed in many different molecular isoforms, however, preserves the possibility that it has more specific functions in neural cell recognition and circuit assembly (Edelman 1983). Although the precise contribution of NCAM to the growth of axons and the formation of neuronal connections remains uncertain, its isolation provided important credibility for the view that cell-adhesive interactions in the nervous system can be dissected in molecular terms. In addition, the realization that NCAM constitutes a divergent member of the immunoglobulin (Ig) domain superfamily (Barthels et al. 1987) brought the study of neural cell adhesion and recognition into the well-worked framework of cell and antigen recognition in the immune system. Since the dis-

A Normal

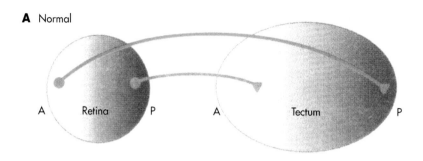

B Overexpression of ephrin A2

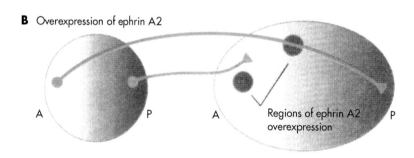

C Mutation of ephrin A5

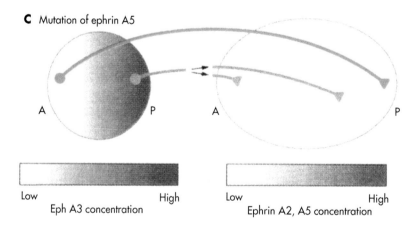

FIGURE 6-12. A role for ephrins and Eph kinases in the formation of the retinotectal map *(opposite page)*.

(A) Members of the Eph kinase class of tyrosine kinase receptors are distributed in gradients in the retina, and some of their ligands, the ephrins, are distributed in gradients in the optic tectum. These two molecular gradients have been proposed to regulate retinotectal topography through the binding of ephrins to kinases and the consequent inhibition of axon growth. The levels of ephrin A2 and ephrin A5 are higher in the posterior tectum than in the anterior tectum, and thus may contribute to the inhibition of extension of posterior retinal axons, which are rich in the kinase eph A3.

(B) Diagram showing the consequences of ephrin A2 expression in portion of the chick optic tectum that normally have low levels of this ligand. Posterior retinal axons avoid sites of ephrin A2 overexpression and terminate in abnormal positions. In contrast, anterior retinal axons, which normally grow into the ephrin-rich posterior tectum, behave normally when they encounter excess ephrin A2.

(C) In mice lacking ephrin A5 function, some posterior retinal axons terminate in inappropriate regions of the tectum.
Source. From the studies of O'Leary, Flanagan, Frisen, Barbacid, and others, as summarized in Kandel ER, Schwartz JH, Jessell T: *Principles of Neural Science,* 4th Edition. New York, McGraw-Hill, 2000.

covery of NCAM, over a hundred Ig domain-containing neural adhesion and recognition proteins have been identified, although the function of most of these proteins in vivo remains unclear (Brummendorf and Rathjen 1996).

In parallel, studies by Masatoshi Takeichi and his colleagues isolated the major calcium-dependent adhesive force binding vertebrate cells, the cadherin proteins (Takeichi 1990). Cadherins have been shown to have major roles in the calcium-dependent adhesive interaction of virtually all cells in the vertebrate embryo, and cadherins have also been identified in *Drosophila* and *C. elegans*. The calcium dependence of cadherin function can be mapped to a critical calcium-binding domain required for protein stability. As we discuss below, cadherins, like Ig domain proteins, are now known to represent a very large family.

A third general adhesive system characterized in the 1980s was that involved in the interaction of cells with glycoproteins of the extracellular matrix. At this time, biochemical studies by many groups had identified collagens, fibronectins, and laminins as key adhesive glycoprotein components of the extracellular matrix. The search for cellular receptors for these structurally distinct glycoproteins converged with the identification of integrins, a large family of heterodimeric integral membrane proteins (Hynes 1987; Ruoslahti 1996). Integrins have prominent roles in cell-matrix adhesion within the nervous system and in virtually all other tissue types. Thus, three main classes of neuronal surface membrane proteins—Ig domain proteins, cadherins, and integrins—appear to provide neural cells with the major adhesive systems necessary for the growth of axons, and these proteins may also contribute to more selective forms of neuronal recognition.

Many additional proteins that are expressed more selectively and appear to have selective roles in axonal growth have now been identified. For example, genetic screens in *C. elegans* and biochemical assays of axon growth regulatory factors in vertebrates collided with the characterization of netrins, a small class of secreted proteins with cell context-dependent axonal attractant and repellent activities (see Culotti and Merz 1998). A similar convergence of biochemical and genetic assays led to the isolation of the semaphorin/collapsin class of growth cone collapse-inducing factors (Kolodkin 1998) and to the characterization of a slit signaling pathway that appear to function both to repel axons and to promote axon branching (Guthrie 1999). Independently, in vitro assays to examine the molecular basis of the topographic mapping of retinotectal projections culminated in the identification and functional characterization of ephrins: surface proteins that function as ligands for receptor tyrosine kinases of the Eph class (Drescher et al. 1997). Ephrin-Eph kinase signaling is now thought to have a dominant role in the establishment of the molecular gradients used to form projection maps in the retinotectal system and in other regions of the central nervous system (Figure 6–12)—perhaps corresponding to some of the matching chemical labels postulated earlier by Sperry.

With each of these discoveries, the veils that had previously shrouded the molecular analysis of axon guidance have been progressively stripped away. As a consequence, it is now realistic to begin to consider, at a molecular level, how the guidance of axons is directed by dynamic sets of molecular cues that either entice or deter the growth of axons at successive stages on their path to a final target. Despite these indisputable advances, many aspects of the logic of axon guidance remain unclear. With the multitude of candidate cues now shown to possess repellent or attractant functions, we still need to understand why individual sets of molecules are used in particular cellular contexts. Are there unique and as yet unappreciated functions provided by one but not another class of guidance cue? Or is there simply molecular opportunism? That is, can similar steps in selective axon pathfinding be achieved by any one of a large and structurally unrelated group of guidance molecules?

One route to resolving such issues will be through the dissection of the signal transduction pathways triggered in growth cones by activation of receptors for guidance cues. Already, such studies have begun to lead to the molecular classification of biochemical signaling pathways and their modulators within the growth cone (Mueller 1999). They have also provided dramatic evidence in vitro that the ability of a growth cone to perceive an extrinsic signal as attractant or repellent can be modified by changing the ambient level of cyclic nucleotide activity. Further dissection of transduction mechanisms in the growth cone may thus help to clarify the logic that underlies the apparent selectivity of action of certain axonal growth and guidance factors. Another critical but poorly resolved issue is that of determining which guidance factors genuinely have instructive roles in directing axon growth and which merely provide permissive signals that enable growth cones to respond to other, more critical, signals.

The Selection and Refinement of Neuronal Connections

With the arrival of developing axons in the vicinity of their final position, growth cones are required to select specific target cells with which to form and maintain functional connections. Although this process is critical in establishing the later functional properties of neural circuits, insight into the molecular basis of neuronal target cell selection remains fragmentary. As discussed above, one recurring issue has been the attempt to determine whether the formation of selective connections is the product of genetically determined factors that specify rules of connectivity in a precise manner, or whether the initial pattern of connections can tolerate a degree of inaccuracy that is subsequently resolved through the elimination of some connections and the consolidation of others (Cowan et al. 1984; Shatz 1997). This latter view then represents the reemergence, albeit in a more restricted and comprehensible form, of the ideas originally articulated by Weiss in the 1940s.

A modern consensus view holds that both genetic predetermination and use-dependent refinement of connections are important contributors to the organization of mature circuits. The relative contribution of these two sets of factors are, however, likely to vary considerably with the particular neural circuit under study. One possibility is that circuits constructed early in evolution or at early stages in the development of an organism, as for example the spinal monosynaptic stretch reflex circuit, are established in a predominantly activity-independent manner (Frank and Wenner 1993). In contrast, the more sophisticated cortical circuits associated with the processing of cognitive information, which emerge later in evolution and development, may require functional validation for the establishment of final patterns of connectivity (Shatz 1997).

The pioneering studies of David Hubel and Torsten Wiesel in the 1960s provided the first evidence for a role for visually driven neural activity in the functional organization of the primary visual cortex (Hubel and Wiesel 1998). Hubel and Wiesel deprived one eye of vision for several weeks during an early critical period of postnatal life. After this procedure, they observed that most neurons in layer four of the primary visual cortex could be activated only by input from the eye that had remained open, thus revealing a marked shift in the pattern of ocular dominance columns in the cortex. At an anatomical level, the terminal arbors of the axons of lateral geniculate neurons supplied by the intact eye were found by Simon LeVay, Michael Stryker, and their colleagues to be considerably more extensive than those supplied by the deprived eye (Antonini and Stryker 1993a, 1993b; Hubel et al. 1977). Many subsequent studies have confirmed the essential role of activity in the formation of visual connections and have shown further that the temporal pattern of activity provided by the two eyes is an important parameter in the establishment of ocular dominance columns (Shatz 1997). Under conditions in which visual input is provided to both eyes in a synchronous manner, the formation of ocular dominance columns is again perturbed (Stryker and Harris 1986). Additional studies have shown that the level of activity in postsynaptic cortical neurons is necessary for ocular dominance column formation (Hata and Stryker 1994). Collectively, these findings have begun to focus attention on the possible mechanisms by which the state of activity of postsynaptic cortical neurons could influence the pattern of arborization of presynaptic afferent fibers as they enter the cortex.

One advance in addressing this problem came with the proposal that the activation of the NMDA subclass of glutamate receptors on postsynaptic neurons might be involved in the normal segregation of afferent input to visual centers (Hofer and Constantine-Paton 1994). An extension of this idea is that the NMDA receptor–mediated activation of cortical neurons results in the release of an activity-dependent retrograde signal that influences the growth and maintenance of presynaptic branches and nerve terminals. Several candidate mediators of such a retrograde signal have now been advanced, including nitric oxide and certain peptide growth factors. Much attention has also been directed at testing the possibility that the activity-dependent release of neurotrophins by cortical neurons is a critical step in the establishment of eye-specific projections into the visual cortex. Some support for this idea has been provided with the demonstration by Carla Shatz and colleagues that local infusion of the neurotrophins NT4 or BDNF into the developing cortex prevents the segregation of ocular dominance columns (Cabelli et al. 1995). Similar developmental defects are observed if the ligand-binding domains of neurotrophin receptors are introduced into the cortex, presumably the consequence of sequestering endogenous neurotrophins (Cabelli et al. 1997). Thus, an attractive if still speculative idea is that neurotrophic factors—classes of proteins identified initially on the basis of their critical roles in promoting the survival of neurons—have later and more subtle roles in shaping neuronal connections in the mammalian CNS.

Although the critical role of activity in the formation of neuronal circuits in the visual system and in many other regions of the CNS is well established, the precise nature of its contribution is less well defined. Information encoded by patterns of activity could be sufficient to direct certain connections. It remains possible, however, that for many neuronal circuits, a basal but unpatterned level of activity is all that is required. In this view, activity may simply permit neurons to respond to other signals that have more direct roles in the control of selective connections or may permit the maintenance of connections formed at earlier stages and through separate mechanisms. Evidence supportive of this latter view has come from studies by Michael Stryker and his colleagues on the role of visually driven activity in the formation of orientation and ocular dominance columns in the developing visual cortex (Crair et al. 1998). Neural activity may therefore exert its influence in large part by consolidating connections that have been established earlier through mechanisms which have their basis in molecular recognition between afferent neurons and their cortical target cells (see Crowley and Katz 1999; Weliky and Katz 1999).

Defining the relative contributions of sensory-evoked activity and genetically determined factors remains difficult, first because the molecular basis of target recognition in any circuit is still unknown and second because the pathways by which activity modifies connectivity are poorly understood. Progress in resolving these issues will therefore require additional insight into the molecules that control synaptic specificity. One anticipated feature of molecules that contribute to the selection of neural connections is that of molecular diversity (Serafini 1999). Several classes of proteins that exhibit inordinate molecular variation have recently been identified and, not surprisingly, have been implicated in the formation of selective connections.

The cadherins as discussed above represent one class of cell surface recognition protein that exists in large numbers. Diversity in cadherin structure can be enhanced dramatically through a process in which one of a chromosomally arrayed cluster of variable cadherin domain gene sequences is appended to a nearby constant region sequence (Wu and Maniatis 1999). The molecular mechanism used to assemble such modularly constructed cadherin proteins remains unclear, but the number of these variable domains is high, bringing the total number of predicted cadherins to well over 100. The vast majority of cadherins are known to be expressed by neural cells and studies of the patterns of expression of the classical cadherins have revealed a striking segregation of individual cadherins within functionally interconnected regions of the brain (Takeichi et al. 1997). In addition, cadherins are concentrated at apposing pre- and postsynaptic membranes at central synapses (Shapiro and Colman 1999). Although intriguing, the link between selective cadherin expression and the specificity of synaptic connections remains to be demonstrated functionally.

A second class of proteins with the potential for considerable structural variation is the neurexins. Neurexins are surface proteins identified originally by virtue of their interaction with the neurotoxin α-latrotoxin (Missler and Südhof 1998; Rudenko et al. 1999). Analysis of the potential for alternative splicing of the *neurexin* genes suggests, in principle, that ~1,000 protein isoforms can be generated and at least some of these potential isoforms are known to be expressed by central neurons. In addition, a class of neurexin receptors termed neuroligins has been identified (Song et al. 1999). Again, though, a functional role for neurexin-neuroligin interactions in the formation of synapses remains to be established.

A third highly diverse class of neuronal surface proteins are the seven-pass odorant receptors expressed on primary sensory neurons in the olfactory epithelium. Several major classes of odorant or pheromone receptors have now been identified in vertebrates, and in total this class of receptors is thought to be encoded by over 1,000 distinct genes (Axel 1995; Buck and Axel 1991). This genetic diversity is likely to underlie the remarkable discriminatory capacity of the mammalian olfactory sensory system. The creative manipulation of odorant receptor gene regulatory sequences to map the central projections of olfactory sensory axons through reporter gene expression in transgenic mice has also revealed a precise anatomical convergence of sensory axons linked by common receptor gene expression to individual target glomeruli in the olfactory bulb (Mombaerts et al. 1996). This finding poses the additional question of the mechanisms directing sensory axon targeting to individual glomeruli. Strikingly, manipulation of the pattern of expression of individual odorant receptor genes in transgenic mice results in a predictable change in the central projection pattern of olfactory sensory axons (Wang et al. 1998). An intriguing implication of these findings is that olfactory sensory receptors function not only in peripheral odor discrimination but also in axon targeting, potentially providing a direct link between the sensory receptive properties of a neuron and its central pattern of connectivity.

Determining whether each or any of these classes of proteins have roles in selective synapse formation in the developing central nervous system CNS) is an important goal in itself and may also provide the entry point for a more rigorous examination of the relationship between neuronal activity, gene expression, and synaptic connectivity.

The events that initiate the formation of selective contacts between pre- and postsynaptic partners are, however, unlikely to provide sufficient information to establish the functional properties of synapses necessary for effective neuronal communication. A separate set of molecules and mechanisms appears to promote the maturation of early neuron–target contacts into specialized synaptic structures. Current views of this aspect of neuronal development derive largely from studies of one peripheral synapse, the neuromuscular junction (Sanes and Lichtman 1999). These studies have their origins in many classical physiological studies of synaptic transmission at the neuromuscular junction. In particular, the ability to measure dynamic changes in the pattern of expression of ACh receptors on the surface of muscle fibers as they become innervated (Fischbach et al. 1978) provided many early insights into the cellular mechanisms by which the motor axon organizes the elaborate program of postsynaptic differentiation necessary for efficient synaptic transmission. By the 1980s, powerful in vivo and in vitro assays to examine synaptic organization under conditions of muscle denervation and reinnervation had been developed, and these assays facilitated biochemical efforts to purify neuronally derived factors with synaptic organizing capacities (McMahan 1990; Sanes and Lichtman 1999).

These efforts culminated in the identification of two major pre- to postsynaptic signaling pathways that appear to coordinate many aspects of the synaptic machinery in the postsynaptic muscle membrane. Signals mediated by agrin, a nerve- and muscle-derived proteoglycan, through its tyrosine kinase receptor MuSK have an essential role in the clustering of ACh receptors and also of other synaptically localized proteins at postsynaptic sites located in precise register with the presynaptic zones specialized for transmitter release (see Kleiman and Reichardt 1996; McMahan 1990). A second set of nerve- and muscle-derived factors, the neuregulins which signal through ErbB class tyrosine kinase receptors, appears instead to control the local synthesis of ACh receptor genes in muscle cells (see Sandrock et al. 1997), and perhaps also to direct the local insertion of newly synthesized receptors at synaptic sites.

These dramatic molecular successes have provided the foundations of a comprehensive understanding of the steps involved in the formation and organization of nerve-muscle synapses. The extent to which the principles that have emerged from the study of this synapse peripherally extend also to the organization of central synapses remains uncertain. There has, however, been considerable progress in recent years in defining the structural components of the presynaptic release apparatus at central synapses (Bock and Scheller 1999) and the proteins that concentrate postsynaptic receptors (Sheng and Pak 1999). From the information now emerging, it seems likely that the identity of molecular signals that orchestrate the maturation of central synapses will soon be known, and in the process we will come to recognize principles of central synaptic organization similar to those that operate at the neuromuscular junction.

A Future for Studies of Neural Development

Despite the dramatic advances of the two past decades, several important but unresolved issues cloud our view of the assembly of synaptic connections.

These problems will need to be addressed before any satisfying understanding of neural circuit assembly can be claimed.

One issue stems from the pursuit of mechanisms of neuronal cell fate determination and of the control of axonal pathfinding and connectivity as largely separate disciplines. With the many available details of cell fate specification and of the regulation of axonal growth and guidance, it is still not clear if and how the transcriptional codes that control neuronal identity intersect with the expression of the effector molecules that direct axonal connectivity. For example, in only a few cases have relevant genetic targets of the transcription factors that control early steps in neuronal identity been identified. Indeed, a superficial survey of patterns of expression of transcription factors and axonal receptors for guidance cues reveals little obvious coincidence at the cellular level. Thus, the extent to which the regulated expression of genes that encode receptors for axon guidance cues depends on the sets of determinant factors implicated in earlier aspects of neuronal subtype identity remains unclear. Defining the full complement of transcription factors that specify the identity of an individual neuronal subtype and the molecular sequence of cell-cell interactions that guide the axon of the same neuron to its target is one obvious but laborious route to resolving this issue.

Similarly, the relationship between transcription factor expression and other later aspects of neuronal phenotype—for example neurotransmitter synthesis and chemosensitivity—also remains unclear. In a few instances, cell-specific transcription factors have been linked to the expression of genes that control neurotransmitter synthesis (see Goridis and Brunet 1999). Nevertheless, the general logic linking transcriptional identity and the expression of the neuronal traits that confer specialized synaptic signaling properties and connectivity remains obscure.

Assuming, as seems likely, that these issues can be solved in a relatively rapid fashion, what does the future hold for studies of neural development? Clearly, there will be interesting variation in the strategies used to establish selective connections in different regions of the developing brain and in different circuits. The documentation of these variations will provide a richer and more profound appreciation of the core principles of neuronal circuit assembly. But the reiteration of a few basic themes in different brain regions can sustain excitement in the field only briefly, and in any event will not provide an obvious intellectual bridge between studies of development and of the function of mature neuronal circuits.

Application of neural development to the study of neurological disease

One future area in which studies of neural development are likely to have significant impact is in the application of fundamental information on the specification of cell fate and the guidance of axons to problems posed by neurodegenerative diseases and traumatic injury to the nervous system.

As discussed above, we are beginning to obtain a rather detailed outline of the relationship between inductive signaling and the expression of cell-specific transcription factors that define cell fate. In some cases, details of these pathways have progressed to the point that certain transcription factors expressed by single classes of CNS neurons have been shown to be sufficient to direct neuronal subtype fate in a manner that is largely independent of the prior developmental history of the progenitor cell (Tanabe et al. 1998). If this is the case for the few classes of neurons in which inductive signaling pathways have been particularly well studied, it seems likely that similar dedicated determinant factors will exist for many other classes of neurons in the CNS. The identification of such factors may be of significance in the context of the many ongoing attempts to identify neural progenitor cells and then to drive them along specific pathways of neurogenesis (Doetsch et al. 1999; Johansson et al. 1999; Morrison et al. 1999; Panchision et al. 1998). One outcome of such developmental studies may therefore be to rationalize strategies for reintroduction of fate-restricted neural progenitor cells into the CNS in vivo. In principle, these advances could offer the potential of more efficient cell replacement therapies in a wide variety of neurological degenerative disorders.

Similarly, the wealth of information on molecules that promote or inhibit axonal growth is likely to be of relevance for studies of axonal regeneration and repair. The pioneering studies of Albert Aguayo and colleagues of the regenerative capacity of central neurons in a cellular environment composed of peripheral rather than CNS nerve cells revealed the potential of central neurons to regenerate (see Goldberg and Barres 2000; Richardson et al. 1997). These studies prompted the search for molecules expressed by cells of the mature CNS that inhibit the growth of axons (see Tatagiba et al. 1997) and for molecules expressed in early development that have the capacity to promote the growth of axons of CNS neurons (Tessier-Lavigne and Goodman 1996). The progress in identification of axon growth–promoting and inhibitory factors may therefore eventually permit rational changes to be made in the environment through which regenerating axons in the mature CNS are required to project. Of equal promise are studies to clarify the signal transduction pathways by which axons respond to these environmental cues. The elucidation of these pathways may permit a more general manipulation of axonal responses—for example rendering axons insensitive to broad classes of inhibitory factors, or supersensitive to many distinct axonal growth–promoting factors. It may also be worth considering whether there is a common molecular basis for the marked differences in the regenerative capacity of different vertebrate species evident in studies of both nerve and limb regeneration (see, for example, Brockes 1997).

Establishing a link between the development
and function of neuronal circuits

An additional, and potentially a more far-reaching, contribution of neural development may emerge by taking advantage of the compendium of information now available on cell-specific gene expression in developing neurons and of the ease of genetic manipulation in mammals, notably the mouse. With these methods in hand, it may be possible to modify the function of highly restricted classes of neurons in the adult animal and to assay resultant changes in the function of specific neuronal circuits.

One initial limitation in the application of information about neuronal subtype–specific gene expression during development is that the majority of such genes are transiently expressed. Thus, the normal temporal profile of gene expression does not permit direct tracing of the relationship between embryonic neuronal subtype identity and the physiological properties of the same neuronal subsets in the adult. This problem can now be overcome through the use of genetically based lineage tracing methods. For example, genes encoding yeast or bacterially derived recombinase enzymes can be introduced into specific genetic loci by targeted recombination (Dymecki 1996; Schwenk et al. 1998), to generate mouse strains which can then be crossed with other genetically modified mice in which recombinase-driven DNA rearrangement results in the irreversible activation of reporter gene expression at all subsequent stages in the life of a neuron (Lee et al. 2000; Zinyk et al. 1998). This relatively simple methodology offers the immediate promise of providing a direct link between subsets of neurons defined at embryonic stages and the location, and functional identity of these neurons within the mature CNS.

With the compilation of such lineage information, variants of this same basic genetic strategy can be used to modify the function of neuronal subsets at predefined times. One drastic method for eliminating neuronal function involves the activation of toxins in a neuron-specific manner, under precise temporal control (see, for example, Grieshammer et al. 1998; Watanabe et al. 1998), thus permitting the physical ablation of predefined populations of CNS neurons with a specificity unattainable by conventional lesioning methods. More subtly, specific populations of neurons could, in principle, be activated or inactivated reversibly in the adult animal through temporally regulated expression of ion channels that change the threshold for neuronal excitability (Johns et al. 1999). In addition, the development of transgenic mice methods for anterograde or retrograde transynaptic transport of foreign marker proteins (Coen et al. 1997; Yoshihara et al. 1999) may be helpful in providing novel information on neuronal connectivity in the CNS that cannot easily be extracted by other anatomical tracing methods.

In this way, the increasingly detailed molecular information that derives from attempts to examine the principles of neural circuit assembly during development should have clear application to the major problems of systems neuroscience discussed in the following sections of this review. At present, the routine application of these genetic methodologies is feasible only in the mouse, and thus the issue of linking studies obtained in lower mammals with information obtained in primates and ideally in man still needs to be addressed. Nevertheless, with advances in the resolution of functional imaging methods that are outlined later in this review, and in the application of these methods to small mammals, the link between studies in mice and primates can be strengthened. When this is achieved, the information that emerges from studies of the development of neural circuits may assume a more prominent place in the repertoire of experimental strategies that aim to decipher how such circuits function in the adult brain.

Neural Systems: From Neurons to Perception

The individual neurons that make up the brain work together in specialized groups, or systems, each of which serves a distinct function. *Systems neuroscience* is the study of these neural systems, which include those involved in vision, memory, and language. Neural systems possess a number of common properties, not the least of which is the fact that they all process higher-order information about an organism's environment and biological needs. In humans, this information often gains access to consciousness. Systems neuroscience thus places great emphasis on uncovering the neural structures and events associated with the steps in an information-processing hierarchy. How is information encoded (sensation), how is it interpreted to confer meaning (perception), how is it stored or modified (learning and memory), how is it used to predict the future state of the environment and the consequences of action (decision making/emotion), and how is it used to guide behavior (motor control) and to communicate (language)? The twentieth century has seen remarkable progress in understanding these processes. This ascendance of modern systems neuroscience is attributable, in part, to the convergence of five key subdisciplines, each of which contributed major technical or conceptual advances.

Neuropsychology: Localization of the Biological Source of Mental Function

The first question one might ask about an information-processing device concerns its gross structure and the relationship between structural elements and their functions. The simplest approach to this question—and the approach that has best withstood the test of time—is to observe the behavioral or psychological consequences of localized lesions of brain tissue. The modern discipline of neuropsychology was founded on this approach and draws both from human clinical case studies—often provided during the early decades of the twentieth century by brain injuries sustained in battle—and from experimental studies of the effects of targeted destruction of brain tissue in animals. Through these means the functions of specific brain regions, such as those involved in sensation, perception, memory, and language, have been inferred.

Neuroanatomy: Patterns of Connectivity Identify Information Processing Stages

The discipline of neuroanatomy, which blossomed at the turn of the century following the adoption of the neuron doctrine and which has benefited from many subsequent technical advances, has revealed much about the fine structure of the brain's components and the manner in which they are connected to one another. As we have seen, one of the earliest and most influential technical developments was the discovery by Camillo Golgi of a method for selective staining of individual neurons, which permitted their visualization by light microscopy. By such methods, it became possible to use differences in the morphology of cells in different brain regions as markers for functional diversity. This procedure, known as cytoarchitectonics, was promoted vigorously in the early decades of the twentieth century by the anatomists Korbinian Brodmann and Oscar and Cecile Vogt. Brodmann's cytoarchitectonic map of the human cerebral cortex, which was published in 1909 and charted the positions of some 50 distinct cortical zones, has served as a guidebook for generations of scientists and clinicians, and as a catalyst for innumerable studies of cortical functional organization.

Arguably the most important outcome of the means to label neurons, however, was the ability it provided to trace connections between different brain regions. To this end, cell labeling techniques have undergone enormous refinement over the past three decades. Small quantities of fluorescent or radioactive substances, for example, can now be injected with precision into one brain region and subsequently detected in other regions, which provides evidence for connectivity. The products of anatomical tract tracing are wiring diagrams of major brain systems, which are continuously evolving in their precision and completeness, and have been indispensable to the analysis of information flow through the brain and for understanding the hierarchy of processing stages.

Neurophysiology: Uncovering Cellular Representations of the World

Adoption of the neuron doctrine and recognition of the electrical nature of nervous tissue paved the way to an understanding of the information represented by neurons via their electrical properties. Techniques for amplification and recording of small electrical potentials were developed in the 1920s by Edgar Adrian. This new technology enabled neurobiologists to relate a neuronal signal directly to a specific event, such as the presentation of a sensory stimulus, and became a cornerstone of systems neuroscience. By the 1930s, electrophysiological methods were sufficiently refined to enable recordings to be made from individual neurons. Sensory processing and motor control emerged as natural targets for study. The great successes of single-neuron electrophysiology are most evident from the work of Vernon Mountcastle in the somatosensory system, and David Hubel and Torsten Wiesel in the visual cortex, whose investigations, beginning in the late 1950s, profoundly shaped our understanding of the relationship between neuronal and sensory events.

Psychophysics: The Objective Study of Behavior

Historically, quantitation of behavior has been the province of experimental psychology, which emerged in the nineteenth century from deep-rooted philosophical traditions to become a distinct scientific discipline and a key component of modern systems neuroscience. Among the most notable steps in this emergence was the development by the German physicist and philosopher Gustav Fechner of a systematic scientific methodology for assessing the relationship between behavior and internal states. Fechner's *Elements of Psychophysics,* published in 1860, founded an "exact science of the functional relationship...between body and mind," based on the assumption that the relationship between brain and perception could be measured experimentally as the relationship between a stimulus and the sensation it gives rise to (Fechner 1860/1966). In practice, Fechner's psychophysics is applied by varying a sensory stimulus along some physical dimension—such as the intensity or wavelength of light—and obtaining reports from an observer regarding the sensations experienced. In this manner, one can identify the function that relates the physical dimension of the stimulus to an internal sensory dimension, and from that relationship infer the rules by which the sensory information is processed.

Throughout the twentieth century, the tools of psychophysics have been extremely useful in identifying the information-processing strategies of sensory, perceptual, and motor systems of the brain. Beginning with the work of Mountcastle in the 1960s (Mountcastle et al. 1969), psychophysics has frequently been paired directly with electrophysiological methods to extra-ordinary effect in identifying the neuronal events that give rise to specific sensory and perceptual processes.

Computation: Divining the Mechanisms of Information Processing

Large neural systems such as those involved in vision, combine and analyze incoming signals to "interpret" their causes and generate appropriate out-puts. The logical steps in these neuronal mechanisms have become accessi-ble to quantitative and theoretical treatment. The goal has been to extract generic computational principles that can account for existing data and have predictive value. Some of the earliest work along these lines was directed at sensory and motor processing and was founded on engineering techniques and principles designed for the study of simple linear systems. One of the most successful examples of this approach is Georg von Bekesy's (1960) in-vestigation of the cochlea and its relation to the frequency encoding of sound. von Bekesy began by investigating the patterns of vibration of the various components of the inner ear, and the relationship of these patterns to the characteristics of sound waves. From these observations he concluded that this system analyzes sound by a linear frequency decomposition—that is, the mechanical properties of the cochlea allow specific frequency compo-nents of sound to be independently isolated and detected by the sensory ep-ithelium. Considerable gains have also been made using similar theoretical approaches to understand early stages of visual processing and the control of movements of the eyes.

Many levels of processing in neural systems deviate from linear forms of computation. The search for alternative computational principles, which was fueled in part by the rise of cognitive science in the 1980s and an un-precedented richness of physiological and anatomical data, has led to a number of novel and sophisticated theoretical approaches, such as those em-bodied by neural networks (Rumelhart et al. 1987). These networks operate on the biologically plausible principle that information can be represented in a distributed fashion across a large population of "units," or modeled neu-rons. Moreover, this information may be combined in many different ways to yield complex cellular representations, simply by changing the strength of "synaptic" connections between modeled neurons.

Vision as a Model System

Collectively, these five areas of neuroscience—neuropsychology, neuroanat-omy, neurophysiology, psychophysics, and computation—constitute an experimental arsenal, which has already revealed in outline the structure, operational mechanisms, and functions of large neural systems, such as those involved in vision, memory, and language. Although the range of successes is broad, and many general principles of system organization and function have been discovered, the visual system has emerged as the model for experimental investigation and is consequently the area in which we have the greatest understanding.

Setting the stage: early explorations of
visual perception and brain

Two early developments presaged the extraordinary progress in understanding visual function that is now a legacy of the twentieth century. The first of these occurred within the field of experimental psychology. Hermann von Helmholtz (1860/1924) and Wilhelm Wundt (1902), two of Fechner's nineteenth-century contemporaries, attempted to identify how different visual stimuli lead to different subjective experiences. Their method was initially observational and introspective, but later they exploited the objective methodology of psychophysics. The lasting outcome of these efforts was a quantitative appreciation of the elements of visual experience—color, brightness, motion, distance—and an initial set of ideas about how they might be represented by the brain. A second early development occurred within the field of neuropsychology. With mounting experimental evidence for localization of function within specific brain regions, Hermann Munk (1881) and Edward Schafer (1888) each used the method of focal ablation of brain tissue at the end of the nineteenth century to identify brain regions that serve visual function. They found that the occipital lobe of the cerebral cortex plays an essential role in the processing of visual information.

The Golden Era of Single-Neuron Electrophysiology

Perhaps the single greatest technical advance in vision science was the application of the electrophysiological methods that had emerged in the late 1920s and 1930s. In a pioneering series of studies begun in the early 1930s, Keefer Hartline recorded from single cells in the eye of the horseshoe crab (*Limulus*) and examined the relationship between the properties of the incoming sensory stimulus—which in this case happened to be a small spot of light projected onto the eye—and the neuronal response (i.e., the frequency of action potentials). Through these (Hartline and Graham 1932) and subsequent experiments of a similar nature that involved recordings from single axons in the frog optic nerve (Hartline 1938), Hartline made two important discoveries. First, he found that individual neurons respond to light only within a well-defined region of visual space, which Hartline termed the visual *receptive field.* Operationally defined, the receptive field is the portion of the sensory epithelium (the sheet of photoreceptors, in the case of vision) that when stimulated elicits a change in the frequency of action potentials for a given neuron. In anatomical terms, the receptive field describes all of the receptor and subsequent cells that converge upon and influence the firing pattern of the neuron under study. The concept of the receptive field has proven to be an extremely useful and general concept in systems neuroscience.

Hartline's second major discovery was that the visual responses to light were dependent upon contrast. Specifically, the amplitude of the neuronal response to a light in one region of visual space was greatest when there was no light in an adjacent region of space. Thus, rather than simply conveying the presence or absence of light, neurons in the visual system communicate information about the spatial structure or pattern of the incoming stimulus. Because these observations paralleled the well-known perceptual enhancement of brightness at contrast boundaries—exploited for centuries by artists wishing to enhance the range of light intensities perceived in their paintings (Leonardo da Vinci 1956)—they were seized upon as a potential physiological substrate for the perceptual experience of brightness.

The tradition of single-neuron studies of visual processing continued through the 1940s and 1950s with a shift of emphasis to mammals (affording closer ties to human visual perception and made possible by advances in electrophysiological techniques). In 1953, Stephen Kuffler, a student of Eccles and Katz, examined the behavior of neurons in the cat retina. Kuffler focused on the retinal ganglion cells, the output cells of the retina, which carry visual information from the photoreceptors through the optic nerve to the lateral geniculate nucleus of the thalamus and other central processing regions. Following Hartline's model, Kuffler described the response characteristics of ganglion cells in terms of their receptive field properties. He discovered that the receptive fields of retinal ganglion cells were round in shape and had distinct concentric excitatory and inhibitory zones, which made them maximally sensitive to spatial contrast. On the basis of the architecture of their receptive field properties, Kuffler divided these cells into two groups. One class of cells had a central excitatory zone and a surrounding inhibitory region ("on-center" cells), whereas the other class of cells had an inhibitory central region and an excitatory surround region ("off-center" cells).

Beyond the retina: visual contours are detected
by neurons in primary visual cortex

Anatomical tracing experiments conducted over the past several decades have shown that the outputs of the retina terminate in several distinct brain regions. One of the largest projections extends from the retina to the lateral geniculate nucleus (LGN) of the thalamus and continues on via the *geniculostriate pathway* to *primary visual cortex*. Otherwise known as *striate cortex* or *area V1*, this latter visual processing stage lies on the occipital pole of the cerebral cortex, and is known from the early neuropsychological studies of Munk, Schafer, Gordon Holmes (1927), and others to be critical for normal visual function.

The electrophysiological approach pioneered by Kuffler in studies of mammalian retinal ganglion cells was carried to these higher processing stages by two of Kuffler's young colleagues, David Hubel and Torsten Wiesel. In the late 1950s, Hubel and Wiesel began to examine the response properties of neurons in the cat and monkey lateral geniculate nucleus. These neurons were found to possess center-surround receptive field properties not unlike those of retinal ganglion cells. By contrast, Hubel and Wiesel (1959) found that the response properties of cells in the primary visual cortex of both cats and monkeys were very much more complicated. Cortical cells could not be effectively stimulated by the simple spots of light that proved so effective in the retina and in the lateral geniculate nucleus. To be effective, a stimulus had to have linear properties; the best stimuli were lines, bars, rectangles, or squares (Figure 6–16). Hubel and Wiesel divided the cortical cells into simple and complex (for a review, see Hubel and Wiesel 1977). We will use simple cells to illustrate in greater detail the types of stimulus selectivities observed (Figure 6–13).

A typical receptive field for a simple cell in primary visual cortex might have a central rectangular excitatory area with its long axis running from twelve to six o'clock, flanked on each side by similarly shaped inhibitory areas. For this type of cortical cell, the most effective excitatory stimulus is a bar of light with a specific axis of orientation—in this case, from twelve o'clock to six o'clock—projected on the central excitatory area of the receptive field. Since this rectangular zone is framed by two rectangular inhibitory areas, the most effective stimulus for inhibition is one that stimulated one or both of the two flanking inhibitory zones. A horizontal or oblique bar of light would stimulate both excitatory and inhibitory areas and would therefore be relatively ineffective. Thus, a stimulus that is highly effective if projected vertically onto a given area of retina so as to be on target for the excitatory zone would become ineffective if held horizontally or obliquely. Other cells had similar receptive field shapes but different axes of orientation (vertical or oblique). For example, the most effective stimulus for a cell with an oblique field would be a bar of light running from ten o'clock to four o'clock or from two o'clock to eight o'clock (Figure 6–13).

The most interesting feature of the simple cortical cells is that they are much more particular in their stimulus requirement than the retinal gan-

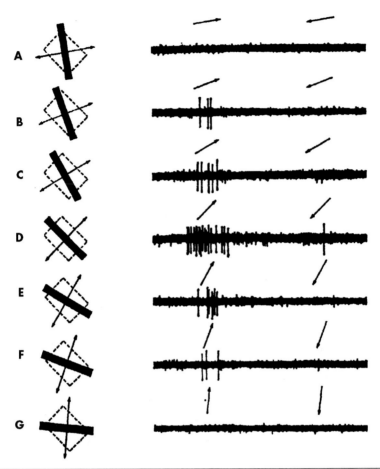

FIGURE 6–13.   Neuronal orientation selectivity, as first observed by Hubel and Wiesel in primary visual cortex (area V1) of a rhesus monkey.

The receptive field of the cell that was recorded is indicated by broken rectangles in the left column. The visual stimulus that was viewed by the monkey consisted of a bar of light that was moved back and forth through the receptive field of the cell in each of seven different orientations (rows A–G). The different directions of motion used for each orientation are indicated by the small arrows. Recorded traces of cellular activity are shown at right, in which the horizontal axis represents time (2 s/trace) and each vertical line represents an action potential. This neuron responded most strongly to a bar of light oriented along the diagonal (stimulus D), particularly when the bar was moved through the receptive field from lower left to upper right. Neurons bearing such properties are common in the visual cortex, and their discovery revolutionized views on the neuronal bases of visual perception.

Source.   From Hubel DH, Wiesel TN: "Receptive Fields and Functional Architecture of Monkey Straite Cortex." *The Journal of Physiology* 195:215–243, 1968.

glion cells or those in the lateral geniculate nucleus in requiring a proper axis of orientation. For a stimulus to be effective for a retinal ganglion or a geniculate cell, it only has to have the proper shape—in general, circular—and the proper retinal position so as to activate appropriate receptors in the retina. Simple cortical cells not only have to represent all retinal positions and several shapes (lines, bars, rectangles), but also for each shape they have to represent all axes of orientation. These findings provided an initial insight as to why the visual cortex (or any cortex) needs so many cells for its normal functions. Cells are required to represent every retinal area in all axes of orientation so as to abstract the information presented to the cortex. Hubel and Wiesel suggested that the simplest explanation for the response properties of a cortical cell with a simple receptive field was that they received innervation from a set of geniculate cells that had appropriate on-center and off-center properties and appropriate retinal positions.

Another feature distinguishes cells in primary visual cortex from those in the lateral geniculate nucleus. Neurons of the lateral geniculate nucleus only respond to stimulation of one or the other eye. In primary visual cortex, one begins to find cells that are activated by stimulation of either eye (Figure 6–14). These cells provide a neural substrate for the integration of information from the two eyes. Binocular properties of this type are essential to the use of stereoscopic cues for depth vision in animals, such as primates with frontally placed eyes.

Beyond V1: specialized functions of higher cortical visual areas

Neuropsychological studies carried out over much of the past century have shown that deficits in visual function follow from damage anywhere within a vast expanse—over 30% in humans—of the cerebral cortex. Moreover, the type of deficit depends upon the site of damage: temporal lobe lesions cause impairments in object recognition (termed "agnosias" by Sigmund Freud) (for a review, see Farah 1995), whereas parietal lobe lesions interfere with use of visual cues for spatially directed actions (for a review, see Mesulam 1999). These early findings suggested that the complex cellular properties discovered in area V1 by Hubel and Wiesel were only the tip of the iceberg.

Motivated by this prospect, in the 1970s there was a dramatic increase in electrophysiological and anatomical studies designed to explore the organization of the *extrastriate visual cortex,* which lies beyond area V1. Two groups—Semir Zeki, and John Allman and John Kaas—noted that not only were extrastriate neurons vastly heterogeneous in their response properties but that the extrastriate visual cortex could be neatly subdivided into a large number of distinct modules on the basis of these properties (for a review, see Van Essen 1985). At present, the visual cortex of monkeys is thought to be

A Normal

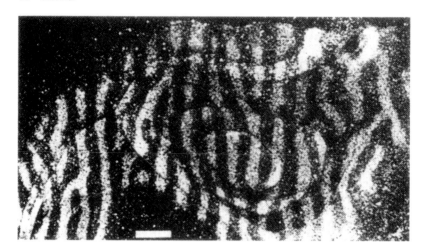

B Reconstruction: normal ocular dominance columns

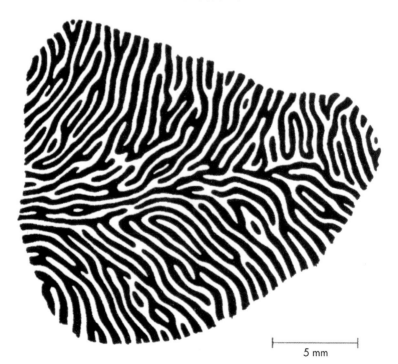

5 mm

FIGURE 6–14. Anatomical representation of ocular dominance columns in primate visual cortex *(opposite page).*
(A) The right eye of a normal adult was injected with the radiolabeled proline and fucose. This dark-field autoradiograph obtained after 10 days shows a tangential section of area 17 of the right hemisphere. Radioactivity can be seen in the form of white stripes, which correspond to thalamic axon terminals in layer 4 that relay input from the injected eye. The alternating dark stripes depict the position to geniculate afferents from the uninjected eye.

(B) Reconstruction of ocular dominance columns in area 17 of the right hemisphere showing the intricate organization of the map.
Source. (A) From Hubel DH, Wiesel TN, LeVay S: "Plasticity of Ocular Dominance Columns in the Monkey Striate Cortex." *Philosophical Transactions of the Royal Society of London. Series B: Biological Sciences* 278:377–409, 1977. (B) From LeVay S, Wiesel TN, Hubel, DH: "The Development of Ocular Dominance Columns in Normal and Visualy Deprived Monkeys." *The Journal of Comparative Neurology* 191:1–51, 1980.

composed of over 30 such modules (Figure 6–16).

These efforts to reveal order in the heterogeneity of visual cortex were a natural extension of the nineteenth-century concept of localization of function. They were, moreover, reinforced by the computational view, advanced by the theorist David Marr (1982), that large system operations (such as seeing) can be subdivided and assigned to task-specific modules. Although little is yet known of the specific tasks "assigned" to the vast majority of extrastriate visual areas, there are some noteworthy exceptions. Of particular interest is the middle temporal (MT) visual area, which appears to be specialized for motion.

Area MT lies near the junction of the occipital, parietal, and temporal lobes and is known to possess a high proportion of neurons that represent the trajectory of a moving visual stimulus, suggesting an important role in visual motion processing (for a review, see Albright 1993). This idea has been supported by three related findings by William Newsome and his colleagues that imply a causal link between the activity of MT neurons and perceived motion. In the first experiment, Newsome and his colleagues found that focal destruction of area MT in monkeys results in motion blindness, demonstrating that MT is *necessary* for motion perception (Newsome and Paré 1988) (Figure 6–15). In a second experiment, Newsome and Anthony Movshon obtained psychophysical measurements of a monkey's ability to discriminate direction of motion, simultaneously with electrophysiological measurements of the motion sensitivity of MT neurons in the monkey's brain. They found that the sensitivity of individual neurons correlated extremely well with performance on the behavioral task, demonstrating that the direction information encoded by neurons of MT is *sufficient* to account for the monkey's judgment of motion (Newsome et al. 1989). Newsome and colleagues reasoned that if this logic were correct, then it should be possible to alter the monkey's perception of motion by artificially modifying the firing rate of MT neurons. In a third experiment, these investigators did exactly that. Small electrical currents were used to stimulate clusters of neurons sensitive to a common direction of motion. Remarkably, doing so was found to bias the monkey's judgment toward that direction of motion (Salzman et al. 1990). Electrical stimulation of this sort thus has the effect of adding a fixed motion signal to the signal received by MT from the retina. Not only do these results strongly support the hypothesized role of area MT in motion processing, but also they imply that perceptual decisions can be based on the activity of relatively small populations of neurons.

Detection of behaviorally significant visual features

The initial discovery that visual neurons integrate complex spatial information led to speculation about the role of such neurons in detection of behaviorally significant visual features. Experiments conducted in the 1950s by Horace Barlow (1953) and by Jerome Lettvin and Humberto Maturana reinforced this view with the finding that ganglion cells in the frog retina respond optimally to patterns of light that resemble the silhouette and flight of an insect (Lettvin et al. 1959), which is, of course, the primary food source for a frog. This finding led naturally to the concept that single visual cells were *feature detectors*. According to this view, receptive fields may operate as highly specialized templates for detection of significant features ("trigger features") in the visual image. The concept was expanded upon in the 1960s by the Polish psychologist Jerzy Konorski (1967), who proposed the existence of "gnostic units"—cellular representations of visual features, such as faces, that convey highly meaningful information to the observer. Horace Barlow (1972) expressed similar views in his "cardinal cell" hypothesis. The possibility that cells specialized for detection of faces might exist at higher levels of processing was a logical extension of the findings of Hubel and Wiesel (1977), which documented increasingly abstract cellular representations as one ascended the hierarchy of processing stages. In addition, the feature detector/cardinal cell hypothesis followed naturally from several decades of neuropsychological research demonstrating that damage to the inferior temporal lobe of the cerebral cortex compromised a human or monkey's ability to identify complex objects (Farah 1995; Gross 1973). Most importantly, damage to a small subregion of this cortex in humans results in an inability to recognize faces, a syndrome called *prosopagnosia*. Patients with prosopagnosia can identify a face as a face, its parts, and even specific emotions ex-

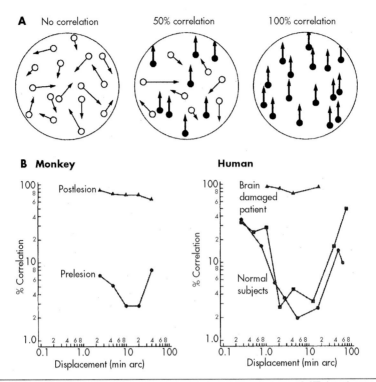

FIGURE 6–15. The involvement of cortical area MT in the perception of visual motion.

A monkey with an MT lesion and a human patient with damage to extrastriate visual cortex have similar deficits in motion perception.

(A) Displays used to study the perception of motion. In the display on the left, there is no correlation between the directions of movement of several dots, and thus no net motion in the display. In the display on the right, all the dots move in the same direction (100% correlation). An intermediate case is in the center; 50% of the dots move in the same direction while the other 50% move in random directions (and are perceived as noise added to the signal).

(B) The perception of visual motion by a monkey before and after a lesion of MT (left). The performance of a human subject with bilateral brain damage is compared to two normal subjects (right). The ordinate of the graph shows the percent correlation in the directions of all moving dots (as in part [A]) required for the monkey to select the common direction. The abscissa indicates the size of the displacement of the dot and thus the degree of apparent motion. Note the general similarity between the performance of the human and that of the monkey and the devastation to this performance after the cortical lesions (Baker et al. 1991; Pare 1998).

Source. From experiments of Newsome and others as illustrated in Kandel ER, Schwartz JH, Jessell T: *Principles of Neural Science*, 4th Edition. New York, McGraw-Hill, 2000.

A

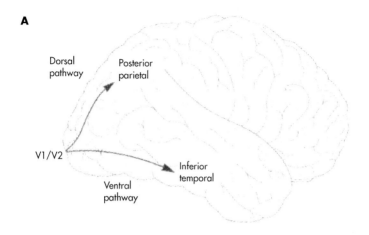

B

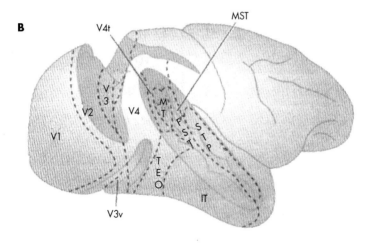

C

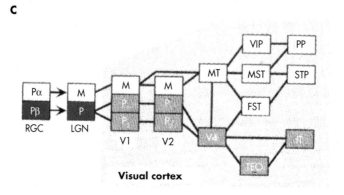

Visual cortex

FIGURE 6–16. Visual information is processed by divergent cortical pathways *(opposite page).*

One of the most important discoveries of the past century in the field of sensory biology is the multiplicity of areas in the cerebral cortex that are involved in visual perception and visually guided behavior. These areas are thought to be functionally specialized and hierarchically organized into two parallel processing streams.

(A) The image is a lateral view of the rhesus monkey brain, illustrating these two major pathways, both originating from V1, the striate cortex. There is a dorsal ("where") cortical stream, which takes a dorsal route to the parietal cortex, and a ventral ("what") cortical stream, which takes a ventral route to the temporal cortex.

(B) The image illustrates the primary visual cortex (area V1) located on the occipital pole (left); the "extrastriate" cortical visual areas extend anteriorly (rightward) and are labeled by their commonly used abbreviations (see below). Indicated borders of visual areas (dashed lines) are approximate. Some sulci have been partially opened (shaded regions) to show visual areas that lie buried within these sulci.

(C) The image illustrates some of the anatomical connections known to exist from the retina through visual cortex. Except where indicated by arrows, anatomical connections are known to be reciprocal.

Abbreviations: FST=fundus superior temporal, LGN=lateral geniculate nucleus of the thalamus, M=magnocellular subdivisions, MST=medial superior temporal, MT=middle temporal, P1 and P2=parvocellular subdivisions, PP=posterior parietal cortex, RGC=retinal ganglion cell layer, STP=superior temporal polysensory VIP=ventral intraparietal.

Source. From Albright TD: "Cortical Processing of Visual Motion." *Reviews of Oculomotor Research* 5:177–201, 1993. Used with permission of Elsevier.

pressed by the face, but they are unable to identify a particular face as belonging to a specific person (Farah 1995).

Strong electrophysiological evidence in support of the feature detector/cardinal cell hypothesis came from the work of Charles Gross beginning in the late 1960s. Using the same methods established by Kuffler, Hubel and Wiesel, and others, Gross discovered cells in the inferior temporal cortex of monkeys that respond to specific types of complex stimuli such as hands and faces (Gross et al. 1969). For cells that respond to a hand, individual fingers are particularly critical. For cells that respond to faces, the frontal view of the face is the most effective stimulus for some cells, while for others it is the side view. Whereas some neurons respond preferentially to faces, others respond to facial expressions. It seems likely that such cells contribute directly to the perceptually meaningful experience of face recognition.

General Principles of Visual System Organization and Function

We have here covered only a few highlights of twentieth-century research on the visual system. The complete legacy has led to a number of general principles of visual system organization and function to which we now turn.

The visual system is hierarchically organized

A consistent feature of the visual system is the presence of multiple hierarchically organized processing stages (Figure 6–16), through which information is represented in increasingly complex and abstract forms (for a review, see Van Essen 1985). As first suggested by Hubel and Wiesel (1962), properties present at each stage result, in part, from selective convergence of information from the previous stage. The hierarchy begins with the photoreceptors, which detect the presence of a spot of light shining upon a particular part of the retina. Each retinal ganglion cell—the output cells of the retina—surveys the activity of retinal bipolar cells, which in turn survey activity in a group of receptors. The product of this convergence of information in the retina is a simple abstraction of light intensity, namely a representation of luminance contrast. A neuron in the lateral geniculate nucleus surveys a group of retinal ganglion cells, and activity in the geniculate cell also signals luminance contrast. A simple cell in primary visual cortex surveys a population of geniculate neurons and the firing of that simple cell reflects a still higher level of abstraction: the presence of an oriented contour in the visual image. At successively higher stages of processing, information is combined to form representations of even greater complexity, such that, for example, individual neurons within the pathway for visual pattern processing encode complex behaviorally significant objects, such as faces.

How far does this hierarchy go? Is there a group of cells that observes the hierarchies of simple cells and makes one aware of the total pattern? And if so, is there a still higher group in the hierarchy that looks at combinations of complex patterns as these enter our awareness? These are important questions for the future of systems neuroscience, which we address below in our discussions of visual feature "binding" and consciousness.

The visual system is organized in parallel processing streams

In addition to a hierarchy of processing stages, the visual system is organized in parallel streams (Figure 6–16). Incoming information of different types is channeled through these different streams, such that the output of each stream serves a unique function. This type of channeling occurs on several scales and at different hierarchical levels (for a review, see Van Essen 1985).

One of the most pronounced examples of channeling occurs within the projection from retina to the lateral geniculate nucleus of the thalamus, and beyond. Different types of retinal ganglion cells project selectively to three different laminar subdivisions of the lateral geniculate nucleus, known as parvocellular, magnocellular, and koniocellular laminae. Each of these subdivisions is known to convey a unique spectrum of retinal image information and to maintain that information in a largely segregated form at least as far into the system as primary visual cortex.

Beyond V1, the ascending anatomical projections give rise to two visual pathways, each of which is organized hierarchically (Figure 6–16). One pathway extends dorsally to terminate within the parietal lobe and includes area MT and visual areas of the posterior parietal cortex (Figure 6–19). A second pathway extends ventrally to terminate within the temporal lobe and includes areas V4 and inferior temporal cortex (for a review, see Felleman and Van Essen 1991). On the basis of a large body of neuropsychological data, Leslie Ungerleider and Mortimer Mishkin concluded in the early 1980s that these two cortical pathways serve different functions (Ungerleider and Mishkin 1982). Accordingly, cortical areas of the dorsal pathway are concerned with "where" an object is in visual space. These areas represent motion, distance, and the spatial relations between surfaces in the visual world, and provide a crucial source of information for initiating and guiding movements. By contrast, areas of the ventral pathway are concerned with "what" an object is. These areas represent information about form and the properties of visual surfaces, such as color or texture, and play important roles in object recognition (for a review, see A.D. Milner and Goodale 1995). Electrophysiological studies of the response properties of neurons in areas of the proposed "where" and "what" pathways have provided support for this functional dichotomy (Figure 6–17).

The "what" versus "where" distinction continues at still higher levels, where visual information is stored in memory for later retrieval. The dorsal stream projects to a subdivision of frontal cortex that is known to be critical for visual spatial memory. The ventral stream projects to a different subdivision of frontal cortex, which serves object recognition memory.

Many visual processing stages are topographically organized

Near the turn of the century, Hermann Munk (1881) suggested, in part on the basis of observed effects of cortical lesions, that the visual cortex may contain a spatial map of the retinal surface. This hypothesis was confirmed in the 1910s by Gordon Holmes (1927), who discovered a precise relationship between the site of damage in human visual cortex and the location in visual space where visual sensitivity was lost. Even better evidence came in 1941 from the work of Wade Marshall and Samuel Talbot, who used gross

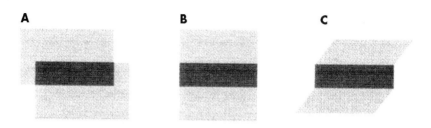

FIGURE 6-17.   The influence of local sensory context on visual perception.

Each of the three images displayed here contains a horizontal dark gray rectangle. Although the rectangles are physically identical, the surrounding features (the contexts) differ in the three images. As a result, the rectangle is attributed perceptually to different environmental causes in the three instances: In the image shown in (A), the rectangle appears to result from the overlap of two surfaces, one of which is transparent (e.g., a piece of tinted glass). In the image shown in (B), the rectangle appears to result from a variation in surface reflectance (e.g., a stripe painted across a large flat canvas). In the image shown in (C), the rectangle appears to result from partial shading of a three-dimensional surface (i.e., variation in the angle of the surface with respect to the source of illumination). These markedly different perceptual interpretations argue for the existence of different neuronal representations of the rectangle in each of the three instances. These representations can only be identified in neurophysiological experiments if the appropriate contextual cues are used for visual stimulation. See the text for details.

Source.   Courtesy of T. Albright and colleagues.

electrophysiological techniques to demonstrate an orderly topographic representation of visual space across the cortical surface, such that neurons with spatially adjacent receptive fields lie adjacent to one another in the brain (Talbot and Marshall 1941). It has since been discovered that neuronal maps of visual space are characteristic of many visual processing stages (for a review, see Van Essen 1985). Such maps may facilitate computations based on comparisons of visual information at adjacent regions of visual space. These maps are commonly distorted relative to the visual field, such that the numbers of neurons representing the center of gaze, which is particularly important for visual object recognition, greatly exceed those representing the visual periphery. These variations in "magnification factor" are thought to underlie variations in the observer's resolving power and sensitivity.

The visual cortex is organized in vertical columns

The existence of a column-like anatomical substructure in the cerebral cortex has been known since the beginning of the twentieth century, following the work of Ramón y Cajal, Constantin von Economo, and Rafael Lorente de Nó. Although Lorente de Nó (1938) first suggested that this characteristic structure might have some functional significance, it was the physiologist Vernon Mountcastle who developed the concept fully and proposed in the 1950s that this may be a general principle of cortical organization. Using single-cell electrophysiological techniques, Mountcastle (1957) obtained the first evidence in support of this proposal through his investigations of the primate somatosensory system.

The best-studied example of columnar organization, however, is that discovered in the 1960s by Hubel and Wiesel. These investigators found that primary visual cortex is arranged in a series of narrow vertical columns, about 100–200 μm in width, running from the surface of the cortex to the white matter (Hubel and Wiesel 1962, 1968). In a given column, cells have similar receptive field positions and generally similar receptive field properties, including preferred orientation. Consistent with Mountcastle's hypothesis, the columns seem to serve as elementary units of cortical organization designed both to bring cells together so that they can be appropriately interconnected and to generate from their interconnections the properties needed for cells with higher-order receptive fields. Additional evidence for functional columns, and for the validity of Mountcastle's proposal, has come from studies of higher visual areas. In the early 1980s, Thomas Albright and colleagues (1984) identified a system of functional columns in area MT. Interestingly, the spatial scale of this columnar system, which represents direction of stimulus motion, is virtually identical to that for orientation in primary visual cortex—consistent with the expression of a general organizational principle. Columnar systems have also been observed in inferior temporal cortex (Fujita et al. 1992).

Why are columns a preferred form of cortical architecture? Mountcastle's original proposal assumed the need for adequate "coverage," such that, for example, the machinery for detecting all contour orientations is available for all parts of the visual field represented in the cortex. There are also computational advantages (Schwartz 1980) afforded by representing similar features adjacent to one another—such as the ability readily to compare and contrast similar orientations. Finally, it may be that columnar structure is derived simply from developmental constraints, such that it is easier and more economical to wire together a cortex that has similar properties in close proximity (Goodhill 1997; Miller 1994; Swindale 1980).

The visual system is modifiable by experience
during early postnatal development

The mammalian brain develops through a complex multistaged process that extends from embryogenesis through early postnatal life. The end product of this developmental sequence is a set of patterned anatomical connections that give rise to the mature system properties of the brain. Once the principal features of mature visual system organization and function became known through the work of Hubel and Wiesel in the 1960s, it was natural to question whether those features reflect a genetically predetermined plan that is implemented during development, or whether they are influenced by the amount and type of visual stimulation that occurs before the developmental process is complete.

These questions have been addressed through experiments in which 1) the properties of the system are assessed at birth or shortly thereafter (precluding the possibility of any significant contribution of experience), or 2) animals are subjected to abnormal visual experience during postnatal development. The earliest experiments of the former type were conducted by Hubel and Wiesel (1963), who reported the visual sensitivities of neurons in the primary visual cortex of newborn kittens to be similar in most respects to those of mature animals. Although these findings suggested a large degree of genetic control, Wiesel and Hubel (1963, 1965) also discovered that abnormal visual experience, such as extended closing of the eyelids or induction of strabismus (both achieved surgically), dramatically altered the visual sensitivities of cortical neurons, provided that the intervention occurred during a "critical period," which extended for several weeks postnatally. Other studies have demonstrated close relationships between such critical periods and their presumed causes and effects—that is, the formation of appropriate anatomical connections (for a review, see L.C. Katz and Shatz 1996) and the development of visual perceptual abilities (for a review, see Teller 1997).

The general view that has emerged from all of these experiments is that the newborn visual system possesses a considerable degree of order, but that visual experience is essential during critical periods to maintain that order and to fine-tune it to achieve optimal performance in adulthood. Hubel and Wiesel (1965) summed up the implications of this view with characteristic prescience and breadth: "All of this makes one wonder whether more subtle types of deprivation may not likewise exert their ill effects through the deterioration of complex central pathways that either were not used or else were used inappropriately." The degree to which this may be true throughout the life of an organism is an issue we address below in the context of perceptual learning.

A Future for the Study of Neural Systems

Despite unprecedented progress, our understanding of visual system organization and function is far from complete. On the contrary, developments and

discoveries of the past century have raised many new and often unanticipated questions regarding the visual system and other large brain systems.
Here we focus on a few of the bigger issues at stake, with some predictions
about where this field of research may be headed in the new millennium.

How do sensory representations lead to perception?

The physiological studies of Hubel and Wiesel and many others over the past
50 years have revealed much about how basic features of the visual image,
such as oriented lines and patches of color, are detected and represented by
cortical neurons. But how do these cellular representations account for our
perceptual experience of the world? The underlying assumption has been
that perception of complex scenes would result from the collective activities
of neurons whose properties we so far have considered and were characterized under reduced stimulus conditions. It is, however, increasingly apparent that this assumption is flawed because it posits that individual neuronal
representations of sensory features are independent of one another, and the
field of sensory physiology is consequently at a turning point in its evolution.

One can fully appreciate the problem and begin to chart a new course for
the future by tracing the origins of current views on the cellular bases of perception. One popular nineteenth-century view was known as "elementism."
According to this influential doctrine, any percept can be explained as a collection of independent internal states (sensations) elicited by individual sensory elements, such as brightness, color, and distance. The undeniable
appeal of elementism rests on the power of reductionism, whereby it should
be possible in principle to dissect out the elemental causes of perceptual experience—a red patch here, a yellow contour there, some motion in the
center, etc.—much in the way one might dismantle a pump. There were,
nonetheless, many early critics of this view, including the physicist Ernst
Mach (1886/1924), and its foundations crumbled at the turn of the century
with the emergence of a Gestalt theory of visual function, which maintained
that the perceptual whole is indeed far greater than the sum of the sensory
parts.

This Gestalt perspective was promoted vigorously by the psychologists
Max Wertheimer (1924/1950), Wolfgang Köhler (1929), and Kurt Koffka
(1935), and its legitimacy was most effectively communicated by simple and
compelling visual demonstrations. Through such demonstrations, it could
easily be seen that the percept elicited by one stimulus element (e.g., brightness) is heavily dependent upon other stimulus attributes (e.g., three-
dimensional form) in the same image. A key feature of Gestalt theory is thus
contextual interaction: the perceptual interpretation of *each* visual image fea-

ture is a function of the context offered by *other* features in the image. *Why* contextual interaction occurs is in itself an interesting question. The answer can be found in the fact that visual images are only ambiguously related to the visual scenes that give rise to them (i.e., there are, in principle, an infinite number of scenes that can lead to the same visual image on the retina). The context in which an image feature appears provides a rich source of information that can be used to resolve its ambiguity—to, in other words, assist the viewer in identifying the "meaning" of the feature, as defined by the content of the scene that led to its appearance.

Although the holistic and eminently functional perspective of the Gestalt tradition bears great validity, the tradition has long lacked momentum, owing in part to its failure to develop mechanistic or neuronal foundations, and it has had surprisingly little influence throughout much of the twentieth century. Indeed, with the rise of single-cell studies of the mammalian visual system, we have witnessed an unwitting return to the principles of element-ism, largely as a matter of investigative convenience. And therein lies the problem. When Hubel and Wiesel stimulated cortical neurons with single oriented lines, they purchased the power to reduce the response to a simple code for oriented image features, which has been of enormous benefit, but at the cost of a lack of generality. From the orientation tuning of a V1 neuron obtained by such means, one learns how the cell encodes the physical properties of the retinal stimulus. If, however, the *meaning* of the stimulus—that is, its environmental cause, *the thing that is perceived*—is only revealed by context, as shown by the Gestalt theorists, then it is frankly impossible to learn what role the cell plays in perception using this experimental approach.

In the search for an alternative approach to carry us into a new millennium, it is useful to return to the principles of Gestalt theory and to develop an operational distinction between candidate neuronal substrates for sensation and for perception. Accordingly, candidate neuronal substrates for sensation—which have been the primary subjects of study over the past 50 years—encode the physical properties of the proximal stimulus (the visual image), such as orientation or direction of motion. Perceptual representations, on the other hand, reflect the world (the visual scene) that likely gave rise to the sensory stimulus. Contextual manipulations make it possible to dissociate local sensory properties from perception, and thereby offer a means to identify neuronal responses that are correlated with perception rather than the proximal sensory stimulus. Francis Crick and Christof Koch (1998) have recently equated perceptual representations of the sort defined here with "neural correlates of consciousness," owing to their belief that perceptual awareness is a legitimate operational definition of consciousness. We take this issue up below in the section on vision and consciousness.

There are many ways in which this research strategy for studying candidate neuronal substrates for perception can be applied. These fall into two complementary categories. First, one can investigate whether neuronal responses to identical receptive field stimuli covary with the different percepts—determined by different contexts—that those stimuli elicit. The set of stimuli illustrated in Figure 6–17 are of this class, and a valid experimental goal would be to identify neuronal responses that vary with the percept elicited, even though the receptive field stimulus (the gray rectangle in Figure 6–17) remains physically unchanged. Second, one can investigate the neuronal responses to different receptive field stimuli—"sensory synonyms"—that elicit the same percept, owing to context. Both of these situations, which are prevalent in normal experience, afford opportunities to experimentally decouple sensation and perception.

The *first* of these two approaches—whether neuronal responses covary with different contexts—was used by Thomas Albright, Gene Stoner, and colleagues to understand the role of cortical visual area MT in motion perception (for a review, see Albright and Stoner 1995). Moving objects in a typical visual scene commonly generate a complex array of moving visual image features. One objective of the visual system is to integrate these moving features to recover the coherent motions of the objects that gave rise to them. That integration process is heavily and necessarily context dependent, such that, for example, the *object* motions that are perceived from two identical collections of *image* motions can vary greatly as a function of the context in which the motions appear. A simple example of this phenomenon can be found in the appearance of two overlapping stripes that move in different directions. This type of visual stimulus, which is known as a "plaid pattern," is a simple laboratory incarnation of a common real-world occurrence, namely two objects moving past one another.

Albright and Stoner proposed that if contextual cues present in the image indicated that the two stripes lay at different distances from the observer, then the two stripes would appear to move in two different directions. On the other hand, if the very same motions were viewed in the presence of contextual cues that indicated no such depth ordering, then the two striped components of the plaid would appear to move coherently in one direction. There are a number of different contextual cues that can be used to elicit a percept in which the stripes are ordered in depth. In their initial experiment, Stoner, Albright, and V.S. Ramachandran used luminance cues for perceptual transparency to achieve this goal. Simply by adjusting the intensities of light coming from different regions of the plaid pattern, it was possible to make the plaids appear as either a single surface or two overlapping surfaces (Stoner et al. 1990). Moreover, as predicted, these contextual manipulations dramatically altered perceived motion, even though the image motion was unchanged. Once this dissociation between image and perceived motion was discovered, the dissociation became a useful tool to investigate whether the responses of cortical motion-sensitive neurons encode image motion or perceived motion.

In a second experiment, Stoner and Albright (1992) pursued this idea and found that when directionally selective neurons in cortical visual area MT were presented with identical image motions that were perceived differently as a function of context, the responses of many neurons covaried with the motion that was perceived. These findings support the view that an important step in the visual processing hierarchy is the use of context to construct cellular representations of visual scene attributes—the stuff of perceptual experience—out of cellular representations of visual image features, such as oriented contours.

The *second* approach to decoupling sensation and perception is the reciprocal of the first and is based upon the phenomenon of perceptual constancy, in which multiple sensory stimuli give rise to the same percept, owing to appropriate contextual cues. Perceptual constancies reflect efforts by the visual system to recover behaviorally significant attributes of the visual scene, in the face of variation along behaviorally irrelevant sensory dimensions. Size constancy—the invariance of perceived size of an object across different retinal sizes—and brightness/color constancy—the invariance of perceived reflectance or color of a surface in the face of illumination changes—are classic examples. Generally speaking, the physiological approach advocated is one in which neuronal responses are evaluated to determine whether they covary with the changing receptive field stimulus, or whether they exhibit an invariance that mirrors the percept. Physiologists have only begun to employ this approach. Several studies, for example, have recently explored a phenomenon termed "form-cue invariance," in which a percept of motion of shape is invariant across different "form cues," such as luminance, chrominance, or texture, that enable the stimulus to be seen. In one such study, Albright (1992) discovered a population of motion-sensitive neurons in area MT that appear to encode the direction of motion of a stimulus independently of the fact that the form cue—an aspect of the retinal stimulus that is, in principle, irrelevant to motion detection—is varied. To paraphrase from Horace Barlow's (1972) neuron doctrine for perceptual psychology, the main function of such cells appears not to be the encoding of specific characteristics of retinal illumination, "but to continue responding invariantly to the same external patterns"—that is, to the *meaningful* attributes of the input.

From the outset, the physiological approach to the operations of the visual system has seen itself as being in the service of perceptual psychology. If the "exploration of psychological territory" is to continue, however, physiologists must advance beyond the acontextual approach that has been the standard of twentieth-century research in this field. New experimental approaches in which contextual influences are exploited as tools for the study of neural substrates of perception are thus likely to be an important feature of future research in this area.

Binding it all together

As we have seen, the representational strategy that the visual system has adopted is one in which the properties of the incoming signal are distributed across many neurons, such that each neuron only conveys a small piece of the larger picture. At the level of the retina, for example, the information represented by single cells is limited to a small circular region of space. At the level of primary visual cortex, cells integrate information from earlier stages in order to convey information about contour orientation, but they remain highly specialized. At still higher levels, visual information is further combined and abstracted to yield even more complex properties and greater specialization of function, as evidenced by the multiplicity of extrastriate visual areas. In view of this strategy, one cannot help but wonder how all of the specialized representations are bound together to render a neuronal signal that conforms to the complex patterns that we perceive. How is it, for example, that the cells representing the edges, the varying orientations, the colors and textures, and the different distances associated with the tree outside my corner window, are linked together to produce my percept of that tree? How are other properties of the same visual image, such as the attributes of a different but nearby tree, "segmented" and bound together as a separate entity from the first? Even more puzzling is the fact that only some of these complex patterns enter my awareness at any point in time. How are those patterns selected, and how are objects that we are simultaneously aware of linked together? What role does visual attention play in this process? Is there a distinguishing feature of the collections of neurons that happen to represent objects or collections of objects that we have become conscious of? Is there, as contemplated by Sherrington (1941) a half century ago, one "pontifical cell" that represents the final outcome of this integration process? The representational problem addressed by these questions has become known as the "binding problem." In a more general form, the problem has preoccupied philosophers and cognitive scientists for decades, and it now stands among the most formidable challenges in modern neuroscience (see, for example, the October 1999 issue of Neuron).

At its most basic level, the binding problem is simply that of representing conjunctions of attributes. There are, in principle, two mechanistic strategies that could accomplish this task, one based on neural space and the other based on time. On the one hand, attributes that must be conjoined—such as the color and the direction of motion of an object—could be represented in that form by selective convergence of information onto single neurons, yielding a neuron, for example, that selectively encodes rightward-moving red objects. The appeal of such a strategy is that it follows naturally from what is already known of the hierarchical properties of the visual system. The neuronal representation of an oriented contour is, after all, nothing more than a product of selective convergence of information from the previous stage. Moreover, long-standing evidence indicates that neurons well up in the hierarchy encode very complex conjunctions of visual attributes, such as those associated with faces (Gross et al. 1969). The problem with this form of binding, however, is one of generality: the variety of unique perceivable conjunctions of visual attributes vastly exceeds the number of available neurons. So while this may be a strategy that the visual system has adopted for early levels of integration and for highly specialized and vital functions like facial recognition, it is simply untenable as a general mechanism for binding.

The alternative strategy is one in which visual attributes are bound in time, rather than by static spatial convergence. The obvious advantage of a dynamic binding mechanism is that, unlike the static design, it places no serious combinatorial limits on the pieces of information that can be conjoined. A form of this mechanism was suggested as early as 1949 by the psychologist Donald Hebb (1949), who hypothesized the existence of "cell assemblies." Each such assembly was conceived as a collection of neurons that are dynamically associated with one another as needed to link the features they independently represent. A key feature of this concept is the ability of each cell to hold membership in multiple overlapping assemblies—such that, for example, a cell that represents upward motion may be assembled with a cell representing the color red on one occasion but assembled with a cell representing the color green on a different occasion. This view of binding was subsequently elaborated upon by Horace Barlow (1972), who noted it to be a particularly efficient form of representation because perceptions commonly "overlap with one another, sharing parts which continue unchanged from one moment to another."

The trick, of course, is identifying a *dynamic binding code* that can be used to transiently link cells into an assembly. One idea, which was implicit in Hebb's original proposal, developed significantly in the early 1980s by the theorist Christoph von der Malsburg (1981), and which has subsequently drawn a great deal of attention (see, for example, reviews in the October 1999 issue of *Neuron*), is that temporal synchrony of neuronal firing patterns may underlie binding. As suggested in 1989 by Charles Gray, Wolf Singer, and colleagues, "synchrony of oscillatory responses in spatially separate regions of the cortex may be used to establish a transient relationship between common but spatially distributed features of a pattern" (Gray et al. 1989). This solution is in effect a dynamic switchboard that binds collections of complex features "on demand" via synchronized firing of the neurons that represent the individual features. Gray and Singer presented provocative data in support of this hypothesis. They found that the temporal spiking patterns of pairs of simultaneously recorded neurons in visual cortex were likely to be correlated if the separate visual stimuli that elicited those patterns appeared (to human observers) to be part of a common object. Other studies, however, have failed to find support for this synchrony hypothesis (e.g., Lamme and Spekreijse 1998; for a review, see Shadlen and Movshon 1999) and the matter remains unsettled.

If we accept the concept of dynamic cell assemblies, and the related proposal for temporal binding by synchronous firing (if only for the sake of argument), we face many critical questions. How, for example, are the transient patterns of synchrony "read out"? Does synchronous firing lead to transient synaptic facilitation of converging inputs onto a multipurpose pontifical cell (or, perhaps more appropriately, given the democratic and ephemeral nature of the hypothesized convergence, a "presidential cell")? Or is the perceptual binding simply implicit in the activity of the synchronized neurons, which constitute flexible cell assemblies for specific percepts? What elicits synchrony to begin with? Is there a top-down supervisory module that identifies attributes that experience tells us are likely to be parts of the same object? Or is the process bottom-up, using a variety of "image segmentation cues" to parse out attributes that belong to the same versus different objects? And what happens when—as is often the case—there are multiple objects perceived simultaneously? Is spike timing sufficiently precise to allow multiple synchrony events to occur simultaneously?

While we await the answers to these and other questions, it nevertheless appears likely that visual integration rests upon a combination of static and dynamic binding mechanisms. Indeed, except for a few clever hypotheses and controversial details, our understanding of these processes has advanced little beyond the view advocated by Barlow in the early 1970s (as a counterpoint to Sherrington's pontifical cell metaphor), according to which a series of "cardinal cells" reside at the top of static convergence hierarchies, but "among the many cardinals only a few speak at once" in the form of dynamic cell assemblies (Barlow 1972). But research now moves rapidly on these fronts and, with vast improvements in technology for monitoring the firing patterns of many neurons simultaneously, the existence and operations of cell assemblies and their role in binding should come into sharper focus in the coming years.

Vision and consciousness

Interestingly, the binding problem and its proposed solution by a dynamic code have also been coupled with the phenomena of visual awareness and consciousness. In the case of visual awareness, the argument is quite natural (if not tautological), as there are good reasons to believe that the perceptual binding of visual attributes is tantamount to their reaching the perceiver's awareness. Indeed, Singer and colleagues have argued that "appropriate synchronization among cortical neurons may be one of the necessary conditions for the…awareness of sensory stimuli" (Engel et al. 1999). Developing the concept of neuronal "metarepresentations," Singer (1998) has furthermore suggested that this code may underlie all of the complex patterns that enter our awareness at any time. The extension of this line of argument to consciousness depends, of course, on how one defines the term. Although there is a long history of confusion about what consciousness actually means, and a plethora of colloquial uses of the term, Francis Crick and Christof Koch (1998) have recently attempted to facilitate scientific progress by arguing for a specific and limited definition that is relevant to vision. (We consider the issue more broadly in "Consciousness: A Challenge for the Next Century" below). Roughly speaking, that definition can be equated with perceptual awareness; it is "enriched" by visual attention, and may fill a window of time, which Gerald Edelman (1983) has termed "the remembered present" (see also James 1890). If we accept that operational and not unreasonable definition, then dynamic representations of the sort proposed by Singer and others may be relevant to consciousness. But this is slick and unstable terrain, newly trodden by neuroscience and lacking the guideposts of established experimental paradigms (deflecting, to paraphrase Bertrand Russell, nearly all but fools and Nobel laureates). It is, nonetheless, one of the most compelling issues facing the future of neuroscience, and is certain to be a focal point of research in the next century.

What are the local cellular mechanisms of vision?

As we have seen, physiological studies have revealed much about the types of visual information carried by neurons at different processing stages. In parallel, anatomical studies have told us a great deal about the gross pattern of connections within and between different stages of processing. Until recently, however, comparatively little has been known about the local circuits that confer neuronal properties and mediate the computations required for perception. This knowledge is an essential starting point for understanding how neurons integrate and store different sources of visual information, as well as alter their sensitivity to compensate for environmental and behavioral changes.

Recent progress in this area has been fueled by new technologies that allow finer-resolution tracing of anatomical connections, in conjunction with methods that allow assessment of the contributions of these connections in their functional state. For example, optical imaging of neuronal activity, combined with cell labeling, is enabling us to determine the relationships between functional architecture and cortical circuits (Malach et al. 1993). Investigation of correlated firing patterns between pairs of neurons, in conjunction with precise measures of receptive field properties, has provided an approach complementary to anatomical tracing of local circuitry (for a review, see Usrey and Reid 1999).

Another promising technique of this sort, known as photostimulation, was applied recently by Edward Callaway and Lawrence Katz. This technique, which enables one to assess the pattern and strength of synaptic connections between neurons in local cortical circuits, exploits a form of the excitatory neurotransmitter glutamate that is inactive ("caged") until illuminated (L. C. Katz and Dalva 1994). Callaway and colleagues have revealed that the pattern of functional connections between different laminae in primary visual cortex provides far more opportunities for cross talk between different visual processing streams than was evident from traditional anatomical studies (Callaway 1998).

Another approach to understanding the relationship between circuitry and function might involve deactivation of individual circuit components, such as specific classes of cells, and assessment of the ensuing loss of function. At a gross level, this approach is recognizable as the lesion method that has been used in neuropsychology for over a century to reveal the functions of large neuronal systems. But is it realistic to expect that these methods can be extended to identify the fine details of circuit organization and function? The fact that circuit components are both anatomically and functionally intermingled—particularly in the cerebral cortex—would seem to preclude this possibility. A resolution, however, can be found in new molecular techniques that enable one to manipulate gene expression and to exploit the genetic distinctiveness of cells that serve different functions.

These new techniques incorporate three key features: 1) the ability to introduce novel genes into neurons, the expressed products of which will alter neuronal function, 2) the ability to regulate expression of these transgenes in a time-dependent manner, and 3) the ability to regulate expression of these transgenes selectively in specific classes of cells. The first of these techniques has been standard fare for some time now in the form of germline transgenic manipulations in mice (for a review, see Picciotto 1999). The same end point is now possible in other species, including primates, using viral vector transfection (Takahashi et al. 1999). In principle, as discussed earlier, it should be possible by these means to introduce novel genes that block neuronal cell firing when expressed, which would effectively remove affected cells from the circuit. Recent evidence suggests that overexpression of K^+ channels may be an effective means to transiently inhibit conduction of action potentials or their propagation into dendrites (Johns et al. 1999). The second technique is also becoming routine using one of a number of inducible systems that promote gene expression only in the presence of exogenous factors, which can be delivered by the experimenter (e.g., Nó et al. 1996). This temporal control permits before and after measures of the contributions of affected cells. The third technique—cell-specific expression—is absolutely critical, of course, if these tools are to provide any greater resolution than standard cell-ablation techniques. As discussed earlier in this review, this technique taps into gene "promoters" that are known to regulate expression of specific genes only in specific cell types. By replacing the genes that are normally regulated with novel genes, one can restrict expression of these transgenes to cells that recognize the promoter. Mark Mayford, Eric Kandel, and colleagues have demonstrated the feasibility of these three basic techniques using germline transgenic manipulations in mice to explore the functions of hippocampal neurons in relation to memory storage (Mayford et al. 1996).

Related techniques might also be used to facilitate anatomical analysis of local circuits. For example, instead of introducing and expressing a gene that disrupts cell firing, one might simply transfect neurons with genes that encode visible proteins, such as GFP. The end result in this case would be selective labeling of a specific class of cells, which could be used, for example, to identify those cells in a brain slice preparation for physiological recording, or simply for analysis of cell morphology and connections using light microscopy. Recent experiments document the feasibility of this general approach using germline transgenic manipulations in mice (Yoshihara et al. 1999).

There are many technical details that will need to be worked out—not the least of which is the identification of additional cell-specific promoters perhaps through the strategies outlined in our earlier discussion of the assembly of neural connections—before these fantasy experiments become a practical means to investigate the organization and function of local circuits in the primate visual system. Nonetheless, the potential benefit afforded by this unprecedented merger of molecular tools and systems approaches to brain function is clearly enormous. They are certain to become a staple of future experiments aimed at understanding the entire realm of brain systems.

How do cellular representations change with visual experience?

As we have seen, a central tenet of modern neuroscience is that stages in brain development correspond to specific stages in the development of perceptual abilities. These stages are known as critical periods, and they are characterized by an extraordinary degree of neuronal and perceptual plasticity. Only recently has it been recognized that the plasticity of the visual system is not restricted to these critical periods early in development but is modifiable throughout the adult life of the organism. The forms of this adult plasticity are many and varied, but all can be viewed as recalibration of incoming signals to compensate for changes in the environment, the fidelity of signal detection (such as that associated with normal aging or trauma to the sensory periphery), or behavioral goals.

One of the most striking and revealing forms of adult neuronal plasticity is that associated with *perceptual learning*, which is an improvement with practice in the ability to discriminate sensory attributes. In humans, these learning phenomena are ubiquitous in everyday life and generally self-evident, and they have been a subject of scientific investigation for decades (for a review, see Karni and Bertini 1997). Consider, for example, the copy editor who over time becomes particularly sensitive to graceless word pairings, or the assembly line worker who can instantly recognize the miswired transistor. Until recently, however, little effort had been made to investigate their neuronal bases. Indeed, the critical period concept had become so widespread and deeply rooted that there seemed little ground for believing that visual representations might be modifiable throughout life.

Thus, it was well before the modern neuroscience community was prepared to embrace the concept of plastic representations in mature animals that Michael Merzenich began to address the degree to which sensory maps could change in response to a variety of manipulations (for a review, see Buonomano and Merzenich 1998). This work began to have a broad impact in the mid-1980s with the demonstration of marked and systematic reorganization of somatosensory cortex in response to a change in the peripheral sensory field (e.g., selective deafferentation). Even more exciting and provocative was the subsequent demonstration that cortical maps reorganize in response to selective use of components of a sensory modality, in a manner that mirrors perceptual learning.

Following in the footsteps of Merzenich, Charles Gilbert has recently begun to investigate the relationship between adult visual perceptual learning and the receptive field properties of cortical neurons. In one set of experiments, Gilbert and colleagues have found that increases in perceptual sensitivity fail to generalize to spatial locations or stimulus configurations beyond those in the set of training stimuli (Crist et al. 1997). This high degree of spatial specificity suggests that the underlying neuronal changes may occur at a very early stage of processing—perhaps V1—where the spatial resolving power of cortical neurons is greatest. Other behavioral observations support this conclusion (e.g., Karni and Sagi 1993). Using a complementary approach, Gilbert and others have demonstrated plasticity of cortical representations more directly (Chino et al. 1992; Gilbert and Wiesel 1992). In this case, the receptive field properties of V1 neurons were found to change following localized interruption of retinal input (caused by small retinal lesions). Similar to previous findings from the somatosensory system, these changes took the form of shifts in the spatial profile of receptive fields, such that cells normally responding to light in the area covered by the retinal lesion become sensitive to stimulation of adjacent regions of visual space. These changes began to occur in a matter of minutes following deafferentation. Although in this case, unlike perceptual learning, the plasticity was not induced by repeated exposure to a sensory stimulus but was rather a response to a marked loss of stimulation, both can be viewed as forms of renormalization and the underlying cellular mechanisms may be similar.

This is among the most exciting areas of systems neuroscience today, bridging as it does the topics of sensory processing and learning. Early contributions to this field have been particularly inspiring and influential because the prevailing wisdom held that sensory representations were largely immutable following critical periods of developmental reorganization. As we have seen, recent observations prove that this is not the case. On the contrary, representational changes occur throughout life as part of a normalization process to compensate for damage or deterioration of the sensory periphery or to meet novel behavioral and perceptual requirements. But these findings naturally raise many new questions that will occupy neuroscientists for years to come. Little is yet known, for example, of the specific neuronal events that give rise to plasticity of the adult visual system, although such processes are sure to include changes in synaptic efficacy, changes in neuronal cell structure, and possibly neurogenesis. Future research is also likely to address the following questions: How are these experience-dependent changes in visual processing mediated? Evidence indicates that higher cortical areas, such as regions of the frontal cortex, contain neurons that represent attributes of memorized stimuli. What role, if any, do these mnemonic representations play in the formation of experience-dependent changes in visual cortex? What are the control signals that initiate such changes? Does representational plasticity occur at all stages of visual processing? Does it occur in the retina? Do such changes constitute the neuronal repository of long-term visual memories?

Beyond vision: exploring links with other brain systems

It is a pedagogical convenience to treat the visual system—as we have done here—as functionally independent and separable from other brain systems. The fact of the matter is that vision is but one cog in the wheel and is in many ways integrated with other major systems, including those responsible for memory, emotion, and motor control. Although we know less—much less, in some cases—about these other systems, it is now clear that the areas of interface between vision and those systems that serve storage, evaluation, and action are among the most important targets for future research in systems neuroscience. Here we consider one of these areas of interface: that associated with motor control.

Visual guidance of behavior: from retina to muscle

A major function of the visual system is to provide sensory input to guide actions, such as moving through the environment. Visual and motor control systems have in common the fact that they both represent space. But the relevant frames of reference—retinal space, in the case of vision, and ultimately muscle space, in the case of action—are radically different. How then does light falling on a particular location on my retina lead to a reaching arm movement (or an eye movement, or a leg movement, etc.) to the source of the light? The problem becomes even more puzzling if we consider that exactly the same arm movement will be executed regardless of what direction I am looking, implying that vastly different retinal signals can lead to the same motor output. In principle, there are a number of different means by which this coordinate transformation could be accomplished.

Perhaps in large part because of the compelling subjective sense that space is stable regardless of the orientations of our sense organs and our muscles, it has often been proposed that the brain contains a unified representation of space. This unified map might represent space in a three-dimensional "world-centered" frame of reference, as opposed to the more specific coordinates of the sensory (retina) and effector (muscle) organs. According to this view, we have an internal neuronal map of the spatial locations of all of the items on my desk in front of us. That map remains coherent and unchanged regardless of which way we are looking—or, for that matter, which way the entire body is oriented. The advantage of such a system is that it provides a generic source of spatial information that can be used to guide all movements, which is independent of the state of the sense organs or muscles. The disadvantage is that, because of its independence from sense organs and muscles, a generic reference frame is extremely difficult to compute.

Numerous studies conducted over the past few decades have evaluated the hypothesis that space is transformed from sensation to action via a unified reference frame. Neuropsychologists have examined the effects of damage to brain regions that lie between early visual processing and motor control—specifically the parietal and premotor areas of the cerebral cortex. The typical consequence of such damage is "neglect," in which subjects ignore stimuli that appear in certain regions of visual space (for a review, see Mesulam 1999) (Figure 6–19). (Neglect is distinguished from blindness by the fact that a neglect patient can clearly see a neglected stimulus if his or her attention is drawn to it.) If there were a single unified map of space, one would expect that neglect would be manifested in the same part of the spatial map—always to the right side of the observer, for example. On the contrary, results indicate that neglect can be present in any of a number of different spatial reference frames (retinal coordinates, body part coordinates, object-based coordinates, world-based coordinates), suggesting that there are multiple spatial maps, which may serve specialized functions (for a review, see Colby and Goldberg 1999).

Neurophysiological data also support the hypothesis that there are multiple types of spatial maps used to transform information from sensory to motor coordinates. This issue has been explored extensively by Richard Andersen and colleagues, who recorded from neurons in the parietal cortex in search of a representation of visual space in head-centered coordinates (which would, in principle, be useful for directing movements of the eyes). Andersen discovered instead that these neurons possess "gain fields," by which the amplitude of the response to a visual stimulus is modulated systematically by the direction of gaze (Andersen et al. 1985). Because these neuronal responses take eye position into account, it is in principle possible to deduce the spatial location of the visual stimulus from the activity of a population of such neurons, regardless of where the eyes are looking (for a review, see Andersen et al. 1993). This information can then be used to guide movements to the stimulus.

Recent physiological experiments by two groups—Carl Olson and Sonya Gettner, and Michael Graziano and Charles Gross—provide fascinating evidence for more explicit but highly specialized spatial maps that could mediate visual-motor control. Olson and Gettner (1995) recorded from individual neurons in premotor cortex of monkeys and found that the neurons responded if an eye movement was made to a particular *part* of an object, regardless of the spatial location of the object. These neurons thus appear to represent space in an object-based coordinate frame. Graziano and Gross studied single neurons in the premotor cortex that possess both visual and tactile receptive fields. The visual receptive field of each neuron was found to be linked to the spatial location of the tactile receptive field, such that, for example, a neuron that was activated by tactile stimulation of the arm was also activated by a visual stimulus in the vicinity of the arm (Graziano et al. 1994). Remarkably, if the arm moved to a new location, the visual receptive field moved along with it. The visual receptive fields of these neurons thus appear to have been transformed from a retinal frame of reference to a reference frame centered on the position of the arm. Graziano and Gross propose that this arm-centered reference frame may be well suited for orchestrating movements of the arm to stimuli that are near the arm. More generally, they speculate that visual-motor transformations of many types may rely upon specialized body-part centered maps of space, rather than upon a single unified spatial map. Although both of these physiological findings suggest promising new approaches to the study of sensorimotor coordinate transformations, they leave many questions of a mechanistic nature unanswered. Perhaps the most nagging question raised by the Graziano and Gross study concerns the apparently profound spatial mobility of the visual receptive field, which dances across retinal space with every movement of the arm. How are retinal signals dynamically rerouted, as it were, to continuously update the visual receptive field of the premotor neuron, using information about arm position as a guide? This and other related questions will be an important focus of research in years to come.

Consciousness: A Challenge for the Next Century

Perhaps the greatest unresolved problem in visual perception, in memory, and, indeed, in all of biology resides in the analysis of consciousness. This is a particularly difficult problem, in part because there is no widespread agreement on exactly what constitutes a successful solution. There is agreement nevertheless that a successful solution will require, at a minimum, insight into two major issues that lie at the heart of the study of consciousness: 1) *awareness* of the sensory world and 2) *volition*, the voluntary control of thoughts and feelings.

In this section, we consider awareness by focusing on two of its components: attentional orienting to sensory signals in the presence and in the absence of stimuli (imagery). We shall then go on to consider volition by focusing on the self-regulation of thoughts, feelings, and actions. These two problems are at once relevant to consciousness, yet tractable, and therefore serve to illustrate how consciousness can be dissected biologically.

As with other problems in biology, there are both reductionist and holistic approaches to these components of consciousness. A reductionist approach would view these aspects of consciousness from a genetic, synaptic, and cellular level. However, in the case of consciousness, it is hard to imagine any solution that would not also require an understanding of the large neural networks that underlie cognition, actions, and emotion. In our view, the appropriate direction in seeking a solution to the problems of consciousness lies in successfully linking understanding at all of these levels, from genes to behavior. In this section, we try to illustrate this integrative approach in relation to orienting to sensory stimuli, imagery, and self-regulation. Finally, we examine how far these scientific approaches will take us in understanding the most subjective aspects of consciousness.

Rigorous top-down approaches to consciousness have been limited by the lack of good methods for resolving the activity of populations of cells. The use of neuroimaging methods during the last decade has made it possible to observe the activity of large numbers of neurons in human subjects while they are studied for their awareness of the sensory world and for their voluntary control of thoughts and feelings (Posner and Raichle 1994, 1998). Cognitive studies using these imaging methods, such as positron emission tomography (PET) and functional magnetic resonance imaging (fMRI), are based upon changes in blood flow and blood oxygenation that occur in localized regions of the brain when neurons increase their activity (Raichle 1998; Rosen et al. 1998). These methods have now been applied with some success to the study of attentional orienting, visual imagery, and regulation of cognitive and emotional states. In each of these domains, the individual functional components have proven to be surprisingly well localized; however, each of the major functions of consciousness—such as attentional orienting to sensory stimuli and volition—involves not one but several functional components. As a result, each function of consciousness appears to involve several networks and these are distributed across a variety of brain areas. Fortunately, the enormous complexity of the problems has been made somewhat more tractable by use of appropriate animal models. In the best case, as with studies of the visual system, it has proven possible to relate neural activity studied at the cellular level in nonhuman primates to the activity of large neural networks studied in the same brain areas but now in human subjects using brain imaging (Tootell et al. 1998). While the results are not definitive, they show that specific aspects of consciousness can even now be analyzed on the cellular level with methods currently available to neuroscience.

Orienting of Attention to Sensory Stimuli

Origins of the modern study of sensory attention

The modern study of attention can be traced to 1958 and the publication by Donald Broadbent of a monograph entitled *Perception and Communication* (Broadbent 1958). Broadbent proposed that when we focus attention on one object to the exclusion of other surrounding objects, the focus of selective attention requires a filter that holds back messages from unattended channels. According to Broadbent's view, attention is a high-level skill that is so developed in some people as to allow them to perform remarkable feats such as simultaneous translation. This skill allows even untrained subjects to have a role in selecting their environment by attending only to certain stimuli while shutting out others. Although there have been challenges to this view in the four decades that have passed since it was proposed, even Broadbent's strongest critics have embraced his general approach. In the next section we begin to explore the neuronal implementation of the type of selective attention studied by Broadbent.

All visual areas, including primary visual cortex (V1), can be biased by a shift of attention

To obtain an idea of how brain areas become involved in selection of a stimulus, consider the task of looking for a file on your computer desktop. If the desktop is cluttered with files, you will have to search for the one you want. Such a search may be accompanied by eye movements, but if the objects are close, the search may involve covert shifts of attention without eye movements. Such visual search tasks involve the coordinated action of the two large-scale brain networks. One network, the ventral visual pathway, which we discussed in the previous section, is concerned with objects and with form recognition, required for obtaining the identity of each file. The second network—located in the posterior parietal cortex of the dorsal visual pathway—is related to the act of shifting attention to the locations *where* the file might be found. Early studies of the dorsal pathway were conducted by Michael Goldberg and Robert Wurtz (1972). They found that cells in the posterior parietal cortex of alert monkeys responded differentially to identical stimuli depending on whether or not the monkey was attending to the stimulus (Figure 6–18). When the monkey attended, the firing of the cell was much more intense than when the monkey ignored the stimulus. These results provided the first data on the cellular level that neurons in the parietal cortex are correlated with attention to the location of visual objects. With the advent of neuroimaging, it became possible to see the distributed network of brain areas involved in attention in human subjects. This network includes the frontal eye fields, the superior colliculus, and the posterior parietal lobe, all of which are also involved in eye movements (Corbetta 1998).

Studies conducted by Robert Desimone and colleagues have addressed the role of the ventral visual pathway (particularly areas V4 and IT) in attentional control. A typical paradigm involves first establishing the stimulus selectively for a cortical neuron. Suppose, for example, that the neuron under study responded well to a red bar of light and poorly to a green bar when these stimuli were individually placed in the neuronal receptive field. At this point, both stimuli—the red and green bars—would be placed in the receptive field of the cell at the *same* time. If the animal was instructed to attend to the "good" stimulus (red bar) then the neuron responded well. If, however, the animal was instructed to attend to the "poor" stimulus (green bar) then the response was correspondingly poor—*despite the fact that the retinal*

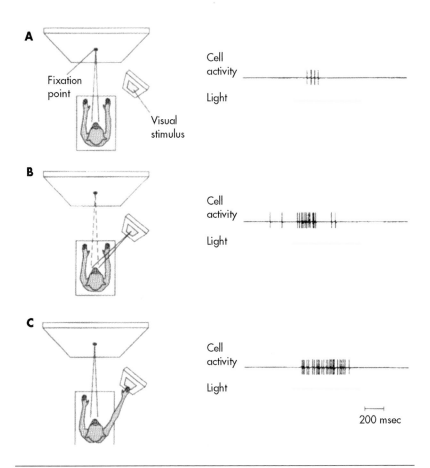

FIGURE 6–18.   The influence of attention on the response of cortical neurons.

Neurons in the posterior parietal cortex of a monkey respond more effectively to a stimulus when the animal is attentive to the stimulus.

(A) A spot of light elicits only a few action potentials in a cell when the animal's gaze is fixed away from the stimulus.

(B) The same cell's activity is enhanced when the animal takes visual notice of the stimulus through a saccadic eye movement.

(C) The cell's activity is enhanced further when the monkey touches the spot, even without eye movement.

Source.   From Wurtz and Goldberg 1989, as illustrated in Kandel ER, Schwartz JH, Jessell T: *Principles of Neural Science,* 4th Edition. New York, McGraw-Hill, 2000.

stimulus was the same in both cases. Desimone and colleagues interpreted these results as evidence that the receptive field shrinks to conform to the attended stimulus, thereby excluding the unattended stimulus and implementing a filtering mechanism of the sort proposed by Broadbent (see Desimone and Duncan 1995 for a review).

The exact visual area that will be biased in the manner revealed by these physiological studies appears to depend upon the task required of the subject (Desimone and Duncan 1995; Kastner et al. 1999; Posner and Gilbert 1999). Imaging studies have shown that if people are asked to attend to target motion, activity is increased in a brain area in the dorsal visual pathway sensitive to movement (area MT). Quite different visual areas become active for attention to other stimulus dimensions such as color or orientation (Corbetta et al. 1991). When attention is shifted to a new location, the neural activity of cells in the ventral, object recognition network is increased even before any target is presented at that location (Kastner et al. 1999).

There is a fundamental distinction between focal and ambient attention

Of course, there is a sense in which, without even trying to attend, you are conscious of all the objects on your desktop. However, when careful tests are made that involve making changes in a complex visual scene, these tests reveal that when attention is focused on one object, other objects within the scene, even large and important ones, can be altered without the subject being aware that a change has taken place (Rensink et al. 1997). Thus, while attention can be summoned efficiently to a novel event, there is surprisingly little awareness of changes that occur at loci that are not attended. It, therefore, is useful to distinguish between *focal attention,* which allows reporting of details of the scene, and *ambient attention,* which forms our general awareness of the scene around us. While both are aspects of consciousness, their underlying neurobiological mechanisms may be quite different. It seems likely that ambient attention may depend primarily upon posterior brain areas. By contrast, focal attention, which is often switched between objects based on instructions, may depend on more anterior areas related to voluntary control of action.

There may be only a small number of networks concerned with attention, and these can be distinguished on the cellular and even on the molecular level

It seems likely that the neuronal computation that occurs in most cortical areas can be influenced by attention. Indeed, there is a surprisingly large number of sites in the brain where attentional influences can be demonstrated. However, the source of those effects is thought to emanate from a small number of networks that perform different functions.

For example, a novel visual stimulus serves both to alert the organism and to orient attention to the location of the stimulus. This distinction can be demonstrated by using separate cues for alerting and orienting. An alerting cue provides the monkey with information about when a target will occur, but not about where that target will be located. An orienting cue provides the monkey with specific information about where the target will be located and thus allows the subject to move attention to the cued location. Two separate brain networks, both located in the parietal lobe, but using distinctly different chemical transmitter systems, are involved in changing the level of alertness and in switching attention toward the stimulus. Thus, the influence of alerting cues is reduced by drugs that block norepinephrine activity, but these drugs do not influence orienting. By contrast, drugs that block cholinergic activity influence orienting to the cue, but do not diminish the alerting effect (Marrocco and Davidson 1998). These pharmacological studies illustrate how one can separate a simple act of attention to a novel event into two distinct components, and pinpoint both the anatomical systems and the modulatory synaptic mechanisms involved.

Findings in patients confirm the importance of neural systems for alerting and orienting to our normal awareness of the world around us. Strokes that interfere with the blood supply to the posterior parietal cortex on the right side produce an inability to orient attention to the left side, the side opposite the lesion. Patients who suffer from these right-sided lesions will show striking deficits in body image and in their perception of spatial relations. Although their somatic sensations are intact, these patients may ignore (*neglect*) the spatial aspects of all sensory input from the left side of their body as well as of external space, and they will ignore the left half of any visual object with which they are confronted. For example, patients with neglect syndrome will exhibit a severe disturbance in their ability to copy drawn figures. This deficit can be so severe that the patient may draw a flower with only petals on the right side of the plant. When asked to copy a clock, the patient may ignore the digits on the left and try to fill in all the digits on the right, or draw them down the side, running off the clock face (Figure 6–19). These patients also may ignore the left half of the body and fail to dress, undress, and wash the affected side.

Less dramatic but similar difficulties in orienting accompany loss of parietal neurons due to degenerative disorders such as Alzheimer's dementia (Parasuraman and Greenwood 1998). In these cases, stimuli going directly to the lesioned area, that would normally produce orienting of attention, may no longer do so and consequently the person may be completely unaware of these stimuli.

Model   Patient's copy

FIGURE 6–19. The contribution of the posterior parietal cortex to visual attention?

The three drawings on the right were made from the models on the left by patients with unilateral visual neglect following lesion of the right posterior parietal cortex. *Source.* From Bloom and Lazerson 1988, as illustrated in Kandel ER, Schwartz JH, Jessell T: *Principles of Neural Science,* 4th Edition. New York, McGraw-Hill, 2000.

Many researchers agree with Francis Crick's view that sensory awareness probably is the most tractable area for a rigorous understanding of consciousness at a mechanistic level. As a result, much research has recently been focused on the study of orienting to sensory stimuli. The discovery that attention can influence activity within primary visual cortex (area 17) has allowed investigators to explore attentional effects within a visual area whose other cellular and physiological features and anatomical characteristics are extremely well characterized (Posner and Gilbert 1999). The work in area 17 therefore provides an opportunity to specify exactly what cellular structures and functions can be modified by the act of attending.

There also now are methods for exploring the consequences of attention on the natural life of the organism. One direction of current research involves the study of the maturation of orienting mechanisms in the human infant. These studies illustrate that the ability to orient attention to visual stimuli undergoes a substantial maturation during the first year of life. For example, an infant 2–4 months of age has great difficulty in disengaging from a strong visual attractor. If the attractor is a checkerboard, the difficulty in disengaging may cause the infant to become distressed; if the attractor is the eyes of the mother, the lack of engagement may contribute to the development of parent-infant bonding. As with age, the parietal mechanisms involved in orienting of attention mature; they appear to allow the infant to disengage from strong attractors. Thus, by 4 months infants can begin to move their eyes in anticipation of the occurrence of a visual stimulus. The ability to show anticipatory eye movements demonstrates an influence of learning on orientation, on where infants attend.

The maturation of the ability to orient to visual stimuli has dramatic consequences for an infant's response to novelty and their ability to know where to look. At 4 months and older, a parent can regulate negative affect by use of distraction, by orienting the infant to a novel stimulus (Posner and Rothbart 1998). Much of the development of orienting skills must relate to the maturation of specific pathways between neural areas. The advent of new forms of neuroimaging may allow one to follow the time course of the maturation of the pathways required for the development of orienting and thus give us a new means of exploring the mechanisms of these developmental changes in infants and children (Conturo et al. 1999).

Imagery

Imagery differs from perception in efficiency of coding

Visual images are an excellent example of a purely mental event, and as such, they are a promising entryway for the neurobiological study of consciousness. Images seem to have a sensory character even though no sensory stimulus is presented. In the early part of the twentieth century, visual images could only be studied by the methods of verbal report and by the systematic collection of surveys (Galton 1907). Because behaviorists thought that the subjective nature of the image would never allow a scientific approach, the study of imagery was largely abandoned. In the period following World War II, imagery again became a focus of study in modern cognitive psychology, and objective experimental methods for probing the characteristics of these mental events were successfully developed (see Kosslyn 1980 for a review).

It is now possible to study brain mechanisms of imagery in humans and animals

It is now possible to design objective tests of imagery. If a test subject is given the name of a letter (e.g., R) and an angle of orientation (75 degrees) most people can construct a mental image of the R at the correct angle. If people are then shown visual probes of R that may be a real or mirror-image R, they can quickly construct a mental image of the R when the probe letter is at the same angle as the image. In fact, they are faster to respond to probes that match the angle of the imaged R than to an upright R (Cooper and Shepard 1973).

But if one tests a subject by asking them to develop a visual image of a somewhat lengthy word, such as "pumpkin," most people believe they can do it until required to perform an objective task of this accomplishment like being asked to spell the word backward. The difficulty encountered when asking a subject to spell backward a word like "pumpkin" led Weber and Harnish (1974) to question whether there really was an image of the word "pumpkin." They then proceeded to compare the performance of subjects when a word stimulus like "pumpkin" was physically present (perceptual condition) and when it had to be created as an image (imagery condition). With short three- or five-letter words, the subjects show the same reaction time for an imagery condition as they show for the perceptual condition. But when the word has more than five letters imagery is much slower than perception (Weber and Harnish 1974). Thus, imagery can indeed produce a remarkably efficient representation, but it is also rather limited to only about three to five separate items. By contrast, most people can hold in memory about seven to eight separate letter names.

It also takes longer to develop a visual image than to provide a name. If you are asked to image each lowercase letter of the alphabet in turn, it will take you 10–20 seconds to go through the alphabet, much longer than if you were to name the letters silently. Creating mental images is thus a complex task that consists of many mental operations. Accordingly, many parts of the brain are involved (Kosslyn 1994).

The ability to study the brain while people construct and inspect visual images has greatly enhanced the field of mental imagery (Kosslyn 1994). These studies have revealed that most of the visual areas that are involved in pattern recognition and in orienting of attention also are recruited during visual imagination. The overlap between areas recruited for the perception of real as opposed to imagined visual objects is substantial. It is clear that visual imagination uses the same apparatus of visual perception and the same systems of visual attention as would be involved if the stimulus were *actually* presented to the sense organ.

This finding also is supported by clinical evidence. Studies of patients with lesions of the right posterior parietal cortex that produce visual neglect in viewing *real objects* also disrupt the experience of imaging the left side of a visual image (Bisiach and Luzzatti 1978). This defect in imagery was first described in a fascinating study by Bisiach and Luzzatti of a group of patients in Milan, all of whom had injury to the *right* parietal lobe. As the patients were sitting in the hospital's examining room, they were asked to imagine that they were standing in the city's main square, the Piazza del Duomo, facing the cathedral, and to describe from memory the key buildings around the square. These subjects identified from memory all the buildings on the right side of the square (ipsilateral to the lesion) but could not recall the buildings on the left, even though these buildings were thoroughly familiar to them.

The patients were next asked to imagine that they were standing on the opposite side of the square, on the steps of the cathedral, so that right and left were reversed. In this imagined position, the patients were now able to name the buildings they previously had been unable to identify but failed to identify or name the buildings they had previously listed. The patients now described what they previously neglected, and neglected what they had previously described, suggesting that they retained in memory full knowledge of the space. What these patients lost was *access* to memories associated with the side of the body opposite the lesion, no matter which way the patient imagined himself facing.

Study of patients with lesions in the right posterior parietal cortex has yielded three insights. First, these lesions commonly lead to a disturbance in orienting of attention. Second, patients who neglect the left side of their body after damage to the right parietal lobe show a disturbance not only in cortical representation of their own body but also in representation of external space. Specifically, they neglect visual stimuli on that side of the body. Third, patients with neglect syndrome also lack access to memories against which perceptions on the neglected side can be compared. Thus, these patients neglect not only real external objects but also objects in memory.

Finally, the lesions not only lead to disorders in perceptive-spatial relationships; they also commonly lead to a disturbance in directed attention (Figure 6–20A).

It has even been possible to develop an animal model of mental rotation of the sort we described previously (Georgopoulos et al. 1989). Monkeys were taught to move a lever in a direction 90 degrees from a target light. Recording from cells in the motor system shows that immediately after the light comes on, the set of active cells has an equivalent vector that would move the limb directly to the light. However, over a 0.25-second interval, the population of cells changes to compute a vector which is at the proper 90-degree angle. Imaging studies of mental rotation in humans have shown that motor areas of the brain as well as the parietal lobe become involved when rotation involves one's own hand (Parsons and Fox 1998).

Future studies of imagery can probe the influence
of learning and individual differences

Traditionally, imagery has been defined by subjective report, but that is no longer necessary. It is now possible to know when the visual system has been activated from the top down, even if the person is unaware that any form of imagery has been used. For example, when reading a vivid description, people often create a visual representation of the scene and individual words may automatically evoke images related to their meaning. Some people are aware of these representations, but others deny having any subjective visual experience. By appropriate imaging studies, we can now determine if the representations differ between people according to their reports or whether the representations are the same, but only some reach threshold for a verbal report of awareness.

Normal people can report whether they have created an image, but in some cases, imagery can be pathological, as in the voices heard by paranoid schizophrenics which are attributed to outside forces, or in the hallucinations of drug states. Lesions of the frontal midline can result in attributing control of one's own hand to alien forces (G. Goldberg 1985). It seems very likely that high-level attentional networks involving this frontal midline area are a source of knowledge that the information has been internally generated. These networks are normally involved in voluntary control of actions. The pathological belief of alien control of one's thoughts that is found in some schizophrenics may arise from abnormalities in the regulation of these networks. In the next section, we examine the operation of these frontal networks as a means of aiding our understanding of the conscious control of behavior.

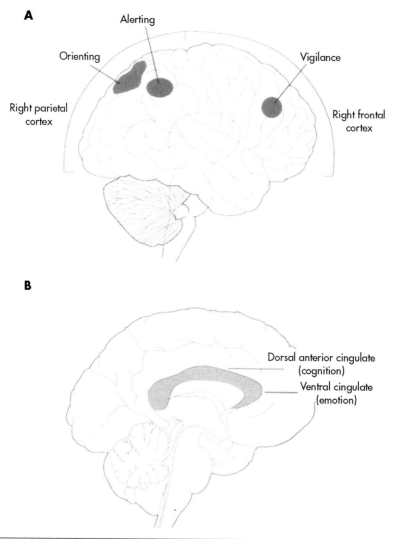

FIGURE 6–20.   Localization of alerting and emotional function.

(A) The localization of alerting and orienting functions in the parietal lobe. A diagram of the lateral surface of the human brain indicating the relation of alerting and orienting mechanisms in the right parietal lobe to the vigilance area in the right frontal cortex.

(B) The localization of cortical regions involved in cognitive and emotional states. A diagram of the medial part of the human brain indicating the dorsal anterior cingulate, which appears to be involved in the monitoring and/or regulation of cognitive activity, and the more ventral portions of the cingulate, which appear to be related to the regulation of emotion.

Source.   (A) Courtesy of M. Posner. (B) Adapted from Bush et al. 1998.

Executive Control Includes Volition as a Major Component

*Areas of the frontal midline appear to be important
in voluntary control of cognition and emotion*

Normal people have a strong subjective feeling of their intentions. They
have a clear sense that they have voluntary control of their own behavior.
These subjective feelings of intentions and voluntary control can be freely
verbalized. Indeed, asking people about their goals or intentions is probably
the single most predictive indicator of their behavior during problem solving
(Newell and Simon 1972). The importance of intention is also illustrated in
patients with frontal lesions (Duncan 1986) or in patients suffering from
with mental disorders (Frith and Dolan 1998), who show disruption in either their voluntary control over behavior or their subjective feelings of control. What are the neural mechanisms of voluntary control?

Norman and Shallice (1986) have argued that an executive attention system is necessary for situations in which routine or automatic processes are
inadequate. These nonroutine, nonautomatic executive functions include
selection among conflicting inputs, resolution of conflict among responses,
and monitoring and correcting errors.

*Priming produced by automatic activation without
attention can facilitate reaction time to the item
when presented consciously*

The existence of executive control was made more concrete in the 1970s and
1980s when cognitive studies first succeeded in separating conscious control of mental events from automatic activation of the same events (Posner
1978). This separation used priming of a target word by a prime word related
in meaning to the target. The method is simple. Subjects are given a task to
perform on a target word. For example, they may be asked to classify
whether the target word is a meaningful word or not, or to categorize it as
representing a living object or not. On some trials, prior to the presentation
of the target word, the subject is given a prime word, a word that is flashed
briefly on the screen without further comment. The prime may be related to
the target (e.g., prime "toy" and target "doll") or unrelated (prime "toy" and
target "stop"). Although the prime provides no direct information on how to
respond to the target, nonetheless, related primes were found to speed up
and unrelated ones to slow down reaction time in comparison to a neutral
warning signal that merely tells the person that a target will be coming.

People do not have to attend to the prime or even be aware that it has
happened to get the priming effect. Priming still takes place—if the prime
word is followed immediately by a visual masking noise (a random visual in-

put) so that subjects are unaware of the identity of the prime word. However, the effects of priming were somewhat different from trials in which the prime word had been carefully attended. Consider a condition when people are given a series of targets that either involve trees or body parts. If ambiguous prime words such as "palm" are followed by a mask so they cannot be reported, they serve to improve the performance on subsequent targets related to both meanings of the word (e.g., "tree" and "hand"). However, when the prime "palm" is presented in the context of trees and unmasked, only the meaning related to the category of the previous trials (e.g., "tree") is primed (Marcel 1983).

James Neely (1977) studied the conscious use of primes by his subjects. In one condition a word from one category (e.g., "animal") was presented as a prime word, and subjects were instructed to associate the category "animal" with the category "building." Target words in the category "building" (e.g., "window") had faster reaction times than targets in a category unrelated category (e.g., "tin"). The subject had voluntarily activated the instructed category. If a specific animal target (e.g., "dog") was presented after a very short interval so that subjects did not have enough time to switch from the prime category "animal" to the instructed category "building," fast reaction times were made to the word "dog." However, if "dog" was delayed until after subjects had a chance to execute the switch to the instructed category, the target "dog" would have a slow reaction time since subjects were now attending to the wrong category. Within a second, Neeley was able to trace the conscious effort involved in switching categories, by its influence on the reaction time to probes.

There is more than one form of priming

These findings give a reality to the difference between the voluntary, conscious control of mental events and the same event when driven unconsciously and automatically by input. Priming can be produced in both ways. First, priming can be produced by automatic activation of a pathway without attention, facilitating reaction times for related items. This form of priming can be totally subconscious. Imaging studies have shown that automatic priming of this sort is produced by a reduction of blood flow within the brain area related to processing the target. For semantic priming, this reduction would be within areas of the brain related to the meaning of the word (Demb et al. 1995). It is as if the prime had tuned the neuron pool so that attention of fewer neurons is required to process the target. Possibly, as a consequence of this reduced overall activity, a primed target, although classified rapidly, is often poorly remembered in a later recall or recognition test.

A second form of priming is produced by directing attention to semantic information. This form of attention appears not to depend on the brain area related to the processing of the word but depends upon different frontal networks, and this information is available to the conscious awareness of the person. Within a second, subjects can voluntarily choose an associated category, and the consequence of that selection is faster processing of related targets and retarded processing of unrelated targets. When a category is attended, items within the category are facilitated in reaction time, but items in other categories are retarded over what they would have been if no priming had taken place. Imaging studies have suggested that attention to a computation increases blood flow within the attended area. Thus, priming may be produced by different brain mechanisms that have quite different consequences for performance. It remains for future studies to tell us how these two priming mechanisms produce what seem like opposite changes at the neural level.

The dorsal anterior cingulate cortex is essential for executive control

Suppose you are asked how many objects you see between the brackets: [two, two, two]. You may at first want to say "two," even though the correct answer is "three." This is because there is a conflict in your mind between the *meaning* of the word as read and the specified task of saying *how many* words are present. This is one form of the Stroop effect. The most frequent form of Stroop effect occurs when a subject has to name the color of ink in which a word is written when the color of the word conflicts with the word name (e.g., when the word "red" is written in blue ink). These conflicting tasks involve focal attention to the critical element of the task when that element must be selected in competition with a more dominant element. Imaging studies of the Stroop effect produced by conflict between elements tend to find very strong activity in the dorsal anterior cingulate gyrus (see Figure 6–20B) often in concert with areas of the basal ganglia and lateral frontal cortex (Bush et al. 1998). For this reason, dorsal anterior cingulate gyrus has been thought to be involved in some aspect of focal or executive attention (Carter et al. 1999).

As is the case with humans, rhesus monkeys trained to associate digits with a quantity show conflict between deciding which of two displays has the greatest number of objects when there is an incompatible relation between the two (e.g., when the larger number of objects is made up of the smaller digit). The monkeys made many more errors on incompatible trials than do humans, despite many hundreds of trials at the task (Washburn 1994). It is as though the monkeys have somewhat less capacity for avoiding interference, despite very extensive training.

In humans, activity in the anterior cingulate gyrus generally is related to the degree of practice or automation of the task. Perhaps the best example is a task where subjects are required to ascribe a use for each noun in a list (e.g., "hammer"→"pound"). There is a conflict between saying the word name aloud and the required task of generating the use of the word. There was strong activation of anterior cingulate when the list was first presented, but with practice on a single list, activity in the cingulate disappears and instead there is activity in the anterior insula, a portion of cortex that lies buried beneath Broca's area (Raichle et al. 1994). Both the anterior insula and Broca's area are closely related to the automatic task of reading the word aloud. Imaging studies have identified two different pathways for producing the use of a word. One pathway is involved when conscious thought is needed to generate a word. This pathway involves the cingulate in conjunction with left lateral areas of the cortex and the right cerebellum. Another, more automatic, pathway is involved when the words are well practiced so that the feeling of conscious search disappears. Now the activity in the cingulate (as well as in the lateral cortex and cerebellum) disappears, and instead one finds activity in Broca's area and in the anterior insula, the structures that are usually involved in the automatic tasks of reading words out loud.

The ventral and dorsal anterior cingulate are concerned with emotion and cognition, respectively

We often think of sensory orienting and memory retrieval as related to focal attention. However, another source of information that frequently engages our attention is emotion. When emotional words are presented in the same conflict tasks described above, a more ventral area of the anterior cingulate becomes active (see Figure 6–20B). In some neuroimaging experiments, the cognitive and emotional areas of the cingulate seem to be mutually inhibitory (Drevets and Raichle 1998). Thus, when strong emotions are involved in the task, the dorsal area is less active than at rest and cognitive conflict tasks tend to reduce activity in the more ventral area of the cingulate.

If the dorsal area of the cingulate is involved in selecting dimensions of a stimulus when there is conflict among competing dimensions, a reasonable idea might be that the dorsal area serves a similar selecting function for emotional conflict. Indeed, we have already discussed the idea that orienting of attention in infancy serves as one means by which caregivers seek to distract their infants from the expression of distress. The control of distress is an important concern of early childhood, and caregivers have the task of first regulating emotions in their infant and later teaching the child to regulate its own emotions. Perhaps, areas of the brain that regulate emotion in infancy have acquired the ability to perform the same functions in response to cognitive challenges. If this idea is correct, children who are well advanced in emotional regulation should be at a specific advantage in regulating cognitive conflict.

We know that children differ in their ability to regulate their emotions. This can be elicited from caregivers when they are asked specific questions about the child's ability to control distress, orient attention and be sensitive to pleasures. The dimension of individual variation in regulation has been called *effortful control*. Studies of 6- to 7-year-olds have found that effortful control can be defined in terms of scales measuring attentional focusing, inhibitory control, low intensity pleasure, and perceptual sensitivity (Rothbart et al. 2000). Effortful control is consistently negatively related to a negative affect in keeping with the notion that attentional skill may help attenuate negative affect. Effortful control also is correlated with the performance of 2-to 4-year-old children in Stroop tasks that require them to handle conflict (Posner and Rothbart 1998). Effortful control is related both to empathy and to the acquisition of conscience, of a sense of moral behavior. Kochanska (1995) has found that individual differences in effortful control have important implications both for the inhibition of antisocial behavior and for the acquisition of prosocial behavior. Children who can effectively employ attention to regulate behavior are better able to inhibit prepotent responses (e.g., striking out, stealing) and are better at taking into consideration the effect of their actions on others.

Empathy and a sense of moral behavior or conscience are at the heart of child socialization. The link between the attentional network of the frontal lobe and conscience might make it possible to at least imagine how aspects of morality might be studied on the neuronal level discussed below in relation to disorders.

Disorders that recruit the cingulate cortex suggest
a connection between cognition and emotion

Attention deficit disorder is defined by a set of cognitive and emotional symptoms. This disorder is usually diagnosed in children but often remains present into adulthood. Neuroimaging of adults who suffer from attention deficit disorder has been carried out under circumstances that require them to do a numerical version of the Stroop effect. Here they are asked to respond to the number of items present. When that number is sometimes in conflict with the quantity indicated by the word (e.g., three copies of the word "two"), adults with attention deficit disorder performed on conflict trials only slightly less efficiently than normals. But unlike the normal controls, adults with attention deficits show no activation of the anterior cingulate.

Instead, they show greater activity on incompatible trials in the anterior insula (Bush et al. 1999). As was suggested in the study of word association discussed above, the insula represents a more automatic pathway than the anterior cingulate, thus allowing for less effortful control over the task.

Another disorder that produces a disruption of voluntary control as well as other emotional and cognitive problems is schizophrenia. Benes (1998) has reported subtle abnormalities of the anterior cingulate in postmortem analyses of schizophrenic brains. She argues that the problem with the anterior cingulate, in the brains of schizophrenics, may be a shift in dopamine regulation from pyramidal to nonpyramidal cells. She has also argued that these changes in the cingulate are related to circuitry involving the amygdala and hippocampus. The schizophrenia studies provide a lead at the cellular level of the possible disregulation of the anterior cingulate in a second abnormality noted for its attentional deficits.

Both schizophrenia and attention deficit disorder have a genetic basis. Studies of attention deficit disorder families have shown that some of them possess a particular allele of the dopamine 4 receptor (LaHoste et al. 1996; Smalley et al. 1998). These studies provide some potential cellular and genetic links between attentional abnormalities found in various pathologies.

A Future for the Study of Consciousness

We have outlined that the problem of consciousness can be considered to consist of two subproblems: awareness and volition. Studies of orienting and of imagery are concerned with the first, and self-regulation with the second. As future research penetrates the organization of attention networks in the frontal cortex, the two functions could prove to be linked. Studies of complex scenes show that presentation of a stimulus does not lead automatically to awareness of even the most central aspects of the scene (Rensink et al. 1997). Even though subjects report that they are aware of a whole scene, they only become aware of a change when their attention is drawn to a change in the scene. We have suggested that posterior brain areas involved in orienting to sensory stimuli may be closely related to ambient awareness and that focal attention might be more associated with the anterior cingulate and other frontal areas related to voluntary control. Perhaps only as we understand the neural basis of the distinction between *focal attention* to limited aspects of the external world and a more *general ambient awareness* of the general scene will we be able to understand the parts of the brain related to consciousness of sensory events (Iwasaki 1993).

The exact functions that the anterior cingulate plays in higher-level attention are not yet clear. We have learned that even very simple acts of attention such as orienting to sensory stimuli involve a network of brain areas that carry out specific functions. During conflict tasks, activation of the anterior cingulate is usually accompanied by activity in lateral frontal cortical areas and in the basal ganglia. It is an important goal to find out what each of these areas contributes.

The functions of attention also relate to issues other than those usually discussed under the term "consciousness". Recently, the contribution of biology to understanding the acquisition of high-level skills such as those learned in schools has been extensively debated (e.g., Bruer 1999). In some areas, such as the neural networks involved in reading and arithmetic, progress has been extensive (Dehaene 1997; Posner et al. 1999). Both skills required attention networks related to those discussed in this section. A better understanding of these mechanisms might help in realizing the goal of a neuroscience-based approach to aspects of education.

Coda

What is the future for neural science in the next millennium? We have seen remarkable and rapid progress in understanding neuronal and synaptic signaling. These advances now invite a structural approach to visualize the static and dynamic structures of ion channels, receptors, and the molecular machinery for signal transduction postsynaptically and for vesicle transport, fusion, and exocytosis presynaptically. We also have made some progress in the analysis of the elementary synaptic mechanisms that contribute to memory storage. These studies have revealed that the different memory systems of brain seem to use similar synaptic mechanisms for the storage of both declarative and nondeclarative knowledge. Similarly, we now have achieved an understanding, at least in broad outline, of the development of the nervous system. Specific inducers, morphogens, attractants, and repellents of process outgrowth and synapse organizing molecules have now been defined, providing a molecular reality to concepts that previously were shrouded in mystery.

Progress in these several areas has in turn made possible a molecular-based neurology, a neurology that will, one hopes, finally be able to address the degenerative diseases of the brain that have for so long eluded our best scientific efforts. In time, advances in these areas may also yield insight into and perhaps solutions for some of the most debilitating diseases confronting medical science—the psychiatric and neurological illnesses of schizophrenia, depression, and Alzheimer's disease. Implicit in this prediction is the expectation that in the future molecular biology will be able to contribute to the system problems of cognitive neural science much as it has recently contributed to signaling, plasticity, and development.

The advances in the cellular understanding of the organization of the somatosensory and visual system by Mountcastle, Hubel and Wiesel, and their followers have helped turn our interest to perception and in the broader sense to cognitive psychology. In turn, contact between cognitive psychology and neural science has given us a new approach to the classical problems of mental function, including attention and consciousness. In both early sensory processes and higher cognitive perceptual and motor systems, we find evidence for the localization of components within a broadly distributed network carrying out complex functions. Indeed, one of the early insights into consciousness is that it shares the properties of other cognitive systems in that like vision and action, it can be dissected into components: attention, imagery, and volition. Each component consists of a set of subcomponents that can be localized within a larger, distributed neural system. Having pointed to that similarity, we must nevertheless acknowledge that of all fields in neural science, in fact of all the fields in all of science, the problems of perception, action, memory, attention, and consciousness provide us with the greatest evidence for our lack of understanding as well as the greatest challenge.

Even if one agreed that the scientific agenda outlined here may be adequate to handle the issues of awareness and volition, there is another aspect of consciousness that needs to be confronted and that is the nature of *subjectivity*. The subjective aspect of consciousness is seen by philosophers of mind such as Searle (1993, 1998) and Nagel (1993) as its defining characteristic and the aspect that poses the greatest scientific challenge. Searle and Nagel argue that each of us experiences a world of private and unique experiences and that these seem much more *real* to us than the experiences of others. We experience our own ideas, moods, and sensations—our successes and disappointments, joys and pains—directly, whereas we can only appreciate other people's ideas, moods, and sensations. Are the purple you see and the jasmine you smell identical to the purple that we see and the jasmine that we smell? The fact that conscious experience is uniquely personal and intensely subjective raises the question whether it is ever possible to determine objectively some common characteristics of experience. We cannot, the argument goes, use those same senses to arrive at an objective understanding of experience.

Clearly, we should be prepared for the possibility that there are aspects of consciousness that will not be solved by the approaches discussed in this review. Some might believe that all that is scientific about the study of life is illuminated at all levels from the molecular to the behavioral by what we know about DNA. But others might believe that there are issues about what it means to be a living being that are really not explained by the most detailed account of DNA. Many issues of awareness and voluntary control are likely to be explained at all levels, from genes to behavior. This might constitute a theory of consciousness in much the same way as DNA serves as the basis for any scientific analysis of what constitutes life. Nonetheless, it is at present hard to imagine how the progress discussed above, even if it continues and intensifies, will solve all the issues of the subjective nature of our experience. We leave it to the readers of *Cell* and *Neuron* in the next millennium to determine how much insight about human consciousness will result from the type of work we have discussed here.

Acknowledgments

We thank G. Gasic, E. Marcus, and L. Pond for helpful comments on the manuscript, Harriet Ayers, Millie Pellan, and Kathy MacArthur for assistance in preparing the text, and Ira Schieren, Sarah Mack, and Charles Lam for help in preparation of the figures. Many of the illustrations used in this article derive from *Principles of Neural Science,* with permission from McGraw-Hill. T.D.A., E.R.K., and T.M.J. are Investigators of the Howard Hughes Medical Institute.

References

Abel T, Nguyen PV, Barad M, et al: Genetic demonstration of a role for PKA in the late phase of LTP and in hippocampus-based long-term memory. Cell 88:615–626, 1997

Adrian ED: The analysis of the nervous system. Sherrington Memorial Lecture. Proc R Soc Med 50:991–998, 1957

Albright TD: Form-cue invariant motion processing in primate visual cortex. Science 255:1141–1143, 1992

Albright TD: Cortical processing of visual motion. Rev Oculomot Res 5:177–201, 1993

Albright TD, Stoner GR: Visual motion perception. Proc Natl Acad Sci USA 92:2433–2440, 1995

Albright TD, Desimone R, Gross CG: Columnar organization of directionally selective cells in visual area MT of the macaque. J Neurophysiol 51:16–31, 1984

Andersen RA, Essick GK, Siegel RM: Encoding of spatial location by posterior parietal neurons. Science 230:456–458, 1985

Andersen RA, Snyder LH, Li CS, et al: Coordinate transformations in the representation of spatial information. Curr Opin Neurobiol 3:171–176, 1993

Antonini A, Stryker M: Development of individual geniculocortical arbors in cat striate cortex and effects of binocular impulse blockade. J Neurosci 13:3549–3573, 1993a

Antonini A, Stryker M: Rapid remodeling of axonal arbors in the visual cortex. Science 260:1819–1821, 1993b

Armstrong CM, Hille B: Voltage gated ion channels and electrical excitability. Neuron 20:371–380, 1998

Attardi DG, Sperry RW: Preferential selection of central pathways by regenerating optic fibers. Exp Neurol 7:46–64, 1963

Axel R: The molecular logic of smell. Sci Am 273:154–159, 1995

Bang AG, Goulding MD: Regulation of vertebrate neural cell fate by transcription factors. Curr Opin Neurobiol 6:25–32, 1996

Barlow HB: Summation and inhibition in the frog's retina. J Physiol 119:69–88, 1953

Barlow HB: Single units and sensation: a neuron doctrine for perceptual psychology? Perception 1:371–394, 1972

Barthels D, Santoni MJ, Wille W, et al: Isolation and nucleotide sequence of mouse NCAM cDNA that codes for a M_r, 79,000 polypeptide without a membrane-spanning region. EMBO J 6:907–914, 1987

Bartsch D, Ghirardi M, Skehel PA, et al: Aplysia CREB-2 represses long-term facilitation: relief of repression converts transient facilitation into long-term functional and structural change. Cell 83:979–992, 1995

Bartsch D, Casadio A, Karl KA, et al: CREB-1 encodes a nuclear activator, a repressor, and a cytoplasmic modulator that form a regulatory unit critical for long-term facilitation. Cell 95:211–223, 1998

Bate CM: Pioneer neurones in an insect embryo. Nature 260:54–56, 1976

Benes FM: Model generation and testing to probe neural circuitry in the cingulate cortex of postmortem schizophrenic brains. Schizophr Bulletin 24:219–230, 1998

Bennett MVL: A comparison of electrically and chemically mediated transmission, in Structure and Function of Synapses. Edited by Pappas GD and Purpura DP. New York, Raven, 1972, pp. 221–256

Bennett MVL: Seeing is relieving: electrical synapses between visualized neurons. Nat Neurosci 3:7–9, 2000

Bisiach E, Luzzatti C: Unilateral neglect of representational space. Cortex 14:129–133, 1978

Bliss TVP, Lømo T: Long-lasting potentiation of synaptic transmission in the dentate area of the anaesthetized rabbit following stimulation of the perforant path. J Physiol 232:331–356, 1973

Bloom F, Lazerson A: Brain, Mind, and Behavior, 2nd Edition. New York, WH Freeman, 1988

Bock JB, Scheller RH: SNARE proteins mediate lipid bilayer fusion. Proc Natl Acad Sci USA 96:12227–12229, 1999

Boistel J, Fatt P: Membrane permeability change during inhibitory transmitter action in crustacean muscle. J Physiol 4:176–191, 1958

Bourtchouladze RA, Franguelli B, Bendy J, et al: Deficient long-term memory in mice with a targeted mutation of the cAMP responsive element binding protein. Cell 79:59–68, 1994

Boyd IA, Martin AR: The end-plate potential in mammalian muscle. J Physiol 132:74–91, 1956

Brackenbury R, Rutishauser U, Edelman GM: Distinct calcium-independent and calcium-dependent adhesion systems of chicken embryo cells. Proc Natl Acad Sci USA 78:387–391, 1981

Briscoe J, Ericson J: The specification of neuronal identity by graded Sonic Hedgehog signalling. Semin Cell Dev Biol 10:353–362, 1999

Broadbent DE: Perception and Communication. London, Pergamon, 1958

Brock LG, Coombs JS, Eccles JC: The recording of potentials from motoneurones with an intracellular electrode. J Physiol 117:431–460, 1952

Brockes JP: Amphibian limb regeneration: rebuilding a complex structure. Science 276:81–87, 1997

Brown RH Jr: Ion channel mutations in periodic paralysis and related myotonic diseases. Ann NY Acad Sci 707:305–316, 1993

Bruer JT: The Myth of the First Three Years: A New Understanding of Early Brain Development and Lifelong Learning. New York, Free Press, 1999

Brummendorf T, Rathjen FG: Structure/function relationships of axon-associated adhesion receptors of the immunoglobulin superfamily. Curr Opin Neurobiol 6:584–593, 1996

Buck L, Axel R: A novel multigene family may encode odorant receptors: a molecular basis for odor recognition. Cell 65:175–187, 1991

Buonomano DV, Merzenich MM: Cortical plasticity: from synapses to maps. Annu Rev Neurosci 21:149–186, 1998

Burns BD: The Mammalian Cerebral Cortex (Monographs of the Physiological Society, No 5). London, Edward Arnold [Publishers] Ltd., 1958

Bush G, Whalen PJ, Rosen BR, et al: The counting Stroop: an interference task specialized for functional neuroimaging—validation study with functional MRI. Hum Brain Map 6:270–282, 1998

Bush G, Frazier JA, Rauch SL, et al: Anterior cingulate cortex dysfunction in attention-deficit/hyperactivity disorder revealed by fMRI and the counting Stroop. Biol Psychiatry 45:1542–1552, 1999

Byrne JH, Kandel ER: Presynaptic facilitation revisited: state and time dependence. J Neurosci 16:425–435, 1996

Cabelli RJ, Hohn A, Shatz CJ: Inhibition of ocular dominance column formation by infusion of NT-4/5 or BDNF. Science 267:1662–1666, 1995

Cabelli RJ, Shelton DL, Segal RA, et al: Blockade of endogenous ligands of trkB inhibits formation of ocular dominance columns. Neuron 19:63–76, 1997

Callaway EM: Local circuits in primary visual cortex of the macaque monkey. Annu Rev Neurosci 21:47–74, 1998

Carew TJ, Hawkins RD, Kandel ER: Differential classical conditioning of a defensive withdrawal reflex in Aplysia californica. Science 219:397–400, 1983

Carter CS, Botvinick MM, Cohen JD: The contribution of the anterior cingulate to executive processes in cognition. Rev Neurosci 10:49–57, 1999

Casadio A, Martin KC, Giustetto M, et al: A transient neuron-wide form of CREB-mediated long-term facilitation can be stabilized at specific synapses by local protein synthesis. Cell 99:221–237, 1999

Castellucci V, Pinsker H, Kupfermann I, et al: Neuronal mechanisms of habituation and dishabituation of the gill-withdrawal reflex in Aplysia. Science 167:1745–1748, 1970

Catterall WA: Structure and function of voltage-sensitive ion channels. Science 242:50–61, 1988

Catterall WA: From ionic currents to molecular mechanisms: the structure and function of voltage-gated sodium channels. Neuron 26:13–25, 2000

Chan YM, Jan YN: Conservation of neurogenic genes and mechanisms. Curr Opin Neurobiol 9:582–588, 1999

Changeux J-P, Galzi JL, Devillers-Thiery A, et al: The functional architecture of the acetylcholine nicotinic receptor explored by affinity labeling and site directed mutagenesis. Q Rev Biophys 25:395–432, 1992

Chen G-Q, Cui C, Mayer ML, et al: Functional characterization of a potassium-selective prokaryotic glutamate receptor. Nature 402:817–821, 1999

Chino YM, Kaas JH, Smith ELD 3rd, et al: Rapid reorganization of cortical maps in adult cats following restricted deafferentation in retina. Vision Res 32:789–796, 1992

Coen L, Osta R, Maury M, et al: Construction of hybrid proteins that migrate retrogradely and transynaptically into the central nervous system. Proc Natl Acad Sci USA 94:9400–9405, 1997

Colby CL, Goldberg ME: Space and attention in parietal cortex. Annu Rev Neurosci 22:319–349, 1999

Cole KS, Curtis HJ: Electrical impedance measurements in the squid giant axon during activity. J Gen Physiol 22:649–670, 1939

Collingridge GL, Bliss TV: Memories of NMDA receptors and LTP. Trends Neurosci 18:54–56, 1995

Colquhoun P, Sakmann B: From muscle endplate to brain synapses: a short history of synapses and agonist-activated ion channels. Neuron 20:381–387, 1998

Conturo TE, Nicolas FL, Cull TS, et al: Tracking fiber pathways in the living human brain. Proc Natl Acad Sci USA 96:10422–10427, 1999

Cooper LA, Shepard RN: Chronometric studies of the rotation of visual images, in Visual Information Processing. Edited by Chase WG. New York, Academic Press, 1973, pp 75–176

Corbetta M: Frontoparietal cortical networks for directing attention and the eye to visual locations: identical, independent, or overlapping neural systems? Proc Natl Acad Sci USA 95:831–838, 1998

Corbetta M, Miezin FM, Dobmeyer S, et al: Selective and divided attention during visual discriminations of shape, color, and speed: functional anatomy by positron emission tomography. J Neurosci 11:2383–2402, 1991

Cowan WM: The emergence of modern neuroanatomy and developmental neurobiology. Neuron 20:413–426, 1998

Cowan WM, Kandel ER: A brief history of synapses and synaptic transmission, in The Synapse. Baltimore, MD, Johns Hopkins University Press, 2000

Cowan WM, Fawcett JW, O'Leary DD, et al: Regressive events in neurogenesis. Science 225:1258–1265, 1984

Cowan WM, Jessell TM, Zipursky SL: Molecular and Cellular Approaches to Neural Development. New York, Oxford University Press, 1997

Cowan WM, Harter DH, Kandel ER: The emergence of modern neuroscience: some implications for neurology and psychiatry. Annu Rev Neurosci 23:343–391, 1999

Crair MC, Gillespie DC, Stryker MP: The role of visual experience in the development of columns in cat visual cortex. Science 279:566–570, 1998

Crick F, Koch C: Consciousness and neuroscience. Cereb Cortex 2:97–107, 1998

Crist RE, Kapadia MK, Westheimer G, et al: Perceptual learning of spatial localization: specificity for orientation, position, and context. J Neurophysiol 78:2889–2894, 1997

Crowley JC, Katz LC: Development of ocular dominance columns in the absence of retinal input. Nat Neurosci 2:1125–1130, 1999

Culotti JG, Merz DC: DCC and netrins. Curr Opin Cell Biol 10:609–613, 1998

Dehaene S: The Number Sense. New York, Oxford University Press, 1997

del Castillo J, Katz B: Quantal components of the endplate potential. J Physiol 124:560–573, 1954

Demb JB, Desmond JE, Wagner AD, et al: Semantic encoding and retrieval in the left inferior prefrontal cortex: a functional MRI study of task difficulty and process specificity. J Neurosci 15:5870–5878, 1995

Desimone R, Duncan J: Neural mechanisms of selective visual attention. Annu Rev Neurosci 18:193–222, 1995

Detwiler SR: Neuroembryology: An Experimental Study. New York, Macmillan, 1936

Doetsch F, Caille I, Lim DA, et al: Subventricular zone astrocytes are neural stem cells in the adult mammalian brain. Cell 97:703–716, 1999

Doyle DA, Morais Cabral J, Pfuetzner RA, et al: The structure of the potassium channel: molecular basis of K^+ conduction and selectivity. Science 280:69–77, 1998

Drescher U, Bonhoeffer F, Muller BK: The Eph family in retinal axon guidance. Curr Opin Neurobiol 7:175–180, 1997

Drevets WC, Raichle ME: Reciprocal suppression of regional blood flow during emotional versus higher cognitive processes: implications for interactions between emotion and cognition. Cogn Emotion 12:353–385, 1998

Dubnau J, Tully T: Gene discovery in Drosophila: new insights for learning and memory. Annu Rev Neurosci 21:407–444, 1998

Dudel J, Kuffler SW: Presynaptic inhibition at the crayfish neuromuscular junction. J Physiol 155:543–562, 1961

Duncan J: Disorganization of behavior after frontal lobe damage. Cognitive Neuropsychology 3:271–290, 1986

Dymecki SM: Flp recombinase promotes site-specific DNA recombination in embryonic stem cells and transgenic mice. Proc Natl Acad Sci USA 93:6191–6196, 1996

Easter SS Jr, Purves D, Rakic P, et al: The changing view of neural specificity. Science 230:507–511, 1985

Eccles JC: The Neurophysiological Basis of Mind: The Principles of Neurophysiology. Oxford: Clarendon Press, 1953

Eccles JC: The Physiology of Synapses. New York, Academic Press, 1964

Edelman GM: Cell adhesion molecules. Science 219:450–457, 1983

Engel AK, Fries P, Konig P, et al: Temporal binding, binocular rivalry, and consciousness. Conscious Cogn 8:128–151, 1999

Farah MJ: Visual Agnosia. Cambridge, MA, MIT Press, 1995

Fatt P, Katz B: An analysis of the end-plate potential recorded with an intra-cellular electrode. J Physiol 115:320–370, 1951

Fatt P, Katz B: Spontaneous subthreshold activity at motor nerve endings. J Physiol 117:109–126, 1952

Fechner G: Elements of Psychophysics, Vol 1 (1860). Translated by Adler HE. New York: Holt, Rinehart and Winston, 1966

Felleman DJ, Van Essen DC: Distributed hierarchical processing in the primate cerebral cortex. Cereb Cortex 1:1–47, 1991

Fernandez-Chacon R, Südhof TC: Genetics of synaptic vesicle function: toward the complete functional anatomy of an organelle. Annu Rev Physiol 61:753–776, 1999

Fischbach GD, Frank E, Jessell TM, et al: Accumulation of acetylcholine receptors and acetylcholinesterase at newly formed nerve-muscle synapses. Pharmacol Rev 30:411–428, 1978

Forbes A: The interpretation of spinal reflexes in terms of present knowledge of nerve conduction. Physiol Rev 2:361–414, 1922

Frank E, Wenner P: Environmental specification of neuronal connectivity. Neuron 10:779–785, 1993

Frey U, Morris RG: Synaptic tagging and long-term potentiation. Nature 385:533–536, 1997

Frith C, Dolan RJ: Images of psychopathology. Curr Opin Neurobiol 8:259–262, 1998

Fujita I, Tanaka K, Ito M, et al: Columns for visual features of objects in monkey inferotemporal cortex. Nature 360:343–346, 1992

Fuortes MGF, Frank K, Becker MC: Steps in the production of motoneuron spikes. J Gen Physiol 40:735–752, 1957

Furshpan EJ, Potter DD: Mechanism of nerve impulse transmission at a crayfish synapse. Nature 180:342–343, 1957

Galton F: Inquiry Into Human Faculty and Its Development. London, JM Dent and Sons, 1907

Galvani L: Commentary on the Effect of Electricity on Muscular Motion (1791). Translated by Green RM. Cambridge, MA, Licht, 1953

Georgopoulos AP, Lurioto JT, Petrides M, et al: Mental rotation of the neuronal population vector. Science 243:234–236, 1989

Gilbert CD, Wiesel TN: Receptive field dynamics in adult primary visual cortex. Nature 356:150–152, 1992

Goldberg G: Supplementary motor area structure and function: review and hypothesis. Behav Brain Sci 8:567–615, 1985

Goldberg JL, Barres AB: Nogo in nerve regeneration. Nature 403:369–370, 2000

Goldberg ME, Wurtz RH: Activity of superior colliculus in behaving monkey, II: effect of attention on neuronal responses. J Neurophysiol 35:560–574, 1972

Goodhill GJ: Stimulating issues in cortical map development. Trends Neurosci 20:375–376, 1997

Goodman CS, Shatz CJ: Developmental mechanisms that generate precise patterns of neuronal connectivity. Cell 72:77–98, 1993

Goridis C, Brunet JF: Transcriptional control of neurotransmitter phenotype. Curr Opin Neurobiol 9:47–53, 1999

Gray CM, Konig P, Engel AK, et al: Synchronization of oscillatory responses in visual cortex: a plausible mechanism for scene segmentation, in Proceedings of the Conference on Synergetics of the Brain, 1989

Graziano MSA, Yap GS, Gross CG: Coding of visual space by premotor neurons. Science 266:1054–1057, 1994

Green T, Heinemann SI, Gusella JF: Molecular neurobiology and genetics: investigation of neural function and dysfunction. Neuron 20:427–444, 1998

Grieshammer U, Lewandoski M, Prevette D, et al: Muscle-specific cell ablation conditional upon Cre-mediated DNA recombination in transgenic mice leads to massive spinal and cranial motoneuron loss. Dev Biol 197:234–247, 1998

Gross CG: Visual functions of inferotemporal cortex, in Handbook of Sensory Physiology, Vol 7. Edited by Autrum H, Jung R, Loewenstein WR, et al. Berlin, Springer-Verlag, 1973, pp 451–482

Gross CG, Bender DB, Rocha-Miranda CE: Visual receptive fields of neurons in inferotemporal cortex of the monkey. Science 166:1303–1306, 1969

Gulbis JM, Mann S, MacKinnon R: Structure of a voltage-dependent K⁺ channel β subunit. Cell 97:943–952, 1999

Gurdon JB, Dyson S, St. Johnston D: Cells' perception of position in a concentration gradient. Cell 95:159–162, 1998

Guthrie S: Axon guidance: starting and stopping with slit. Curr Biol 9:R432–R435, 1999

Hamburger V: Cell death in the development of the lateral motor column of the chick embryo. J Comp Neurol 160:535–546, 1975

Hamburger V: The history of the discovery of the nerve growth factor. J Neurobiol 24:893–897, 1993

Hamburger V, Levi-Montalcini R: Proliferation differentiation and degeneration in the spinal ganglia of the chick embryo under normal and experimental conditions. J Exp Zool 111:457–501, 1949

Han EB, Stevens CI: Of mice and memory. Learn Mem 6:539–541, 1999

Hanson PJ, Roth R, Morisaki H, et al: Structure and conformational changes in NSF and its membrane receptor complexes visualized by quick-freeze/deep-etch electron microscopy. Cell 90:523–535, 1997

Harland R, Gerhart J: Formation and function of Spemann's organizer. Annu Rev Cell Dev Biol 13:611–667, 1997

Hartline HK: The response of single optic nerve fibers of the vertebrate eye to illumination of the retina. Am J Physiol 121:400–415, 1938

Hartline HK, Graham CH: Nerve impulses from single receptors in the eye. J Cell Comp Physiol 1:277–295, 1932

Hata Y, Stryker M: Control of thalamocortical afferent rearrangement by postsynaptic activity in developing visual cortex. Science 265:1732–1735, 1994

Hebb DO: The Organization of Behavior: A Neuropsychological Theory. New York, Wiley, 1949

Hengartner MO, Horvitz HR: Programmed cell death in Caenorhabditis elegans. Curr Opin Genet Dev 4:581–586, 1994

Hertting G, Axelrod J: Fate of tritiated noradrenaline at sympathetic nerve endings. Nature 192:172–173, 1961

Heuser JE: Synaptic vesicle exocytosis revealed in quick-frozen frog neuromuscular junctions treated with 4-amino-pyridine and given a single electric shock, in Approaches to the Cell Biology of Neurons. Edited by Cowan WM, Ferrendelli JA. Washington, DC, Society for Neuroscience, 1977, pp 215–239

Hille B, Armstrong CM, MacKinnon R: Ion channels: from ideas to reality. Nat Med 5:iii–vii, 1999

Hodgkin AL: Evidence for electrical transmission in nerve, parts I and II. J Physiol 90:183–232, 1937

Hodgkin AL, Huxley AF: Action potentials recorded from inside a nerve fiber. Nature 144:710–711, 1939

Hodgkin AL, Katz B: The effect of sodium ions on the electrical activity of the giant axon of the squid. J Physiol 108:37–77, 1949

Hodgkin AL, Huxley AF, Katz B: Measurement of current voltage relations in the membrane of the giant axon of Loligo. J Physiol 116:424–448, 1952

Hofer M, Constantine-Paton M: Regulation of N-methyl-D-aspartate (NMDA) receptor function during the rearrangement of developing neuronal connections. Prog Brain Res 102:277–285, 1994

Hoffman EP, Kunkel LM: Dystrophin abnormalities in Duchenne/Becker muscular dystrophy. Neuron 2:1019–1029, 1989

Hoffman EP, Brown RH Jr, Kunkel LM: Dystrophin: the protein product of the Duchenne muscular dystrophy locus. Cell 51:919–928, 1987

Hokfelt T: Neuropeptides in perspective: the last ten years. Neuron 7:867–879, 1991

Holmes G: Disorders of sensation produced by cortical lesions. Brain 50:413–427, 1927

Hubel DH, Wiesel TN: Receptive fields of single neurones in the cat's striate cortex. J Physiol 148:574–591, 1959

Hubel DH, Wiesel TN: Receptive fields, binocular interaction and functional architecture in the cat's visual cortex. J Physiol 160:106–154, 1962

Hubel DH, Wiesel TN: Receptive fields of cells in striate cortex of very young, visually inexperienced kittens. J Neurophysiol 26:994–1002, 1963

Hubel DH, Wiesel TN: Binocular interaction in striate cortex of kittens reared with artificial squint. J Neurophysiol 28:1041–1059, 1965

Hubel DH, Wiesel TN: Receptive fields and functional architecture of monkey straite cortex. J Physiol 195:215–243, 1968

Hubel DH, Wiesel TN: Ferrier lecture: functional architecture of macaque monkey visual cortex. Proc R Soc Lond B Biol Sci 198:1–59, 1977

Hubel DH, Wiesel TN: Early exploration of the visual cortex. Neuron 20, 401–412, 1998

Hubel DH, Wiesel TN, LeVay S: Plasticity of ocular dominance columns in the monkey striate cortex. Philos Trans R Soc Lond Biol 278:377–409, 1977

Hunt RK, Cowan WM: The chemoaffinity hypothesis: an appreciation of Roger W. Sperry's contributions to developmental biology, in Brain Circuits and Functions of the Mind. Edited by Trevarthen C. New York, Cambridge University Press, 1990, pp 19–74

Huntington's Disease Collaborative Research Group: A novel gene containing a trinucleotide repeat that is expanded and unstable on Huntington's disease chromosomes. Cell 72:971–983, 1993

Hynes RO: Integrins: a family of cell surface receptors. Cell 48:549–554, 1987

Impey S, Mark M, Villacres EC, et al: Induction of CRE-mediated gene expression by stimuli that generate long-lasting LTP in area CA1 of the hippocampus. Neuron 16:973–982, 1996

Impey S, Smith DM, Obrietan K, et al: Stimulation of cAMP response element (CRE)-mediated transcription during contextual learning. Nat Neurosci 1:595–601, 1998

Impey S, Obrietan K, Storm DR: Making new connections: role of ERK/MAP kinase signaling in neuronal plasticity. Neuron 23:11–14, 1999

Isom LL, De Jongh KS, Catterall WA: Auxiliary subunits of voltage-gated ion channels. Neuron 12:1183–1194, 1994

Iversen LL: The Uptake and Storage of Noradrenaline in Sympathetic Nerves. London, Cambridge University Press, 1967

Iwasaki S: Spatial attention and two modes of visual consciousness. Cognition 49:211–233, 1993

Jacobson MD, Weil M, Raff MC: Programmed cell death in animal development. Cell 88:347–354, 1997

James W: Principles of Psychology. New York, Henry Holt, 1890

Johansson CB, Momma S, Clarke DL, et al: Identification of a neural stem cell in the adult mammalian central nervous system. Cell 96:25–34, 1999

Johns DC, Marx R, Mains RE, et al: Inducible genetic suppression of neuronal excitability. J Neurosci 19:1691–1697, 1999

Jones CM, Smith JC: Mesoderm induction assays. Methods Mol Biol 97:341–350, 1999

Kandel ER, Spencer WA: Cellular neurophysiological approaches in the study of learning. Physiol Rev 48:65–134, 1968

Kandel ER, Schwartz JH, Jessell T: Principles of Neural Science, 4th Edition. New York, McGraw-Hill, 2000

Karlin A, Akabas MH: Toward a structural basis for the function of nicotinic acetylcholine receptors and their cousins. Neuron 15:1231–1244, 1995

Karni A, Bertini G: Learning perceptual skills: behavioral probes into adult cortical plasticity. Curr Opin Neurobiol 7:530–535, 1997

Karni A, Sagi D: The time course of learning a visual skill. Nature 365:250–252, 1993

Kastner S, Pinsk MA, De Weerd P, et al: Increased activity in human visual cortex during directed attention in the absence of visual stimulation. Neuron 22:751–761, 1999

Katz B: The release of neural transmitter substances, in The Sherrington Lectures, 10. Springfield, IL, Charles C. Thomas, 1969

Katz LC, Dalva MB: Scanning laser photostimulation: a new approach for analyzing brain circuits. J Neurosci Methods 54:205–218, 1994

Katz LC, Shatz CJ: Synaptic activity and the construction of cortical circuits. Science 274:1133–1138, 1996

Kleiman RJ, Reichardt LF: Testing the agrin hypothesis. Cell 85:461–464, 1996

Kochanska G: Children's temperament, mothers' discipline, and security of attachment: multiple pathways to emerging internalization. Child Dev 66:597–615, 1995

Koffka K: Principles of Gestalt Psychology. New York, Harcourt Brace, 1935

Köhler W: Gestalt Psychology. London, Bell and Sons, 1929

Konorski J: Integrative Activity of the Brain. Chicago, IL, University of Chicago Press, 1967

Kosslyn SM: Image and Mind. Cambridge, MA, Harvard University Press, 1980

Kosslyn SM: Image and Brain. Cambridge, MA, MIT Press, 1994

Kremer EJ, Pritchard M, Lynch M, et al: Mapping of DNA instability at the fragile X to a trinucleotide repeat sequence P(CCG)n. Science 252:1711–1714, 1991

Kuffler SW: Discharge patterns and functional organization of mammalian retina. J Neurophysiol 16:37–68, 1953

LaHoste GJ, Swanson JM, Wigal SB, et al: Dopamine D4 receptor gene polymorphism is associated with attention deficit hyperactivity disorder. Mol Psychiatry 1:121–124, 1996

Lamme VA, Spekreijse H: Neuronal synchrony does not represent texture segregation. Nature 396:362–366, 1998

Lance-Jones C, Landmesser L: Pathway selection by chick lumbosacral motoneurons during normal development. Proc R Soc Lond B Biol Sci 214:1–18, 1981

Langley JN: On the regeneration of pre-ganglionic and post-ganglionic visceral nerve fibers. J Physiol 22:215–230, 1897

Langley JN: On nerve endings and on special excitable substances in cells. Proc R Soc Lond B Biol Sci 78:170–194, 1906

Lee K, Dietrich P, Jessell TM: Genetic ablation reveals the essential role of the roof plate in dorsal interneuron specification. Nature 403:734–740, 2000

Leonardo da Vinci: Treatise on Painting (Codes Urbinus Latinus 1270). Princeton, NJ, Princeton University Press, 1956

Lettvin JY, Maturana HR, McCulloch WS, et al: What the frog's eye tells the frog's brain. Proc Instit Radio Engineers 47:1940–1951, 1959

LeVay S, Wiesel TN, Hubel DH: The development of ocular dominance columns in normal and visually deprived monkeys. J Comp Neurol 191:1–51, 1980

Levi-Montalcini R: The nerve growth factor: its mode of action on sensory and sympathetic nerve cells. Harvey Lect 60:217–259, 1966

Lewis EB: Regulation of the genes of the bithorax complex in Drosophila. Cold Spring Harb Symp Quant Biol 50:155–164, 1985

Liley AW: The quantal components of the mammalian endplate potential. J Physiol 133:571–587, 1956

Lorente de Nó R: Analysis of the activity of the chains of internuncial neurons. J Neurophysiol 1:207–244, 1938

Lumsden A, Krumlauf R: Patterning the vertebrate neuraxis. Science 274:1109–1115, 1996

Mach E: Contributions to the Analysis of Sensations (1886). Translated by Southall JPC. Optical Society of America, 1924

Malach R, Amir Y, Harel M, et al: Relationship between intrinsic connections and functional architecture revealed by optical imaging and in vivo targeted biocytin injections in primate striate cortex. Proc Natl Acad Sci USA 90:10469–10473, 1993

Marcel AJ: Conscious and unconscious perception: experiments on visual masking and word recognition. Cognit Psychol 15:197–237, 1983

Marr D: Vision: A Computational Investigation Into the Human Representation and Processing of Visual Information. San Francisco, CA, WH Freeman, 1982

Marrocco RT, Davidson MC: Neurochemistry of attention, in The Attentive Brain. Edited by Parasuraman R. Cambridge, MA, MIT Press, 1998

Martin KC, Casadio A, Zhu H, et al: Synapse-specific transcription-dependent long-term facilitation of the sensory to motor neuron connection in Aplysia: a function for local protein synthesis in memory storage. Cell 91:927–938, 1998

Mayford M, Kandel ER: Genetic approaches to memory storage. Trends Genet 15:463–470, 1999

Mayford M, Bach ME, Huang Y-Y, et al: Control of memory formation through regulated expression of a CaMKII transgene. Science 274:1678–1683, 1996

McGinnis W, Levine MS, Hafen E, et al: A conserved DNA sequence in homoeotic genes of the Drosophila Antennapedia and bithorax complexes. Nature 308:428–433, 1984

McMahan UJ: The agrin hypothesis. Cold Spring Harb Symp Quant Biol 55:407–418, 1990

Mesulam MM: Spatial attention and neglect: parietal, frontal and cingulate contributions to the mental representation and attentional targeting of salient extrapersonal events. Philos Trans R Soc Lond B Biol Sci 354:1325–1346, 1999

Metzstein MM, Stanfield GM, Horvitz HR: Genetics of programmed cell death in C. elegans: past, present and future. Trends Genet 14:410–416, 1998

Miller KD: A model for the development of simple cell receptive fields and the ordered arrangement of orientation columns through activity-dependent competition between ON- and OFF-center inputs. J Neurosci 14:409–441, 1994

Milner AD, Goodale MA: The Visual Brain in Action. New York, Oxford University Press, 1995

Milner B: Memory disturbance after bilateral hippocampal lesions, in Cognitive Processes and the Brain. Edited by Milner PM, Glickman SE. Princeton, NJ, Van Nostrand, 1965

Milner B, Squire LR, Kandel ER: Cognitive neuroscience and the study of memory. Neuron 20:445–468, 1998

Missler M, Südhof TC: Neurexins: three genes and 1,001 products. Trends Genet 4:20–26, 1998

Mombaerts P, Wang F, Dulac C, et al: Visualizing an olfactory sensory map. Cell 4:675–686, 1996

Montarolo PG, Goelet P, Castellucci VF, et al: A critical period for macromolecular synthesis in long-term heterosynaptic facilitation in Aplysia. Science 234:1249–1254, 1986

Morrison SJ, White PM, Zock C, et al: Prospective identification, isolation by flow cytometry, and in vivo self-renewal of multipotent mammalian neural crest stem cells. Cell 96:737–749, 1999

Mountcastle VB: Modality and topographic properties of single neurons of cat's somatic sensory cortex. J Neurophysiol 20:408–434, 1957

Mountcastle VB, Talbot WH, Sakata H, et al: Cortical neuronal mechanisms in flutter-vibration studied in unanesthetized monkeys. Neuronal periodicity and frequency discrimination. J Neurophysiol 32:452–484, 1969

Mueller BK: Growth cone guidance: first steps towards a deeper understanding. Annu Rev Neurosci 22:351–388, 1999

Munk H: Über die Funktionen der Grosshirnrinde, 3te Mitteilung. Berlin, A Hirschwald, 1881, pp 28–53

Nagel T: What is the mind-brain problem? in Experimental and Theoretical Studies of Consciousness (CIBA Foundation Symposium, 174). New York, Wiley Interscience/CIBA Foundation, 1993, pp 1–13

Neher E, Sakmann B: Single channel membrane of denervated frog muscle fibers. Nature 260:799–802, 1976

Neher E, Sakmann B: Single channel membrane of denervated frog muscle fibers. Nature 260:799–802, 1976

Neely JH: Semantic priming and retrieval from lexical memory: roles of inhibitionless spreading activation and limited-capacity attention. J Exp Psychol 106:226–254

Newell A, Simon HA: Human Problem Solving. Engelwood Cliffs, NJ, Prentice-Hall, 1972

Newsome WT, Paré EB: A selective impairment of motion perception following lesions of the middle temporal visual area (MT). J Neurosci 8:2201–2211, 1988

Newsome WT, Britten KH, Movshon JA: Neuronal correlates of a perceptual decision. Nature 341:52–54, 1989

Nickel W, Weber T, McNew JA, et al: Content mixing and membrane integrity during membrane fusion driven by pairing of isolated v-SNAREs and t-SNAREs. Proc Natl Acad Sci USA 96:12571–12576, 1999

Nieuwkoop PD: Short historical survey of pattern formation in the endo-mesoderm and the neural anlage in the vertebrates: the role of vertical and planar inductive actions. Cell Mol Life Sci 53:305–318, 1997

Nó D, Yao TP, Evans RM: Ecdysone-inducible gene expression in mammalian cells and transgenic mice. Proc Natl Acad Sci USA 93:3346–3351, 1996

Norman DA, Shallice T: Attention to action: willed and automatic control of behavior, in Consciousness and Self-Regulation. Edited by Davidson RJ, Schwartz GE, Shapiro D. New York, Plenum, 1986, pp 1–18

Numa S: A molecular view of neurotransmitter regions and ion channels. Harvey Lect 63:121–165, 1989

Nüsslein-Volhard C, Wieschaus E: Mutations affecting segment number and polarity in Drosophila. Nature 287:795–801, 1980

Olson CR, Gettner SN: Object-centered direction selectivity in the macaque supplementary eye field. Science 269:985–988, 1995

Oppenheim RW: Neuronal cell death and some related regressive phenomena during neurogenesis: a selective historical review and progress report, in Studies in Developmental Neurobiology: Essays in Honor of Viktor Hamburger. Edited by Cowan WM. New York, Oxford University Press, 1981

Panchision D, Hazel T, McKay R: Plasticity and stem cells in the vertebrate nervous system. Curr Opin Cell Biol 10:727–733, 1998

Parasuraman R, Greenwood PM: Selective attention in aging and dementia, in The Attentive Brain. Edited by Parasuraman R. Cambridge, MA, MIT Press, 1998

Parlati F, Weber T, McNew JA, et al: Rapid and efficient fusion of phospholipid vesicles by the α-helical core of a SNARE complex in the absence of an N-terminal regulatory domain. Proc Natl Acad Sci USA 96:12565–12570, 1999

Parsons LM, Fox PT: The neural basis of implicit movements used in recognizing hand shape. Cognitive Neuropsychology 15:583–615, 1998

Paulson HL, Fischbeck KH: Trinucleotide repeats in neurogenetic disorders. Annu Rev Neurosci 19:79–107, 1996

Perutz MF, Johnson T, Suzuki M, et al: Glutamine repeats as polar zippers: their possible role in inherited neurodegenerative diseases. Proc Natl Acad Sci USA 91:5355–5358, 1994

Pettmann B, Henderson CE: Neuronal cell death. Neuron 20:633–647, 1998

Picciotto MR: Knock-out mouse models used to study neurobiological systems. Crit Rev Neurobiol 13:103–149, 1999

Posner MI: Chronometric Explorations of Mind. Hillsdale, NJ, Erlbaum, 1978

Posner MI, Gilbert CD: Attention and primary visual cortex. Proc Natl Acad Sci USA 96:2585–2587, 1999

Posner MI, Raichle ME: Images of Mind. New York, Scientific American Books, 1994

Posner MI, Raichle ME: The neuroimaging of human brain function. Proc Natl Acad Sci USA 95:763–764, 1998

Posner MI, Rothbart MK: Attention, self-regulation and consciousness. Philos Trans R Soc Lond B Biol Sci 353:1915–1927, 1998

Posner MI, Abdullaev YG, McCandliss BD, et al: Neuroanatomy, circuitry and plasticity of word reading. Neuroreport 10:R12–R23, 1999

Ptácek LJ: Channelopathies: ion channel disorders of muscle as a paradigm for paroxysmal disorders of the nervous system. Neuromuscul Disord 7:250–255, 1997

Ptácek LJ: The familial periodic paralyses and nondystrophic myotonias. Am J Med 104:58–70, 1998

Purves D, Lichtman JW: Principles of Neural Development. Sunderland, MA, Sinauer, 1985

Raichle ME: Behind the scenes of functional brain imaging: a historical and physiological perspective. Proc Natl Acad Sci USA 95:765–772, 1998

Raichle ME, Fiez JA, Videen TO, et al: Practice-related changes in the human brain: functional anatomy during non-motor learning. Cereb Cortex 4:8–26, 1994

Ramón y Cajal S: The Croonian Lecture: la fine structure des centres nerveux. Proc R Soc Biol Lond B Sci 55:444–467, 1894

Ramón y Cajal S: The structure and connexions of neurons (1906), in Nobel Lectures: Physiology or Medicine (1901–1921). Amsterdam, Elsevier, 1967, pp 220–253

Ramón y Cajal S: Histologie du système nerveux de l'homme et des vertébrés, Vols 1 and 2 (1911). A Maloine, Paris. Reprinted by Consejo Superior de Investigaciones Cientificas, Inst. Ramón y Cajal, Madrid, 1955

Reddy PS, Housman DE: The complex pathology of trinucleotide repeats. Curr Opin Cell Biol 9:364–372, 1997

Reichardt LF, Fariñas I: Neurotrophic factors and their receptors, in Molecular and Cellular Approaches to Neural Development. Edited by Cowan WM, Jessell TM, Zipursky SL. New York, Oxford University Press, 1997, pp 220–263

Rensink RA, O'Regan JK, Clark JJ: To see or not to see: the need for attention to perceive changes in scenes. Psychol Sci 8:368–373, 1997

Richardson PM, McGuinness UM, Aguayo AJ: Axons from CNS neurons regenerate into PNS grafts. Nature 284:264–265, 1997

Rosen BR, Buckner RL, Dale AM: Event related functional MRI: past, present, and future. Proc Natl Acad Sci USA 95:773–780, 1998

Ross CA: Intranuclear neuronal inclusions: a common pathogenic mechanism for glutamine-repeat neurogenerative diseases? Neuron 19:1147–1150, 1997

Rothbart MK, Ahadi SA, Evans DW: Temperament and personality: origins and outcomes. J Per Soc Psychol 78:122–135, 2000

Rudenko G, Nguyen T, Chelliah Y, et al: The structure of the ligand-binding domain of neurexin Iβ: regulation of LNS domain function by alternative splicing. Cell 99:93–101, 1999

Rumelhart DE, McClelland JL, Group PR: Parallel Distributed Processing. Cambridge, MA, MIT Press, 1987

Ruoslahti E: RGD and other recognition sequences for integrins. Annu Rev Cell Dev Biol 12:697–715, 1996

Rutishauser U, Hoffman S, Edelman GM: Binding properties of a cell adhesion molecule from neural tissue. Proc Natl Acad Sci USA 79:685–689, 1982

Ryalls J, Lecours AR: Broca's first two cases: from bumps on the head to cortical convolutions, in Classic Cases in Neuropsychology. Edited by Code C, Wallesch C W, Joanelle Y, et al. New York, Psychology Press, 1996

Salzman CD, Britten KH, Newsome WT: Microstimulation of visual area MT influences perceived direction of motion. Invest Ophthalmol Vis Sci 31:238, 1990

Sandrock AW Jr, Dryer SE, Rosen KM, et al: Maintenance of acetylcholine receptor number by neuregulins at the neuromuscular junction in vivo. Science 276: 599–603, 1997

Sanes JR, Lichtman JW: Development of the vertebrate neuromuscular junction. Annu Rev Neurosci 22:389–442, 1999

Schafer EA: Experiments on special sense localisations in the cortex cerebri of the monkey. Brain 10:362–380, 1888

Schiller F: Paul Broca: Explorer of the Brain. New York, Oxford University Press, 1992

Schwartz EL: Computational anatomy and functional architecture of striate cortex: spatial mapping approach to perceptual coding. Vision Res 20:645–669, 1980

Schwenk F, Kuhn R, Angrand PO, et al: Temporally and spatially regulated somatic mutagenesis in mice. Nucleic Acids Res 26:1427–1432, 1998

Scott MP, Weiner AJ: Structural relationships among genes that control development: sequence homology between the Antennapedia, Ultrabithorax, and fushi tarazu loci of Drosophila. Proc Natl Acad Sci USA 81:4115–4119, 1984

Searle JR: The problem of consciousness, in Experimental and Theoretical Studies of Consciousness (CIBA Foundation Symposium, 174). New York, Wiley Interscience/CIBA Foundation, 1993, pp 61–80

Searle JR: How to study consciousness scientifically, in Towards an Understanding of Integrative Brain Function. Edited by Fuxe K, Grillner S, Hokfelt T, et al. Amsterdam, Elsevier, 1998, pp 379–387

Serafini T: Finding a partner in a crowd: neuronal diversity and synaptogenesis. Cell 98:133–136, 1999

Shadlen MN, Movshon JA: Synchrony unbound: a critical evaluation of the temporal binding hypothesis. Neuron 24:67–77, 1999

Shapiro L, Colman DR: The diversity of cadherins and implications for a synaptic adhesive code in the CNS. Neuron 23:427–430, 1999

Shatz CJ: Neurotrophins and visual system plasticity, in Molecular and Cellular Approaches to Neural Development. Edited by Cowan WM, Jessell TM, Zipursky SL. New York, Oxford University Press, 1997

Sheng M, Pak DT: Glutamate receptor anchoring proteins and the molecular organization of excitatory synapses. Ann NY Acad Sci 868:483–493, 1999

Shepherd GM: Foundations of the Neuron Doctrine. New York, Oxford University Press, 1991

Sherrington CS: The Central Nervous System, Vol 3, in A Textbook of Physiology, 7th Edition. Edited by Foster M. London, Macmillan, 1897

Sherrington CS: The Integrative Action of the Nervous System, 2nd Edition. New Haven, NJ, Yale University Press, 1906

Sherrington CS: Flexor-reflex of the limb, crossed extension reflex, and reflex stepping and standing (cat and dog). J Physiol 40:28–116, 1910

Sherrington CS: Inhibition as a Coordinative Factor. Nobel Lecture. Stockholm, PA Norstedt, 1932

Sherrington CS: Man on His Nature. New York, Macmillan, 1941

Silva AJ, Kogan JH, Frankland PW, et al: CREB and memory. Annu Rev Neurosci 21:127–148, 1998

Singer W: Consciousness and the structure of neuronal representations. Philos Trans R Soc Lond B Biol Sci 353:1829–1840, 1998

Smalley SL, Bailey JN, Palmer CG, et al: Evidence that the D4 receptor is a susceptibility gene in attention deficit hyperactivity disorder. Mol Psychiatry 3:427–430, 1998

Smith JC: Induction and early amphibian development. Curr Opin Cell Biol 1:1061–1070, 1989

Söllner T, Whiteheart SW, Brunner M, et al: SNAP receptors implicated in vesicle targeting and fusion. Nature 362:318–324, 1993

Song JY, Ichtchenko K, Südhof TC, et al: Neuroligin 1 is a postsynaptic cell-adhesion molecule of excitatory synapses. Proc Natl Acad Sci USA 96:1100–1105, 1999

Speidel CC: Studies of living nerves. II. Activities of ameboid growth cones, sheath cells, and myelin segments, as revealed by prolonged observation of individual nerve fibers in frog tadpoles. Am J Anat 52:1–75, 1933

Spemann H, Mangold H: Ü Induktion von Embryonalanlagen durch Implantation artfremder Organisatoren. Arch Mikrosk Anat Entwicklungsmech 100:599–638, 1924

Sperry RW: Effect of 180° rotation of the retinal field on visuomotor coordination. J Exp Zool 92:263–279, 1943

Sperry RW: Chemoaffinity in the orderly growth of nerve fiber patterns and connections. Proc Natl Acad Sci USA 50:703–710, 1963

Squire LR, Kandel ER: Memory: From Mind to Molecules. New York, Scientific American Books, 1999

Squire LR, Zola-Morgan S: The medial temporal lobe memory system. Science 253:1380–1386, 1991

Stevens CF: Making a submicroscopic hole in one. Nature 349:657–658, 1991

Stevens CF: Neuronal diversity: too many cell types for comfort? Curr Biol 8:R708–R710, 1998

St. Johnston D, Nüsslein-Volhard C: The origin of pattern and polarity in the Drosophila embryo. Cell 68:201–219, 1992

Stoner GR, Albright TD: Neural correlates of perceptual motion coherence. Nature 358:412–414, 1992

Stoner GR, Albright TD, Ramachandran VS: Transparency and coherence in human motion perception. Nature 344:153–155, 1990

Straub V, Campbell KP: Muscular dystrophies and the dystrophin-glycoprotein complex. Curr Opin Neurol 10:168–175, 1997

Strittmatter WJ, Roses AD: Apolipoprotein E and Alzheimer's disease. Annu Rev Neurosci 19:53–77, 1996

Stryker MP, Harris W: Binocular impulse blockade prevents the formation of ocular dominance columns in cat visual cortex. J Neurosci 6:2117–2133, 1986

Sutton RB, Fasshauer D, Jahn R, et al: Crystal structure of a SNARE complex involved in synaptic exocytosis at 2.4 Å resolution. Nature 395:347–353, 1998

Swindale NV: A model for the formation of ocular dominance stripes. Proc R Soc Lond B Biol Sci 208:243–264, 1980

Takahashi M, Miyoshi H, Verma IM, et al: Rescue from photoreceptor degeneration in the rd mouse by human immunodeficiency virus vector-mediated gene transfer. J Virol 73:7812–7816, 1999

Takeichi M: Cadherins: a molecular family important in selective cell-cell adhesion. Annu Rev Biochem 59:237–252, 1990

Takeichi M, Uemura T, Iwai Y, et al: Cadherins in brain patterning and neural network formation. Cold Spring Harb Symp Quant Biol 62:505–510, 1997

Takeuchi A, Takeuchi N: On the permeability of the end-plate membrane during the action of transmitter. J Physiol (Paris) 154:52–67, 1960

Talbot SA, Marshall WH: Physiological studies on neural mechanisms of visual localization and discrimination. Am J Ophthalmol 24:1255–1264, 1941

Tanabe Y, Jessell TM: Diversity and pattern in the developing spinal cord. Science 274:1115–1123, 1996

Tanabe Y, William C, Jessell TM: Specification of motor neuron identity by the MNR2 homeodomain protein. Cell 95:67–80, 1998

Tang YP, Shimizu E, Duber R, et al: Genetic enhancement of learning and memory in mice. Nature 401:63–69, 1999

Tatagiba M, Brosamle C, Schwab ME: Regeneration of injured axons in the adult mammalian central nervous system. Neurosurgery 40:541–547, 1997

Teller DY: First glances: the vision of infants. The Friedenwald Lecture. Invest Ophthalmol Vis Sci 38:2183–2203, 1997

Tempel BL, Papazian PM, Schwartz TL, et al: Sequence of probable potassium channels at Shaker locus of Drosophila. Science 237:770–775, 1987

Tessier-Lavigne M, Goodman CS: The molecular biology of axon guidance. Science 274:1123–1133, 1996

Thomas JB, Bastiani MJ, Bate M, et al: From grasshopper to Drosophila: a common plan for neuronal development. Nature 310:203–207, 1984

Tootell RBH, Nouchine K, Hadjikhani WV, et al: Functional analysis of primary-visual cortex in humans. Proc Natl Acad Sci USA 95:811–817, 1998

Tsien JZ, Heurta PT, Tonegawa S: The essential role of hippocampal CA1 NMDA receptor-dependent synaptic plasticity in spatial memory. Cell 87:1327–1338, 1996

Ungerleider LG, Mishkin M: Two cortical visual systems, in Analysis of Visual Behavior. Edited by Ingle DJ, Goodale MA, Mansfield RJW. Cambridge, MA, MIT Press, 1982, pp 549–586

Usrey WM, Reid RC: Synchronous activity in the visual system. Annu Rev Physiol 61:435–456, 1999

Van Essen DC: Functional organization of primate visual cortex, in Cerebral Cortex, Vol 3. Edited by Peters A, Jones EG. New York, Plenum, 1985, pp 259–327

Verkerk AJ, Pieretti M, Sutcliffe JS, et al: Identification of a gene (FMR-1) containing a CGG repeat coincident with a breakpoint cluster region exhibiting length variations in fragile X syndrome. Cell 65:905–914, 1991 von Bekesy G: Experiments in Hearing. Edited and translated by Wever EG. New York, McGraw-Hill, 1960

von der Malsburg C: The correlation theory of brain function. MPI Biophysical Chemistry, Internal Report 81–2. MPI Biophysical Chemistry, 1981

von Helmholtz H: Treatise on Psychological Optics, Vol 2 (1860). Translated by Southall JPC. [Rochester, NY], Optical Society of America, 1924

Wang F, Nemes A, Mendelsohn M, et al: Odorant receptors govern the formation of a precise topographic map. Cell 93:47–60, 1998

Warren ST, Ashley CT Jr: Triplet repeat expansion mutations: the example of fragile X syndrome. Annu Rev Neurosci 18:77–99, 1995

Washburn DA: Stroop-like effects for monkeys and humans: processing speed or strength of association? Psychol Sci 5:375–379, 1994

Watanabe D, Inokawa H, Hashimoto K, et al: Ablation of cerebellar Golgi cells disrupts synaptic integration involving GABA inhibition and NMDA receptor activation in motor coordination. Cell 95:17–27, 1998

Watkins JC, Evans RH: Excitatory amino acid transmitters. Annu Rev Pharmacol Toxicol 21:165–204, 1981

Weber RJ, Harnish R: Visual imagery for words: the Hebb test. J Exp Psychol 102:409–414, 1974

Weber T, Zemelman BV, McNew JA, et al: SNAREpins: minimal machinery for membrane fusion. Cell 92:759–772, 1998

Weiss P: Self-differentiation of the basic patterns of coordination. Comparative Psychology Monographs 17:1–96, 1941

Weliky M, Katz LC: Correlational structure of spontaneous neuronal activity in the developing lateral geniculate nucleus in vivo. Science 285:599–604, 1999

Wertheimer M: Gestalt Theory (1924). New York, Humanities Press, 1950

Wessells NK: Tissue Interactions and Development. Menlo Park, CA, Benjamin/Cummings, 1977

Wiesel TN, Hubel DH: Single-cell responses in striate cortex of kittens deprived of vision in one eye. J Neurophysiol 26:1003–1017, 1963

Wiesel TN, Hubel DH: Comparison of the effects of unilateral and bilateral eye closure on cortical unit responses in kittens. J Neurophysiol 28:1029–1040, 1965

Wolpert L: Positional information and the spatial pattern of cellular differentiation. J Theor Biol 25:1–47, 1969

Wu Q, Maniatis T: A striking organization of a large family of human neural cadherin-like cell adhesion genes. Cell 97:779–790, 1999

Wundt WM-G: Principles of Physiological Psychology. Translated from the 5th German Edition. New York, Macmillan, 1902

Wurtz RH, Goldberg ME (eds): The Neurobiology of Saccadic Eye Movements (Reviews of Oculomotor Research, Vol 3). Amsterdam, Elsevier, 1989

Yin JC, Tully T: CREB and the formation of long-term memory. Curr Opin Neurobiol 6:264–268, 1996

Yin JCP, Del Vecchio M, Zhou H, et al: CREB as a memory modulator: induced expression of a dCREB-2 activator isoform enhances long-term memory in Drosophila. Cell 81:107–115, 1995

Yoshihara Y, Mizuno T, Nakahira M, et al: A genetic approach to visualization of mul-
tisynaptic neural pathways using plant lectin transgene. Neuron 22:33–41, 1999

Zamanillo D, Sprengel R, Hvalby O, et al: Importance of AMPA receptors for hippocampal LTP but not for spatial learning. Science 284:1805–1811, 1999

Zinyk DL, Mercer EH, Harris E, et al: Fate mapping of the mouse midbrain-hindbrain
constriction using a site-specific recombination system. Curr Biol 8:665–668,
1998

"THE MOLECULAR BIOLOGY OF MEMORY STORAGE"

Charles F. Zorumski, M.D.

The ability of an organism to modify its behavior based on experience is arguably the most important and fascinating property of the nervous system. For serious students of behavior, it is a given that the neural changes underlying learning and memory involve alterations in the strength of connections between neurons. It is important to remember that this was not the case even 50 years ago. In the late 1950s, when Eric Kandel embarked on a scientific journey that radically changed how we now think about complex behaviors, there were several plausible and competing hypotheses under consideration, none of which had particularly strong experimental support. Using what he called a "radically reductionist approach" based on studying relatively simple behaviors in the giant marine snail *Aplysia californica*, Kandel systematically changed the field. By focusing on a simpler organism with a simple nervous system, Kandel and colleagues were able to map in detail the neural pathways underlying specific reflex behaviors in the snail to determine ways in which these behaviors could be modified by experience in a laboratory setting. In a series of dramatic and definitive electrophysiological studies, Kandel demonstrated that the changes underlying several forms of learning result from persistent changes in the function of specific synapses used to generate the behaviors of interest. Furthermore, learning in *Aplysia*, like learning in more complex mammalian systems, has several components with differing degrees of long-term stability. Over several decades of study, the Kandel group dissected the physiological, biochemical, and molecular details of learning, including the identification of specific changes in synapses, ion channels, second messenger systems, protein synthesis, and gene expression.

Of great importance, the spectacular insights gained from work in *Aplysia* did not end with the invertebrate system. Rather, Kandel and colleagues extended these insights into mouse models, where they coupled the power of synaptic physiology and sophisticated behavioral analyses with the ability to manipulate gene expression to examine specific molecules that contribute to mammalian declarative memory. While the molecular and biophysical mechanisms underlying synaptic plasticity in the mouse do not always exactly mimic the events in *Aplysia*, the overall conservation of guiding principles and similarity in processes, including some of the key molecular components, is staggering and a bit humbling for those who see mammals, particularly humans, as so much more complex than mollusks.

The following essay, "The Molecular Biology of Memory Storage: A Dialogue Between Genes and Synapses," was delivered on December 8, 2000, as Kandel's Nobel lecture, and highlights the paths that Kandel and his colleagues pursued in developing what are now textbook insights into learning and memory. These studies unequivocally set the stage for the next generation of studies.

What have we learned from this work? As Kandel summarizes, there are at least four major sets of insights. First, the mechanisms underlying memory reside in changes in synaptic transmission and are conserved across species. These changes can be both pre- and postsynaptic, and, importantly, the same sets of synapses can be altered in different and completely opposite ways by different types of learning. Second, learning affects neuronal excitability as well as synaptic transmission. Thus, the firing properties of neurons are also subject to modification and contribute to synaptic and network changes. Third, synaptic modifications underlying learning can have different temporal components, contributing specifically to short- and long-term forms of memory; overlapping, but not necessarily identical, cellular processes contribute to the different time courses. Fourth, changes underlying memory not only affect synaptic function but also involve structural changes in synaptic contacts. While short-term information storage involves covalent modifications of already existing proteins, long-term storage requires changes in gene expression and protein synthesis.

These four points are profound insights into the workings of nervous systems across species. It is already clear that these principles have a much broader impact than the conditioned reflexes in a mollusk or declarative memory in a rodent. The insights we are now gaining into the properties of emotional and motivational systems, including the actions of therapeutic and abused psychoactive drugs, reflect many of the points outlined above. While the specific molecular details may differ in different systems, the principles developed by the Kandel group serve as a guiding light for future work.

In closing, there is one additional concept that this work teaches us. This concerns the importance of working in simple but definable systems for gaining knowledge of more complex processes. The brilliance of Kandel's foresight in choosing to work with *Aplysia* cannot be overemphasized. While studies of the cellular basis of learning and memory remain a work in progress, it is safe to say that we would not have the insights into mammalian learning and the scientific directions that we have today had it not been for Kandel's determined decision to employ a "radically reductionist approach." This is a lesson that may trump all others as the field of psychiatry anticipates its scientific future.

CHAPTER 7

THE MOLECULAR BIOLOGY OF MEMORY STORAGE

A Dialogue Between Genes and Synapses

Eric R. Kandel, M.D.

One of the most remarkable aspects of an animal's behavior is the ability to modify that behavior by learning, an ability that reaches its highest form in human beings. For me, learning and memory have proven to be endlessly fascinating mental processes because they address one of the fundamental features of human activity: our ability to acquire new ideas from experience and to retain these ideas over time in memory. Moreover, unlike other mental processes such as thought, language, and consciousness, learning seemed from the outset to be readily accessible to cellular and molecular analysis.

This article was originally published in *Science*, Volume 294, Number 5544, pp. 1030–1038.

Howard Hughes Medical Institute, Center for Neurobiology and Behavior, College of Physicians and Surgeons of Columbia University, New York State Psychiatric Institute, 1051 Riverside Drive, New York, NY 10032, USA.

This essay is adapted from the author's address to the Nobel Foundation, December 2000.

I, therefore, have been curious to know, What changes in the brain when we learn? And, once something is learned, how is that information retained in the brain? I have tried to address these questions through a reductionist approach that would allow me to investigate elementary forms of learning and memory at a cellular molecular level—as specific molecular activities within identified nerve cells.

I first became interested in the study of memory in 1950 as a result of my readings in psychoanalysis while still an undergraduate at Harvard College. Later, during medical training, I began to find the psychoanalytic approach limiting because it tended to treat the brain, the organ that generates behavior, as a black box. In the mid-1950s, while still in medical school, I began to appreciate that during my lifetime the black box of the brain would be opened and that the problems of memory storage, once the exclusive domain of psychologists and psychoanalysts, could be investigated with the methods of modern biology. As a result, my interest in memory shifted from a psychoanalytic to a biological approach. As a postdoctoral fellow at the National Institutes of Health (NIH) in Bethesda from 1957 to 1960, I focused on learning more about the biology of the brain and became interested in knowing how learning produces changes in the neural networks of the brain.

My purpose in translating questions about the psychology of learning into the empirical language of biology was not to replace the logic of psychology or psychoanalysis with the logic of cellular molecular biology, but to try to join these two disciplines and to contribute to a new synthesis that would combine the mentalistic psychology of memory storage with the biology of neuronal signaling. I hoped further that the biological analysis of memory might carry with it an extra bonus, that the study of memory storage might reveal new aspects of neuronal signaling. Indeed, this has proven true.

A Radical Reductionist Strategy to Learning and Memory

At first thought, someone interested in learning and memory might be tempted to tackle the problem in its most complex and interesting form. This was the approach that Alden Spencer and I took when we joined forces at NIH in 1958 to study the cellular properties of the hippocampus, the part of the mammalian brain thought to be most directly involved in aspects of complex memory (Kandel and Spencer 1961). We initially asked, rather naively: Are the electrophysiological properties of the pyramidal cells of the hippocampus, which were thought to be the key hippocampal cells involved in memory storage, fundamentally different from other neurons in the brain? With study, it became clear to us that all nerve cells, including the pyramidal cells of the hippocampus, have similar signaling properties. Therefore, the intrinsic signaling properties of neurons would themselves not give us key insights into memory storage (Kandel and Spencer 1968). The unique functions of the hippocampus had to arise not so much from the intrinsic properties of pyramidal neurons but from the pattern of functional interconnections of these cells, and how those interconnections are affected by learning. To tackle that problem, we needed to know how sensory information about a learning task reaches the hippocampus and how information processed by the hippocampus influences behavioral output. This was a formidable challenge, since the hippocampus has a large number of neurons and an immense number of interconnections. It seemed unlikely that we would be able to work out in any reasonable period of time how the neural networks, in which the hippocampus was embedded, participate in behavior and how those networks are affected by learning.

To bring the power of modern biology to bear on the study of learning, it seemed necessary to take a very different approach—a radically reductionist approach. We needed to study not the most complex but the simplest instances of memory storage, and to study them in animals that were most tractable experimentally. Such a reductionist approach was hardly new in twentieth-century biology. One need only think of the use of *Drosophila* in genetics, of bacteria and bacteriophages in molecular biology, and of the squid giant axon in the study of the conduction of nerve impulses. Nevertheless, when it came to the study of behavior, many investigators were reluctant to use a reductionist strategy. In the 1950s and 1960s, many biologists and most psychologists believed that learning was the one area of biology in which the use of simple animal models, particularly invertebrate ones, was least likely to succeed. They argued that only higher animals exhibit interesting forms of learning and that these forms require neuronal organizations and neuronal mechanisms qualitatively different from those found in simple animals.

It was my belief, however, that concerns about the use of a simple experimental system to study learning were misplaced. If elementary forms of learning are common to all animals with an evolved nervous system, there must be conserved features in the mechanisms of learning at the cell and molecular level that can be studied effectively even in simple invertebrate animals.

A Simple Learned Behavior in an Invertebrate

After an extensive search for a suitable experimental animal, I settled on the giant marine snail *Aplysia* (Figure 7–1A) because it offers three important

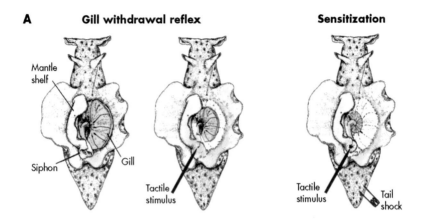

A   **Gill withdrawal reflex**   **Sensitization**

Mantle shelf

Siphon

Gill

Tactile stimulus

Tactile stimulus

Tail shock

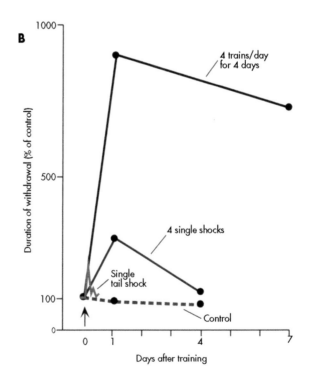

B

4 trains/day for 4 days

4 single shocks

Single tail shock

Control

Duration of withdrawal (% of control)

Days after training

FIGURE 7-1.   A simple learned behavior *(opposite page)*.

(A) A dorsal view of *Aplysia* showing the gill, the animal's respiratory organ. A light touch to the siphon with a fine probe causes the siphon to contract and the gill to withdraw. Here, the mantle shelf is retracted for a better view of the gill. Sensitization of the gill-withdrawal reflex, by applying a noxious stimulus to another part of the body, such as the tail, enhances the withdrawal reflex of both the siphon and the gill.

(B) Spaced repetition converts short-term memory into long-term memory in *Aplysia*. Before sensitization training, a weak touch to the siphon causes only a weak, brief siphon- and gill- withdrawal reflex. Following a single noxious, sensitizing shock to the tail, that same weak touch produces a much larger siphon and gill reflex withdrawal response, an enhancement that lasts about 1 hour. More tail shocks increase the size and duration of the response.

Source.   Modified from Frost WN, Castellucci VF, Hawkins RD, et al: "Monosynaptic Connections Made by the Sensory Neurons of the Gill- and Siphon-Withdrawal Reflex in *Aplysia* Participate in the Storage of Long-Term Memory for Sensitization." *Proceedings of the National Academy of Sciences of the United States of America* 82:8266–8269, 1985.

advantages: its nervous system is made up of a small number of nerve cells; many of these are gigantic; and (as became evident to me later) many are uniquely identifiable (Frazier et al. 1967; Kandel 1976). Whereas the mammalian brain has a trillion central nerve cells, *Aplysia* has only 20,000, and the simplest behaviors that can be modified by learning may directly involve less than 100 central nerve cells. In addition to being few in number, these cells are the largest nerve cells in the animal kingdom, reaching up to 1,000 μm in diameter, large enough to be seen with the naked eye. One can record from these large cells for many hours without any difficulty, and the same cell can be returned to and recorded from over a period of days. The cells can easily be dissected out for biochemical studies, so that from a single cell one can obtain sufficient mRNA to make a cDNA library. Finally, these identified cells can readily be injected with labeled compounds, antibodies, or genetic constructs, procedures that opened up the molecular study of signal transduction within individual nerve cells.

Irving Kupfermann and I soon delineated a very simple defensive reflex: the withdrawal of the gill upon stimulation of the siphon, an action that is like the quick withdrawal of a hand from a hot object. When a weak tactile stimulus is applied to the siphon, both the siphon and gill are withdrawn into the mantle cavity for protection under the mantle shelf (Pinsker et al. 1970) (Figure 7–1A). Kupfermann, Harold Pinsker, and later Tom Carew, Robert Hawkins, and I found that this simple reflex could be modified by three different forms of learning: habituation, sensitization, and classical conditioning (Carew et al. 1972; Pinsker et al. 1970, 1973). As we examined these three forms of learning, we were struck by the resemblance each had to corresponding forms of learning in higher vertebrates and humans. As with vertebrate learning, memory storage for each type of learning in *Aplysia* has two phases: a transient memory that lasts minutes and an enduring memory that lasts days. Conversion of short-term to long-term memory storage requires spaced repetition—practice makes perfect, even in snails (Figure 7–1B) (Carew et al. 1972; Frost et al. 1985; Pinsker et al. 1973).

We focused initially on one type of learning. *Sensitization* is a form of learned fear in which a person or an experimental animal learns to respond strongly to an otherwise neutral stimulus (Frost et al. 1985; Pinsker et al. 1970, 1973). For example, if a person is suddenly exposed to an aversive stimulus, such as a gunshot going off nearby, that person will be sensitized by the unexpected noise. As a result, that person will be frightened and will now startle to an otherwise innocuous stimulus like a tap on the shoulder. Similarly, on receiving an aversive shock to a part of the body such as the tail, an *Aplysia* recognizes the stimulus as aversive and learns to enhance its defensive reflex responses to a variety of subsequent stimuli applied to the siphon, even innocuous stimuli (Castellucci et al. 1989) (Figure 7–1A). The animal remembers the shock, and the duration of this memory is a function of the number of repetitions of the noxious experience (Figure 7–1B). A single shock gives rise to a memory lasting only minutes; this short-term memory does not require the synthesis of new protein. In contrast, four or five spaced shocks to the tail give rise to a memory lasting several days; this long-term memory does require the synthesis of new protein. Further training, four brief trainings a day for 4 days, gives rise to an even more enduring memory lasting weeks, which also requires new protein synthesis. Thus, just as in complex learning in mammals (Ebbinghaus 1885/1963; Flexner and Flexner 1966), the long-term memory for sensitization differs from the short-term memory in requiring the synthesis of new proteins. This was our first clear evidence for the conservation of biochemical mechanisms between *Aplysia* and vertebrates.

Kupfermann, Castellucci, Carew, Hawkins, John Byrne, and I worked out significant components of the neural circuit gill-withdrawal reflex (Figure 7–2). The circuit is located in the abdominal ganglion and has 24 central mechanoreceptor sensory neurons that innervate the siphon skin and make direct monosynaptic connections with 6 gill motor cells (Byrne et al. 1974, 1978a; Castellucci et al. 1970) (Figure 7–2C). The sensory neurons also made indirect connections with the motor cells through small groups of excitatory and inhibitory interneurons (Hawkins et al. 1981a,b). In addition to being identifiable, individual cells also have surprisingly large effects on behavior (Figure 7–2B) (Byrne et al. 1978a,b; Kandel 1976). As we examined the neural circuit of this reflex, we were struck by its invariance. In every animal we examined, each cell connected only to certain target cells and not to others (Figure 7–2C). This also was true in the neural circuitry of other behaviors in *Aplysia* including inking, control of the circulation, and locomotion (Kandel 1976, 1979). This raised a key question in the cell-biological study of learning: How can learning occur in a neural circuit that is so precisely wired?

In 1894, Santiago Ramón y Cajal (1894) proposed a theory of memory storage according to which memory is stored in the growth of new connections. This prescient idea was neglected in good part for half a century as students of learning fought over newer competing ideas. First, Karl Lashley, Wolfgang Köhler, and a number of Gestalt psychologists proposed that learning leads to changes in electric fields or chemical gradients, which they postulated surround neuronal populations and are produced by the aggregate activity of cells recruited by the learning process. Second, Alexander Forbes and Lorente de Nó proposed that memory is stored dynamically by a self-reexciting chain of neurons. Donald Hebb later championed this idea as a mechanism for short-term memory. Finally, Holger Hyden proposed that learning led to changes in the base composition of DNA or RNA. Even though there was much discussion about the merits of each of these ideas, there was no direct evidence to support any of them (Kandel 1968).

Kupfermann, Castellucci, Carew, Hawkins, and I addressed these alternative ideas directly by confronting the question of how learning can occur in a circuit with fixed neuronal elements. To address this question, we examined the neural circuit of the gill-withdrawal reflex while the animal underwent sensitization, classical conditioning, or habituation. Our studies provided clear evidence for the idea proposed by Ramón y Cajal, that learning results from changes in the strength of the synaptic connections between precisely interconnected cells (Castellucci et al. 1970; Kupfermann et al. 1970). Thus, while the organism's developmental program assures that the connections between cells are invariant, it does not specify their precise strength. Rather, experience alters the strength and effectiveness of these preexisting chemical connections. Seen in the perspective of these three forms of learning, synaptic plasticity emerged as a fundamental mechanism for information storage by the nervous system, a mechanism that is built into the very molecular architecture of chemical synapses (Milner et al. 1998).

Molecular Biology of Short- and Long-Term Memory Storage

What are the molecular mechanisms whereby short-term memory is established, and how is it converted to long-term memory? Initially, we focused

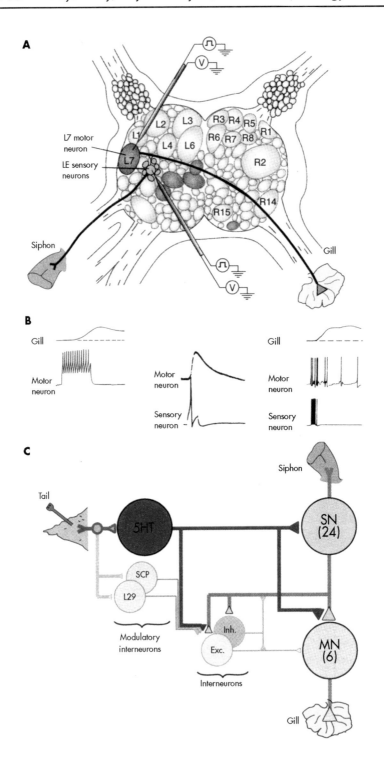

FIGURE 7-2. The neural circuit of the *Aplysia* gill-withdrawal reflex *(opposite page)*.

(A) In this dorsal view of the abdominal ganglion, the six identified motor cells to the gill are brown and the seven sensory neurons are blue. A sensory neuron that synapses on gill motor neuron L7 is stimulated electrically with an intracellular electrode and a microelectrode in the motor neuron records the synaptic potential produced by the action potential in the sensory neuron [see middle trace in (B)]. The sensory neuron carries the input from the siphon skin; the motor neuron makes direct connections onto the gill.

(B) Individual cells make significant contributions to the reflex. Stimulating a single motor neuron (traces on the left) produces a detectable change in the gill and stimulating a single sensory neuron produces a large synaptic potential in the motor neuron (traces in the middle). Repeated stimulation of a single sensory neuron increases the frequency of firing in the motor neuron, leading to a visible reflex contraction of the gill (traces on the right). A single tactile stimulus to the skin normally activates 6–8 of the 24 sensory neurons, causing each to fire 1–2 action potentials. The repetitive firing of 10 action potentials in a single sensory neuron, designed to simulate the firing of the total population (trace on the right) simulates the reflex behavior reasonably well.

(C) Diagram of the circuit of the gill-withdrawal reflex. The siphon is innervated by 24 sensory neurons that connect directly with the 6 motor neurons. The sensory neurons also connect to populations of excitatory and inhibitory interneurons that in turn connect with the motor neurons. Stimulating the tail activates three classes of modulatory interneurons (serotonergic neurons, neurons that release the small cardioactive peptide, and the L29 cells) that act on the terminals of the sensory neurons as well as on those of the excitatory interneurons. The serotonergic modulatory action is the most important; blocking the action of these cells blocks the effects of sensitizing stimuli.

Source. From Glanzman DL, Mackey SL, Hawkins RD, et al: "Depletion of Serotonin in the Nervous System of *Aplysia* Reduces the Behavioral Enhancement of Gill Withdrawal as Well as the Heterosynaptic Facilitation Produced by Tail Shock." *The Journal of Neuroscience* 9:4200–4213, 1989.

on short-term sensitization. In collaboration with James H. Schwartz, we found that the synaptic changes, like short-term behavior, were expressed even when protein synthesis was inhibited. This finding first suggested to us that short-term synaptic plasticity might be mediated by a second-messenger system such as cyclic AMP (cAMP) (Schwartz et al. 1971). Following up on this idea, Schwartz, Howard Cedar, and I found in 1972 that stimulation of the modulatory pathways recruited during heterosynaptic facilitation led to an increase in cAMP in the abdominal ganglion (Cedar et al. 1972). Cedar and Schwartz (1972) found that the neurotransmitter candidates serotonin and dopamine could simulate this action of electrical stimulation and increase levels of cAMP. Later, Hawkins, Castellucci, David Glanzman, and I delineated the modulatory system activated by a sensitizing stimulus to the tail (Glanzman et al. 1989; Hawkins et al. 1981b; Mackey et al. 1989), and confirmed that it contains serotonergic interneurons.

We next found that serotonin acts on specific receptors in the presynaptic terminals of the sensory neuron to enhance transmitter release. In 1976, Marcello Brunelli, Castellucci, and I injected cAMP directly into the presynaptic cells and found that it too produced presynaptic facilitation (Brunelli et al. 1976; Kandel et al. 1976). This provided the most compelling evidence then available that cAMP is involved in controlling synaptic strength and gave us our first insight into one molecular mechanism of short-term memory—the regulation of transmitter release (Figure 7–3).

How does cAMP enhance transmitter release? Serotonin, or injected cAMP, leads to increased excitability and a broadening of the action potential by reducing specific K^+ currents, allowing greater Ca^{2+} influx into the presynaptic terminal with each action potential (Klein and Kandel 1980). The greater Ca^{2+} influx could contribute to the enhanced transmitter release. Following the lead of Paul Greengard, who had proposed that cAMP produces its action in the brain through the cAMP-dependent protein kinase (PKA), Marc Klein and I suggested that cAMP may cause phosphorylation of this K^+ channel by activating PKA (Klein and Kandel 1980). In collaborative experiments with Paul Greengard in 1980, Castellucci, Schwartz, and I found that the active catalytic subunit of PKA by itself produced broadening of the action potential and enhancement of glutamate release (Castellucci et al. 1980). Conversely, the specific peptide inhibitor of PKA (PKI) blocked the actions of serotonin. These findings provided direct evidence for the role of PKA in short-term presynaptic facilitation (Byrne and Kandel 1995; Castellucci et al. 1982).

In an elegant series of experiments, Steven Siegelbaum, Joseph Camardo, and Michael Shuster identified a novel K^+ channel, the S-type K^+ channel, and showed that it too could be modulated by cAMP (Siegelbaum et al. 1982) and that PKA could act on the S-type K^+ channel directly (Shuster et al. 1985). Later, Byrne showed that serotonin also modulates a delayed-rectifier K^+ (Byrne and Kandel 1995). The S-type channel mediated the increase in excitability with a minor contribution to broadening, whereas the delayed-rectifier K^+ channel contributed little to excitability but had a major role in spike broadening. Finally, Hochner, Klein, and I—and independently, Jack Byrne and his colleagues—showed that, in addition to spike broadening, serotonin also enhanced release by an as-yet-unspecified action on the release machinery. Thus, serotonin leads to an increase in presynaptic cAMP, which activates PKA and leads to synaptic strengthening through enhanced

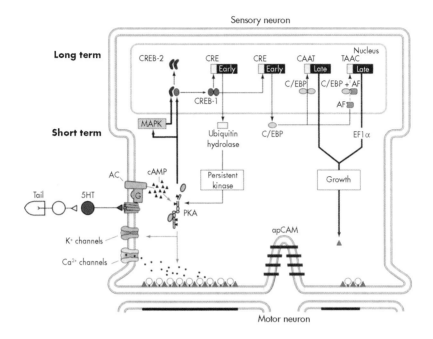

FIGURE 7–3. Effects of short- and long-term sensitization on the monosynaptic component of the gill-withdrawal reflex of *Aplysia*.

In short-term sensitization (lasting minutes to hours), a single tail shock causes a transient release of serotonin that leads to covalent modification of preexisting proteins. The serotonin acts on a transmembrane serotonin receptor to activate the enzyme adenylyl cyclase (AC), which converts ATP to the second-messenger cAMP. In turn, cAMP recruits the cAMP-dependent protein kinase A (PKA) by binding to the regulatory subunits (spindles), causing them to dissociate from and free the catalytic subunits (ovals). These subunits can then phosphorylate substrates (channels and exocytosis machinery) in the presynaptic terminals, leading to enhanced transmitter availability and release. In long-term sensitization, repeated stimulation causes the level of cAMP to rise and persist for several minutes. The catalytic subunits can then translocate to the nucleus, and recruit the mitogen-activated protein kinase (MAPK). In the nucleus, PKA and MAPK phosphorylate and activate the cAMP response element-binding (CREB) protein and remove the repressive action of CREB-2, an inhibitor of CREB-1. CREB-1 in turn activates several immediate-response genes, including a ubiquitin hydrolase necessary for regulated proteolysis of the regulatory subunit of PKA. Cleavage of the (inhibitory) regulatory subunit results in persistent activity of PKA, leading to persistent phosphorylation of the substrate proteins of PKA. A second immediate-response gene activated by CREB-1 is C/EBP, which acts both as a homodimer and as a heterodimer with activating factor (AF) to activate downstream genes [including elongation factor 1α (EF1α)] that lead to the growth of new synaptic connections.

transmitter release produced by a combination of mechanisms (Figure 7–3) (Byrne and Kandel 1995).

CREB-1 Mediated Transcription

By substituting puffs of serotonin (the transmitter released by tail shocks), for the tail shocks themselves, Samuel Schacher, Pier Giorgio Montarolo, Philip Goelet, and I modeled sensitization in a culture dish consisting of a single sensory cell making synaptic connections with a single motor cell (Montarolo et al. 1986). We were able to induce both short- and long-term facilitation and found, as with the intact animal, that the long-term process differed from the short-term process in requiring the synthesis of new proteins.

We used this cell culture to ask: What genes are activated to convert the short-term to the long-term process, and what genes are essential for the maintenance of the long-term process? We found that five spaced puffs of serotonin (simulating five spaced shocks to the tail) activate PKA, which in turn recruits the mitogen-activated protein kinase (MAPK). Both translocate to the nucleus, where they activate a transcriptional cascade beginning with the transcription factor CREB-1, the cAMP response element-binding protein–1, so called because it binds to a cAMP response element (CRE) in the promoters of target genes (Figure 7–3). The first clue to the importance of CREB in long-term memory was provided in 1990 by Pramod Dash and Binyamin Hochner (Dash et al. 1990). They injected, into the nucleus of a sensory neuron in culture, oligonucleotides carrying the CRE DNA element, thereby titrating out CREB. This treatment blocked long-term but not short-term facilitation (Figure 7–3). Later, Dusan Bartsch cloned *Aplysia* CREB-1a (ApCREB-1a) and showed that injection of the phosphorylated form of this transcription factor by itself could initiate the long-term memory process. Downstream from ApCREB (Bartsch et al. 1998), Cristina Alberini and Bartsch found two additional positive transcription regulators, the CAAT box enhancer binding protein (ApC/EBP) and activation factor (Ap/AF) (Alberini et al. 1994; Bartsch et al. 2000). CREB-1 activates this set of immediate response genes, which in turn act on downstream genes, to give rise to the growth of new synaptic connections (Bacskai et al. 1993; Bailey and Kandel 1993; Bailey et al. 1992; Dash et al. 1990; Glanzman et al. 1990; Kaange et al. 1993; Martin et al. 1997; Schacher et al. 1988) (Figure 7–3). As first shown by Craig Bailey and Mary Chen, long-term memory endures by virtue of the growth of new synaptic connections, a structural change that parallels the duration of the behavioral memory (Bailey and Chen 1988, 1989; Bailey and Kandel 1993; Bailey et al. 1992). As the memory fades, the connections retract over time. A typical sensory neuron in the intact *Aplysia* has about

1,200 synaptic varicosities. Following long-term sensitization, the number more than doubles to about 2,600; with time the number returns to about 1,500.

Inhibitory Constraints

In 1995, Bartsch found that positive regulators are only half the story—there are also inhibitory constraints on memory. Long-term synaptic facilitation requires not only activation of memory-enhancer genes but also inactivation of memory-suppressor genes (Figure 7–3). One of these, the transcription factor ApCREB-2, can repress ApCREB-1a mediated transcription; relieving this repression lowers the threshold for the long-term process.

Thus, during long-term memory storage, a tightly controlled cascade of gene activation is switched on, with memory-suppressor genes providing a threshold or checkpoint for memory storage, presumably to ensure that only salient features are learned. Memory suppressors may allow for the modulation of memory storage by emotional stimuli, as occurs in "flashbulb memories," memories of emotionally charged events that are recalled in detail, as if a complete picture had been instantly and powerfully etched in the brain.

Synapse Specificity of Long-Term Facilitation

The finding of a transcriptional cascade explained why long-term memory requires new protein synthesis immediately after training, but it posed a new cell-biological problem. A single neuron makes hundreds of contacts on many different target cells. Short-term synaptic changes are synapse-specific. Since long-lasting synaptic changes require transcription and thus the nucleus, is long-term memory storage a cell-wide process, or are there cellbiological mechanisms that maintain the synapse specificity of long-term facilitation?

To examine these questions, Kelsey Martin cultured one *Aplysia* sensory cell with a bifurcating axon with two motor neurons, forming two widely separated synapses (Figure 7–4A). In this culture system, a single puff of serotonin applied to one synapse produces transient facilitation at that synapse only, as expected (Casadio et al. 1999; Martin et al. 1998). Five puffs of serotonin applied to one branch produces long-lasting facilitation (72 hours), also restricted to the stimulated synapse (Figure 7–4B). This long-lasting synapse-specific facilitation requires CREB and also leads to structural changes. Thus, despite recruitment of nuclear processes, long-term changes in synaptic function and structure are confined only to those synapses stimulated by serotonin.

How does this come about? Martin, Andrea Casadio, Bailey, and I found

A

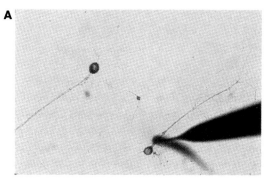

B

Initiation   Capture

(A)   (B)

5 × 5HT   1 × 5HT

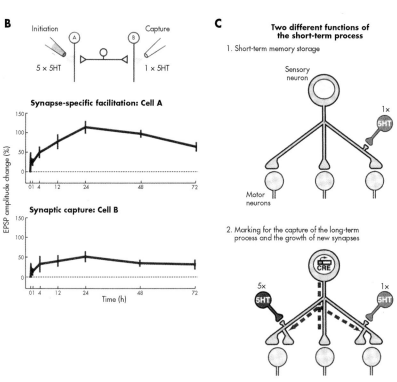

Synapse-specific facilitation: Cell A

EPSP amplitude change (%)

Synaptic capture: Cell B

Time (h)

C

**Two different functions of
the short-term process**

1. Short-term memory storage

Sensory
neuron

1×
5HT

Motor
neurons

2. Marking for the capture of the long-term
process and the growth of new synapses

CRE

5×   1×
5HT   5HT

FIGURE 7–4. A single sensory neuron connects to many target cells *(opposite page)*.

The requirement of a transcriptional mechanism for long-term memory raises the question, What is the unit of long-term information storage? Is it a single synapse, as with short-term facilitation, or the entire neuron? Is there a mechanism for restricting synaptic facilitation to some synaptic connections?

(A) This photomicrograph shows a culture system developed to examine the action of two independent branches of a single in *Aplysia* sensory neuron (the small neuron in the middle) on two different motor neurons (large neurons). Serotonin can be selectively applied to one and not the other of the two branches. The flow of the serotonin can be monitored with the dye, fast green.

(B) Long-term facilitation is synapse-specific and can be captured at another branch by the stimulus that initiates the short-term process. Five puffs of serotonin applied at the initiation site (cell A) produce a synapse-specific facilitation shown in (B). This synapse-specific facilitation is not evident at the synapse of cell B unless that synapse is itself primed with a single puff of serotonin.

(C) Two effects of short-term facilitation: short-term memory storage when acting by itself and marking of the specific synapse to which it is applied for subsequent capture of the proteins necessary for long-term facilitation and growth when applied in conjunction with five pulses to another set of terminals.

Source. (A) and (B) From Martin KC, Casadio A, Zhu H, et al: "Synapse-Specific Transcription-Dependent Long-Term Facilitation of the Sensory to Motor Neuron Connection in *Aplysia:* A Function for Local Protein Synthesis in Memory Storage." *Cell* 91:927–938, 1998. Used with permission of Elsevier.

that five puffs of serotonin send a signal to the nucleus to activate CREB-1, which then appears to send proteins to all terminals; however, only those terminals that have been marked by serotonin can use these proteins productively for synaptic growth. Indeed, one puff of serotonin to the previously unstimulated synapse is sufficient to mark that synapse so that it can capture a reduced form of the long-term facilitation induced at the other site by five puffs of serotonin (Figure 7–4B).

These results gave us a new and surprising insight into short-term facilitation. The stimulus that produces the short-term process has two functions (Figure 7–4C). When acting alone, it provides a selective, synapse-specific enhancement of synaptic strength, which contributes to short-term memory, lasting minutes. When acting in conjunction with the activation of CREB initiated by a long-term process in either that synapse or in any other synapse on the same neuron, the stimulus locally marks those synapses at which it occurs. The marked synapse can then utilize the proteins activated by CREB for synaptic growth to produce a persistent change in synaptic strength. Thus, the logic for the long-term process involves a long-range integration that is different from the short-term process. In the long term, the function of a synapse is not only determined by the history of usage of that synapse. It is also determined by the state of the transcriptional machinery in the nucleus.

How does one puff of serotonin mark a synapse for long-term change? For structural changes to persist, local protein synthesis is required (Casadio et al. 1999). Oswald Steward's important work in the early 1980s had shown that dendrites contain ribosomes, and that specific mRNAs are transported to the dendrites and translated there (Steward 1997). Our experiments showed that one function of these locally translated mRNAs was to stabilize the synapse-specific long-term functional and structural changes.

Neurotransmitter Regulation of Local Protein Synthesis

These studies thus revealed a new, fourth type of synaptic action mediated by neurotransmitter signaling (Figure 7–5). Three of these four have emerged, at least in part, from the study of learning and memory. First, in 1951, Katz and Fatt opened up the modern study of chemical transmission with their discovery of ionotropic receptors that regulate ion flux through transmitter-gated ion channels to produce fast synaptic actions, lasting milliseconds (Fatt and Katz 1951). Second, in the 1970s, metabotropic receptors were found to activate second-messenger pathways, such as the cAMP-PKA pathway, to produce slow synaptic activity lasting minutes (Greengard 1976). As we have seen in *Aplysia,* this slow synaptic action can regulate transmitter release, thereby contributing to short-term memory for sensitization. Third, an even more persistent synaptic action, lasting days, results from repeated action of a modulatory transmitter such as serotonin. With repeated applications of serotonin, second-messenger kinases translocate to the nucleus, where they activate a cascade of gene induction leading to the growth of new synaptic connections. This of course raises the problem of synapse specificity that we have considered above. Our experiments, in the bifurcated culture system, revealed a novel fourth action of neurotransmitters, the marking of the synapse and the regulation of local protein synthesis, which contributes to the establishment of synapse-specific long-term facilitation.

Explicit Memory

I have so far considered only the simplest cases of memory storage—those involving reflexes—a form called implicit or procedural memory. Implicit memory is memory for perceptual and motor skills and is expressed through performance, without conscious recall of past episodes. In contrast, the memories we hold near and dear are called explicit (or declarative) memories. These memories require conscious recall and are concerned with peo-

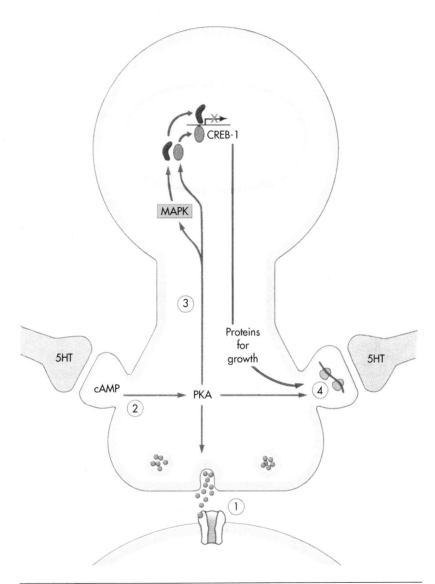

FIGURE 7–5. A dialog between genes and synapses.

Four consequences of the action of neurotransmitters. 1) Transmitter activation of a ligand-gated ion channel leads to a rapid synaptic action lasting milliseconds. 2) Transmitter activation of a seven-transmembrane receptor and a second-messenger kinase leads to a more enduring synaptic action lasting minutes. 3) Repeated transmitter activation of a seven transmembrane receptor leads to the translocation of the kinase to the nucleus and to activation of transcription, producing a persistent synaptic action. 4) Transmitter activation of local protein synthesis to stabilize the synapse-specific facilitation.

A

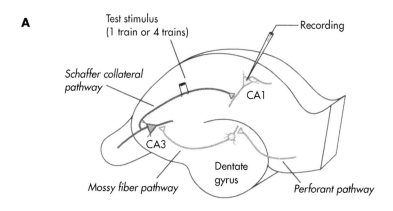

B

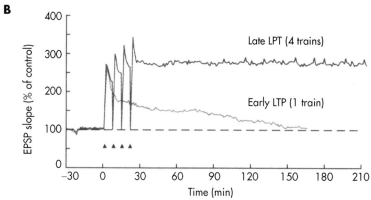

C

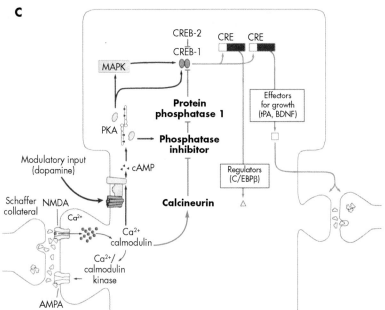

FIGURE 7–6. Long-term potentiation (LTP) in the hippocampus *(opposite page)*.

(A) Three major pathways, each of which gives rise to LTP. The *perforant pathway* from the subiculum forms excitatory connections with the granule cells of the dentate gyrus. The *mossy fiber pathway*, formed by the axons of the granule cells of the dentate gyrus, connects the granule cells with the pyramidal cells in area CA3 of the hippocampus. The *Schaffer collateral pathway* connects the pyramidal cells of the CA3 region with the pyramidal cells in the CA1 region of the hippocampus.

(B) The early and late phases of LTP in the Schaffer collateral pathway. A single train of stimuli for 1 second at 100 Hz elicits an early LTP, and four trains at 10-minute intervals elicit the late phase of LTP. The early LTP lasts about 2 hours, the late LTP more than 24 hours.

(C) A model for the late phase of LTP in the Schaffer collateral pathway. A single train of action potentials initiates early LTP by activating NMDA receptors, Ca^{2+} influx into the postsynaptic cell, and the activation of a set of second messengers. With repeated trains of action potentials (illustrated here), the Ca^{2+} influx also recruits an adenylyl cyclase (AC), which activates the cAMP-dependent protein kinase. The kinase is transported to the nucleus, where it phosphorylates CREB. CREB in turn activates targets (C/EBPβ, tPA, BDNF) that are thought to lead to structural changes. Mutations in mice that block PKA or CREB reduce or eliminate the late phase of LTP. The adenylyl cyclase can also be modulated by dopamine signals and perhaps other modulatory inputs. In addition, there are constraints (bold lines) that inhibit L-LTP and memory storage. Removal of these constraints lowers the threshold for L-LTP and enhances memory storage.

ple, places, objects, and events. Explicit memory involves a specialized anatomical system in the medial temporal lobe, and a structure deep to it, the hippocampus (Bacskai et al. 1993; Castellucci et al. 1978; Milner et al. 1998) (Figure 7–6A). How is explicit memory stored? Louis Flexner, Bernard Agranoff, Sam Barondes, and Larry Squire had shown that explicit memory, like implicit memory, has a short-term phase that does not require protein synthesis and a long-term phase that does (Bacskai et al. 1993). Are these two components of memory storage also represented at the cellular level? What rules govern explicit memory storage?

A decade ago, when I reached my sixtieth birthday, I gathered up my courage and returned to the hippocampus. Mario Capecchi and Oliver Smithies, by achieving targeted gene ablation in mouse embryonic stem cells, provided a superb genetic system for relating individual genes to synaptic plasticity, on the one hand, and to complex explicit memory storage on the other. Mice have a medial temporal lobe system, including a hippocampus, that resembles that of humans, and they use their hippocampus much as we do to store memory of places and objects (Figure 7–6A).

Although we still do not know much about how information is transformed as it gets into and out of the hippocampus, it is well established that the hippocampus contains a cellular representation of extrapersonal space—a cognitive map of space—and lesions of the hippocampus interfere with spatial tasks (Grant et al. 1992). Moreover, in 1972, Terje Lømo and Tim Bliss discovered that the perforant path, a major pathway within the hippocampus, exhibits activity-dependent plasticity, a change now called long-term potentiation (LTP) (Figure 7–6B). In the CA1 region of the hippocampus, LTP is induced postsynaptically by activation of an NMDA receptor to glutamate. In the late 1980s, Richard Morris found that blocking the NMDA receptor pharmacologically not only interfered with LTP but also blocked memory storage (Bliss and Lømo 1973; Morris et al. 1986).

This earlier work on LTP in hippocampal slices had focused on the response to one or two trains of electrical stimuli. But in *Aplysia* we had found that long-term memory emerges most effectively with repeated stimuli (Figure 7–1B). So when Uwe Frey, Yan-You Huang, Peter Nguyen, and I turned to the hippocampus, we examined whether LTP changed with repeated stimulation (Frey et al. 1993; Nguyen et al. 1994; Nicoll and Malenka 1999) and found that hippocampal LTP has phases, much like facilitation in *Aplysia*. The early phase of LTP, produced by a single train of stimuli, lasts only 1–3 hours and does not require new protein synthesis (Nguyen et al. 1994); it involves covalent modifications of preexisting proteins that lead to the strengthening of preexisting connections, similar in principle to short-term facilitation in *Aplysia*. By contrast, repeated trains of electrical stimuli produce a late phase of LTP, which has properties quite different from early LTP and similar to long-term facilitation in *Aplysia* (Figure 7–6B). The late phase of LTP persists for at least a day and requires both translation and transcription. The late phase of LTP, like long-term storage of implicit memory, requires PKA, MAPK, and CREB and appears to lead to the growth of new synaptic connections (Bolshakov et al. 1997; Bourtchouladze et al. 1994; Engert and Bonhoeffer 1999; Frey et al. 1993; Impey et al. 1998; Ma et al. 1999; Muller 1997; Nguyen et al. 1994; Nicoll and Malenka 1999; Yin and Tully 1996) (Figure 7–6C).

The Late Phase of LTP and Explicit Memory

To explore further the specific role of PKA and late LTP in memory storage, Ted Abel, Mark Barad, Rusiko Bourtchouladze, Peter Nguyen, and I generated transgenic mice that express R(AB), a mutant form of the regulatory subunit of PKA that inhibits enzyme activity (Abel et al. 1997). In these R(AB) transgenic mice, the reduction in hippocampal PKA activity was paralleled by a significant decrease in late LTP, while basal synaptic transmission and early LTP remained unchanged. Most interesting, this deficit in the late phase of LTP was paralleled by behavioral deficits in hippocampus-dependent long-term memory for extrapersonal space, whereas learning, and short-term memory, are unimpaired (Figure 7–7A and B). Thus, in the storage of explicit memory of extrapersonal space in the mammalian hippocampus, PKA plays a critical role in the transformation of short-term memory into long-term memory, much as it does in the storage of implicit memory in *Aplysia* and *Drosophila*.

Using the R(AB) mice we could now ask, Why do animals with compromised PKA signaling have difficulty with space (Abel et al. 1997)? We were influenced by the classic studies of John O'Keefe and John Dostrovsky, who in 1971 discovered that the pyramidal cells of the hippocampus—the cells one examines artificially by electrically stimulating the Schaffer collateral pathway while studying LTP—are "place cells;" they actually encode extrapersonal space in the animal (O'Keefe and Nadel 1978). A given pyramidal cell will fire only when the head of the mouse is in a certain part of an enclosed space—the cell's place field. When placed in a new environment, within minutes an animal develops an internal representation of the space (by the coordinated firing of a population of place cells), which is normally stable for days. The same cell will have the same firing field each time the animal is reintroduced to that environment. When now placed in a second environment, a new map is formed—again in minutes—in part from some of the cells that made up the map of the first environment and in part from pyramidal cells that had been silent previously (O'Keefe and Nadel 1978).

It struck me that the formation of a new map resembled a learning process. The map develops with time as the animal familiarizes itself with the space, and once learned, the map of space is retained for days and weeks. To first test whether the molecular pathways underlying the late phase of LTP were important for the long-term stabilization of this map, Cliff Kentros, Robert Muller, Hawkins, and I simply blocked LTP pharmacologically with an NMDA receptor antagonist (Kentros et al. 1998). When placed in a new environment, the animals with blocked NMDA receptors formed a good spatial map that was still stable 1 hour later. However, by 24 hours, most pyramidal cells no longer retained the representation of the field they had initially. This suggested that activation of NMDA receptors—perhaps a step in modifying the strength of the synapse—is required for the long-term stabilization of a place cell map, a result consistent with the role for the late phase of LTP in the stabilization of a place cell map.

We next asked whether a selective deficit that affects only the late phase of LTP causes a selective abnormality in the long-term stability of place cells. Since only the late phase of LTP requires PKA, Alex Rotenberg, Muller, Abel, Hawkins, and I returned to the R(AB) transgenic mice with diminished PKA

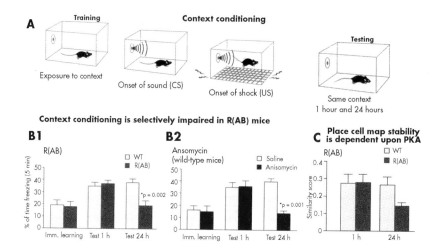

FIGURE 7–7.   Contextual learning and the stability of place cells.

(A) The protocol for context conditioning consists of exposure to the context followed by a tone and then a shock. The animals are then tested 1 hour and 24 hours after training.

(B1) Mutant mice that express the R(AB) gene in the hippocampus, blocking the action of PKA, have a selective defect for long-term contextual memory. Mice that express *R(AB)* were conditioned to freeze to the context. After becoming familiar with the context, the mice heard a sound and received a shock through the electrified grid in the floor. As a result, the animals learned to associate the context of the space with shock and to freeze when placed in the box at a future time. These mice had good short-term memory at 1 hour for freezing to context, but at 24 hours they no longer froze to context, indicating a defect in a form of long-term explicit (declarative) memory that requires the hippocampus. (B2) Wild-type mice exposed to anisomycin, an inhibitor of protein synthesis, during training show a similar defect for long-term memory when tested 24 hours after conditioning.

(C) Place cell stability for *R(AB)* and wild-type mice. *R(AB)* mice with a defect in PKA and late LTP form place fields that are stable at 1 hour. These fields are not stable at 24 hours.

Source.   (A) and (B) From Abel T, Nguyen PV, Barad M, et al: "Genetic Demonstration of a Role for PKA in the Late Phase of LTP and in Hippocampus-Based Long-Term Memory." *Cell* 88:615–626, 1997. (C) From Rotenberg A, Abel T, Hawkins RD, et al: "Parallel Instabilities of Long-Term Potentiation, Place Cells, and Learning Caused by Decreased Protein Kinase A Activity." *The Journal of Neuroscience* 20:8096–8102, 2000; and Agnihotri N, Hawkins RD, Kandel ER, et al: "Protein Synthesis Inhibition Selectively Abolishes Long-Term Stability of Hippocampal Place Cell Maps." *Abstracts—Society for Neuroscience* 27:316.14, 2001. Used with permission of Elsevier.

activity and a diminished form of late LTP (Rotenberg et al. 2000). If reduced activity of PKA affected the stability of place cells, R(AB) mice should be able to form a stable map of space in a novel environment, as in normal animals, that is stable for at least 1 hour. However, the cell field should be unstable when recorded 24 hours later. This is precisely what we found (Figure 7–7C). The fact that long-term instability in the spatial map and the deficit in long-term memory paralleled the deficit in the late phase of LTP suggested that PKA-mediated gene activation and the synthesis of new protein might be essential for the stabilization of the spatial map. Naveen Agnihotri, Kentros, Hawkins, and I tested this idea and found that inhibiting protein synthesis indeed destabilized the place fields in the long term much as does inhibiting PKA (Kentros et al. 2001).

In the course of this work, Kentros and Agnihotri found, remarkably, that, as is the case with explicit memories in humans, a key feature in the stabilization of PKA and protein synthesis–dependent phase of memory is attention. When a mouse does not attend to the space it walks through, the map forms but is unstable after 3–6 hours. When the mouse is forced to attend to the space, however, the map invariably is stable for days!

Inhibitory Constraints on Explicit Memory

Recently we (Malleret et al. 2001) and others (Blitzer et al. 1998) have found that the threshold for hippocampal synaptic plasticity and memory storage is determined by the balance between protein phosphorylation governed by PKA and dephosphorylation (Malleret et al. 2001; Mansuy et al. 1998). To determine whether the endogenous Ca^{2+}-sensitive phosphatase calcineurin acts as a constraint on this balance, we inhibited calcineurin and examined the effects on synaptic plasticity and memory storage. Isabelle Mansuy, Gael Malleret, Danny Winder, Tim Bliss, and I found that a transient reduction of calcineurin activity resulted in facilitation of LTP both in vitro and in vivo (Malleret et al. 2001). This facilitation persisted for several days in the intact animal and was accompanied by enhanced learning and strengthening of short- and long-term memory on several spatial and nonspatial tasks requiring the hippocampus. These results, together with previous findings by Winder and Mansuy showing that overexpression of calcineurin impairs PKA-dependent components of LTP and memory (Mansuy et al. 1998; Winder et al. 1998), demonstrate that endogenous calcineurin can act as a negative regulator of synaptic plasticity, learning, and memory (Figure 7–6C).

An Overall View

Our studies of the storage component of memory, the molecular mechanism whereby information is stored, have led to two general conclusions.

First, our research suggests that the cellular and molecular strategies used in *Aplysia* for storing short- and long-term memory are conserved in mammals and that the same molecular strategies are employed in both implicit and explicit memory storage. With both implicit and explicit memory there are stages in memory that are encoded as changes in synaptic strength and that correlate with the behavioral phases of short- and long-term memory. The short-term synaptic changes involve covalent modification of preexisting proteins, leading to modification of preexisting synaptic connections, whereas the long-term synaptic changes involve activation of gene expression, new protein synthesis, and the formation of new connections. Whereas short-term memory storage for implicit and explicit memory requires different signaling, long-term storage of both implicit and explicit memory uses as a core signaling pathway PKA, MAPK, and CREB-1. At least in the mouse, additional components are likely recruited. In both implicit and explicit memory the switch from short-term to long-term memory is regulated by inhibitory constraints.

Second, the study of learning has revealed new features of synaptic transmission and new cell-biological functions of synaptic signaling. For example, different forms of learning recruit different modulatory transmitters, which then act in one of three ways: 1) They activate second-messenger kinases that are transported to the nucleus, where they initiate processes required for neuronal growth and long-term memory; 2) they mark the specific synapses for capture of the long-term process and regulate local protein synthesis for stabilization; and 3) they mediate, in ways we are just beginning to understand, attentional processes required for memory formation and recall.

Most important, the study of long-term memory has made us aware of the extensive dialog between the synapse and the nucleus, and the nucleus and the synapse (Figure 7–5). In the long-term process, the response of a synapse is not determined simply by its own history of activity (as in short-term plasticity) but also by the history of transcriptional activation in the nucleus.

I started this essay by pointing out that 40 years ago, at the beginning of my career, I thought that a reductionist approach based on the use of a simple experimental system such as *Aplysia* might allow us to address fundamental questions in learning and memory. That was a leap of faith for which I have been rewarded beyond my fondest hopes. Still, the complexity of explicit memory is formidable, and we have only begun to explore it. We as yet know little about the molecular mechanisms that initiate or stabilize the synaptic growth associated with long-term memory. What signaling molecules lead to the cytoskeletal rearrangements during synaptic remodeling? How do they relate to the molecules that control synapse formation during development?

In addition, we have here only considered the molecular mechanisms of memory storage. The more difficult part of memory—especially explicit memory—is a systems problem. We still need to seek answers to a family of important questions. How do different regions of the hippocampus and the medial temporal lobe—the subiculum, the entorhinal, parahippocampal, and perirhinal cortices—interact in the storage of explicit memory? How is information in any of these regions transferred for ultimate consolidation in the neocortex? We do not, for example, understand why the initial storage of long-term memory requires the hippocampus, whereas the hippocampus is not required once a memory has been stored for weeks or months (Milner et al. 1998; Squire and Zola-Morgan 1991). What critical information does the hippocampus convey to the neocortex? We also know very little about the nature of recall of explicit (declarative) memory, a recall that requires conscious effort. These systems problems will require more than the bottoms-up approach of molecular biology. They will also require the top-down approaches of cognitive psychology, neurology, and psychiatry. Ultimately, we will need syntheses that bridge the two approaches.

Despite these complexities, these and other questions in the biology of learning no doubt will be vigorously addressed in the near future. For the biology of the mind has now captured the imagination of the scientific community of the twenty-first century, much as the biology of the gene fascinated the scientists of the twentieth century. As the biological study of the mind assumes the central position within biology and medicine, we have every reason to expect that a succession of brain scientists will be called to Stockholm and honored for their own leaps of faith (Kentros et al. 2001).

Acknowledgments

I have had the privilege to work with and to learn from many gifted students, fellows, and collaborators, and I have tried throughout this lecture to acknowledge their contributions. My science has benefited enormously from the interactive environment created by the Center for Neurobiology and Behavior at the College of Physicians and Surgeons of Columbia University. It would be hard to find a more ideal environment in which to mature as a scientist. Specifically, I have benefited greatly from my long-standing friendship with R. Axel, C. Bailey, J. Dodd, R. Hawkins, J. Koester, T. Jessell, J.H. Schwartz, S. Siegelbaum, and G. Fischbach, the current dean of the College of Physicians and Surgeons. I am further grateful to J. Koester for his excellent leadership of the Center for Neurobiology and Behavior, and to D. Hirsh, S. Silverstein, and J. Oldham, chairs of the three departments to which I belong. Finally, I am indebted to H. Pardes who, until recently, served as dean of the College of Physicians and Surgeons. My research has been generously supported by the Howard Hughes Medical Institute, the NIH, the Mathers Foundation, FRAXA, and the Lieber Trust. I am particularly indebted to the Howard Hughes Medical Institute and its leadership, D. Fredrickson, G. Cahill, P. Chopin, M. Cowan, D. Harter, and more recently T. Cech and G. Rubin, whose farsighted vision has encouraged Hughes investigators to take a long-term perspective so as to be able to tackle challenging problems. Research on learning and memory certainly meets both of these criteria!

References

Abel T, Nguyen PV, Barad M, et al: Genetic demonstration of a role for PKA in the late phase of LTP and in hippocampus-based long-term memory. Cell 88:615–626, 1997

Agnihotri N, Hawkins RD, Kandel ER, et al: Protein synthesis inhibition selectively abolishes long-term stability of hippocampal place cell maps. Abstr Soc Neurosci 27:316.14, 2001

Alberini C, Ghirardi M, Metz R, et al: C/EBP is an immediate-early gene required for the consolidation of long-term facilitation in Aplysia. Cell 76:1099–1114, 1994

Bacskai BJ, Hochner B, Mahaut-Smith M, et al: Spatially resolved dynamics of cAMP and protein kinase A subunits in Aplysia sensory neurons. Science 260:222–226, 1993

Bailey CH, Chen M: Long-term memory in Aplysia modulates the total number of varicosities of single identified sensory neurons. Proc Natl Acad Sci USA 85:2373–2377, 1988

Bailey CH, Chen M: Time course of structural changes at identified sensory neuron synapses during long-term sensitization in Aplysia. J Neurosci 9:1774–1780, 1989

Bailey CH, Kandel ER: Structural changes accompanying memory storage. Annu Rev Physiol 55:397–426, 1993

Bailey CH, Montarolo P, Chen M, et al: Inhibitors of protein and RNA synthesis block structural changes that accompany long-term heterosynaptic plasticity in Aplysia. Neuron 9:749–758, 1992

Bartsch D, Ghirardi M, Skehel PA, et al: Aplysia CREB2 represses long-term facilitation: relief of repression converts transient facilitation into long-term functional and structural change. Cell 83:979–992, 1995

Bartsch D, Casadio A, Karl KA, et al: CREB1 encodes a nuclear activator, a repressor, and a cytoplasmic modulator that form a regulatory unit critical for long-term facilitation. Cell 95:211–223, 1998

Bartsch D, Ghirardi M, Casadio A, et al: Enhancement of memory-related long-term facilitation by ApAF, a novel transcription factor that acts downstream from both CREB1 and CREB2. Cell 103:595–608, 2000

Bliss TV, Lømo T: Long-lasting potentiation of synaptic transmission in the denate area of the anaesthetized rabbit following stimulation of the perforant path. J Physiol 232:331–356, 1973

Blitzer RD, Connor JH, Brown GP, et al: Gating of CaMKII by cAMP-regulated protein phosphatase activity during LTP. Science 280:1940–1942, 1998

Bolshakov VY, Golan H, Kandel ER, et al: Recruitment of new sites of synaptic transmission during the cAMP-dependent late phase of LTP at CA3-CA1 synapses in the hippocampus. Neuron 19:635–651, 1997

Bourtchouladze R, Franguelli B, Blendy J, et al: Deficient long-term memory in mice with a targeted mutation of the cAMP-responsive element-binding protein. Cell 79:59–68, 1994

Brunelli M, Castellucci V, Kandel ER: Synaptic facilitation and behavioral sensitization in Aplysia: possible role of serotonin and cyclic AMP. Science 194:1178–1181, 1976

Byrne JH, Kandel ER: Presynaptic facilitation revisited: state and time dependence. J Neurosci 16:425–435, 1995

Byrne J, Castellucci V, Kandel ER: Receptive fields and response properties of mechanoreceptor neurons innervating skin and mantle shelf of Aplysia. J Neurophysiol 37:1041–1064, 1974

Byrne JH, Castellucci VF, Carew TJ, et al: Stimulus-response relations and stability of mechanoreceptor and motor neurons mediating defensive gill-withdrawal reflex in Aplysia. J Neurophysiol 41:402–417, 1978a

Byrne JH, Castellucci VF, Kandel ER: Contribution of individual mechanoreceptor sensory neurons to defensive gill-withdrawal reflex in Aplysia. J Neurophysiol 41:418–431, 1978b

Carew TJ, Pinsker HM, Kandel ER: Long-term habituation of a defensive withdrawal reflex in Aplysia. Science 175:451–454, 1972

Casadio A, Martin KC, Giustetto M, et al: A transient neuron-wide form of CREB-mediated long-term facilitation can be stabilized at specific synapses by local protein synthesis. Cell 99:221–237, 1999

Castellucci VF, Pinsker H, Kupfermann I, et al: Neuronal mechanisms of habituation and dishabituation of the gill-withdrawal reflex in Aplysia. Science 167:1745–1748, 1970

Castellucci VF, Carew TJ, Kandel ER: Cellular analysis of long-term habituation of the gill-withdrawal reflex in Aplysia. Science 202:1306–1308, 1978

Castellucci VF, Kandel ER, Schwartz JH, et al: Intracellular injection of the catalytic subunit of cyclic AMP–dependent protein kinase simulates facilitation of transmitter release underlying behavioral sensitization in Aplysia. Proc Natl Acad Sci USA 77:7492–7496, 1980

Castellucci VF, Nairn A, Greengard P, et al: Inhibitor of adenosine 3′:5′-monophosphate-dependent protein kinase blocks presynaptic facilitation in Aplysia. J Neurosci 2:1673–1681, 1982

Castellucci VF, Blumenfeld H, Goelet P, et al: Inhibitor of protein synthesis blocks long-term behavioral sensitization in the isolated gill-withdrawal reflex of Aplysia. J Neurobiol 20:1–9, 1989

Cedar H, Schwartz JH: Cyclic adenosine monophosphate in the nervous system of Aplysia californica, II: effect of serotonin and dopamine. J Gen Physiol 60:570–587, 1972

Cedar H, Kandel ER, Schwartz JH: Cyclic adenosine monophosphate in the nervous system of Aplysia californica, I: increased synthesis in response to synaptic stimulation. J Gen Physiol 60:558–569, 1972

Dash PK, Hochner B, Kandel ER: Injection of cAMP-responsive element into the nucleus of Aplysia sensory neurons blocks long-term facilitation. Nature 345:718–721, 1990

Ebbinghaus H: Memory: A Contribution to Experimental Psychology (1885). New York, Dover, 1963

Engert F, Bonhoeffer T: Dendritic spine changes associated with hippocampal long-term synaptic plasticity. Nature 399:66–70, 1999

Fatt P, Katz B: An analysis of the end-plate potential recorded with an intracellular electrode. J Physiol 115:320–370, 1951

Flexner LB, Flexner JB: Effect of acetoxycycloheximide and of an acetoxycycloheximide-puromycin mixture on cerebral protein synthesis and memory in mice. Proc Natl Acad Sci USA 55:369–374, 1966

Frazier WT, Kandel ER, Kupfermann I, et al: Morphological and functional properties of identified neurons in the abdominal ganglion of Aplysia californica. J Neurophysiol 30:1288–1351, 1967

Frey U, Huang Y-Y, Kandel ER: Effects of cAMP simulate a late stage of LTP in hippocampal CA1 neurons. Science 260:1661–1664, 1993

Frost WN, Castellucci VF, Hawkins RD, et al: Monosynaptic connections made by the sensory neurons of the gill- and siphon-withdrawal reflex in Aplysia participate in the storage of long-term memory for sensitization. Proc Natl Acad Sci USA 82:8266–8269, 1985

Glanzman DL, Mackey SL, Hawkins RD, et al: Depletion of serotonin in the nervous system of Aplysia reduces the behavioral enhancement of gill withdrawal as well as the heterosynaptic facilitation produced by tail shock. J Neurosci 9:4200–4213, 1989

Glanzman DL, Kandel ER, Schacher S: Target-dependent structural changes accompanying long-term synaptic facilitation in Aplysia neurons. Science 249:799–802, 1990

Grant SGN, O'Dell TJ, Karl KA, et al: Impaired long-term potentiation, spatial learning, and hippocampal development in fyn mutant mice. Science 258:1903–1910, 1992

Greengard P: Possible role for cyclic nucleotides and phosphorylated membrane proteins in postsynaptic actions of neurotransmitters. Nature 260:101–108, 1976

Hawkins RD, Castellucci VF, Kandel ER: Interneurons involved in mediation and modulation of the gill-withdrawal reflex in Aplysia, I: identification and characterization. J Neurophysiol 45:304–314, 1981a

Hawkins RD, Castellucci VF, Kandel ER: Interneurons involved in mediation and modulation of the gill-withdrawal reflex in Aplysia, II: identified neurons produce hetero-synaptic facilitation contributing to behavioral sensitization. J Neurophysiol 45:315–326, 1981b

Impey S, Obrietan K, Wong ST, et al: Cross talk between ERK and PKA is required for Ca^{2+} stimulation of CEB-dependent transcription and ERK nuclear translocation. Neuron 21:869–883, 1998

Kaang B-K, Kandel ER, Grant SGN: Activation of cAMP-responsive genes by stimuli that produce long-term facilitation in Aplysia sensory neurons. Neuron 10:427–435, 1993

Kandel ER: Cellular Basis of Behavior: An Introduction to Behavioral Biology. San Francisco, CA, Freeman, 1976

Kandel ER: The Behavioral Biology of Aplysia: A Contribution to the Comparative Study of Opisthobranch Molluscs. San Francisco, CA, Freeman, 1979

Kandel ER, Spencer WA: Electrophysiological properties of an archicortical neuron. Ann NY Acad Sci 94:570–603, 1961

Kandel ER, Spencer WA: Cellular neurophysiological approaches in the study of learning. Physiol Rev 48:65–134, 1968

Kandel ER, Brunelli M, Byrne J, et al: A common presynaptic locus for the synaptic changes underlying short-term habituation and sensitization of the gill-withdrawal reflex in Aplysia. Cold Spring Harb Symp Quant Biol 40:465–482, 1976

Kentros C, Hargreaves E, Hawkins RD, et al: Abolition of long-term stability of new hippocampal place cell maps by NMDA receptor blockage. Science 280:2121–2126, 1998

Kentros C, Agnihotri N, Hawkins R, et al: Stabilization of the hippocampal representation of space in mice requires attention. Abstr Soc Neurosci 27:316.15, 2001

Klein M, Kandel ER: Mechanism of calcium current modulation underlying presynaptic facilitation and behavioral sensitization in Aplysia. Proc Natl Acad Sci Abstr 77:6912–6916, 1980

Kupfermann I, Castellucci V, Pinsker H, et al: Neuronal correlates of habituation and dishabituation of the gill-withdrawal reflex in Aplysia. Science 167:1743–1745, 1970

Ma L, Zablow L, Kandel ER, et al: Cyclic AMP induces functional presynaptic boutons in hippocampal CA3–CA1 neuronal cultures. Nat Neurosci 2:24–30, 1999

Mackey SL, Kandel ER, Hawkins RD: Identified serotonergic neurons LCB1 and RCB1 in the cerebral ganglia of Aplysia produce presynaptic facilitation of siphon sensory neurons. J Neurosci 9:4227–4235, 1989

Malleret G, Haditsch U, Genoux D, et al: Inducible and reversible enhancement of learning, memory, and long-term potentiation by genetic inhibition of calcineurin. Cell 104:675–686, 2001

Mansuy IM, Mayford M, Jacob B, et al: Restricted and regulated overexpression reveals calcineurin as a key component in the transition from short-term to long-term memory. Cell 92:39–49, 1998

Martin KC, Michael D, Rose JC, et al: MAP kinase translocates into the nucleus of the presynaptic cell and is required for long-term facilitation in Aplysia. Neuron 18:899–912, 1997

Martin KC, Casadio A, Zhu H, et al: Synapse-specific transcription-dependent long-term facilitation of the sensory to motor neuron connection in Aplysia: a function for local protein synthesis in memory storage. Cell 91:927–938, 1998

Milner B, Squire LR, Kandel ER: Cognitive neuroscience and the study of memory. Neuron 20:445–468, 1998

Montarolo PG, Goelet P, Castellucci VF, et al: A critical period for macromolecular synthesis in long-term heterosynaptic facilitation in Aplysia. Science 234:1249–1254, 1986

Morris RGM, Anderson E, Lynch GS, et al: Selective impairment of learning and blockade of long-term potentiation by an N-methyl-D-aspartate receptor antagonist, AP5. Nature 319:774–776, 1986

Muller D: Ultrastructural plasticity of excitatory synapses. Rev Neurosci 8:77–93, 1997

Nguyen PV, Abel T, Kandel ER: Requirement of a critical period of transcription for induction of a late phase of LTP. Science 265:1104–1107, 1994

Nicoll RA, Malenka RC: Expression mechanisms underlying NMDA receptor-dependent long-term potentiation. Ann NY Acad Sci 868:515–525, 1999

O'Keefe J, Nadel L: The Hippocampus as a Cognitive Map. Oxford, Clarendon, 1978

Pinsker HM, Kupfermann I, Castellucci V, et al: Habituation and dishabituation of the gill-withdrawal reflex in Aplysia. Science 167:1740–1742, 1970

Pinsker HM, Hening WA, Carew TJ, et al: Long-term sensitization of a defenseive withdrawal reflex in Aplysia. Science 182:1039–1042, 1973

Ramón y Cajal S: The Croonian Lecture: la fine structure des centres nerveux. Proc R Soc London B Bio Sci 55:444–468, 1894

Rotenberg A, Abel T, Hawkins RD, et al: Parallel instabilities of long-term potentiation, place cells, and learning caused by decreased protein kinase A activity. J Neurosci 20:8096–8102, 2000

Schacher S, Castellucci VF, Kandel ER: cAMP evokes long-term facilitation in Aplysia sensory neurons that requires new protein synthesis. Science 240:1667–1669, 1988

Schwartz JH, Castellucci VF, Kandel ER: Functioning of identified neurons and synapses in abdominal ganglion of Aplysia in absence of protein synthesis. J Neurophysiol 34:939–953, 1971

Shuster MJ, Camardo JS, Siegelbaum SA, et al: Cyclic AMP–dependent protein kinase closes the serotonin-sensitive K⁺ channels of Aplysia sensory neurones in cell-free membrane patches. Nature 313:392–395, 1985

Siegelbaum S, Camardo JS, Kandel ER: Serotonin and cAMP close single K⁺ channels in Aplysia sensory neurones. Nature 299:413–417, 1982

Squire LR, Zola-Morgan S: The medial temporal lobe memory system. Science 253:1380–1386, 1991

Steward O: mRNA localization in neurons: a multipurpose mechanism? Neuron 18:9–12, 1997

Winder DG, Mansuy IM, Osman M, et al: Genetic and pharmacological evidence for a novel, intermediate phase of long-term potentiation (I-LTP) suppressed by calcineurin. Cell 92:25–37, 1998

Yin J, Tully T: CREB and the formation of long-term memory. Curr Opin Neurobiol 6:264–268, 1996s

"GENES, BRAINS, AND SELF-UNDERSTANDING"

John M. Oldham, M.D.

Kandel's essay "Genes, Brains, and Self-Understanding: Biology's Aspirations for a New Humanism" is eloquent, integrative, and visionary. With characteristic prescience, Kandel outlines the rapidly changing face and breathtaking potential of the science and practice of medicine. Indeed, it takes only a moment of reflection to envy the graduating medical students at Columbia University College of Physicians and Surgeons as they embarked upon their careers, launched with this inspirational, scientifically derived prophecy.

The potential for the new knowledge of the human genome to move us from a focus on populations at risk to the specific genetic vulnerabilities of an individual is exciting and increasingly real, paving the way for renewed and individualized emphasis on protective mechanisms and prevention. However, how effective this new individualized information will be remains unclear, and herein lies a fundamental challenge that, with all of our new knowledge, we must not overlook. We already know from studies of clinical populations, for example, that a healthy diet and regular exercise are protective factors for individuals at risk for coronary artery disease, or that careful adherence to antihypertensives and minimization of stress are protective against dangerous hypertensive episodes, or that major depressive disorder is a serious medical condition that should be treated quickly (which is usually quite effective) when it occurs. And we know that countless numbers of people who repeatedly receive this type of advice do not follow it. Human behavior is, to understate it, complicated.

Kandel ranks at the very top of "the family of deep problems that confront the study of the mind," the need to understand "the biology of consciousness," which is so clear and urgent that surely there is no controversy here. But I would extend this challenge to emphasize the need to understand the biology of the unconscious—to continue our efforts to understand what motivates human behavior at all levels. We already know a great deal about certain behaviors that are self-injurious, such as the molecular neurobiology of addiction, but even so, addiction is a complex mix of biology and behavior. We have every reason to expect that advances in our knowledge of genetics will help us sort out those individuals at highest genetic risk for a given type of addictive behavior—who may need our greatest attention—from those who, with help, may have better odds to leave the addiction behind. But what about patterns of self-injurious behavior that may go unrecognized by the individual in question? For example, gambling to the brink of bankruptcy while being the sole support of a loving family; being driven by extreme, narcissistic personal ambition while leaving an unnoticed trail of used and discarded co-workers behind; making repeated bad choices in relationships without knowing why, leading to escalating frustration and bewilderment, or even, for those at risk, the emergence of severe depression.

Kandel's three laws emphasize pleasures and obligations that are wonderfully relevant to his listening audience of graduating students. There are, however, many future visitors to the consulting room who have obligations and burdens, yet little or no access to pleasure or relief—those illustrated above who create and perpetuate their own unhappiness, and a large universe of others who may be at or near the poverty level, without education, or ill equipped to keep their balance in tough parts of the world. It may well be here, in the realm of negligible resources that the pleasures of a high-fat meal trump the self-discipline of a healthy diet, especially if living longer looks like prolonged misery.

So while science moves at warp speed and takes medicine to a new humanism that is individualized, we must not forget to look for ways to help each individual's world become a better one. Otherwise, that stress-filled environment doubles right back to attack the health of the individual—the very health that we're trying hard to sustain and improve.

CHAPTER 8

GENES, BRAINS, AND SELF-UNDERSTANDING

Biology's Aspirations for a New Humanism

Eric R. Kandel, M.D.

Revolution in Genomics

Members of the graduating class of the year 2001, relatives and friends of the graduates, Dean Gerald Fischbach, colleagues, ladies and gentlemen: I am extremely pleased to be asked to participate in the commencement today because it gives me the opportunity of celebrating the academic achievements of this college and of this class. I owe this college an enormous personal and scientific debt! I have found this medical school to be the very best place in the world to do scientific research and I have benefited greatly from the interactive and supportive environment engendered by the faculty and the students of this school. In addition, throughout my 27 years on this faculty, I have always enjoyed teaching medical students, including, of course, the privilege of teaching the distinguished class we celebrate today. In fact, *Principles of Neuroscience*, the textbook that your class has come to know and love—which is now universally acknowledged to be the heaviest and most expensive book of its kind—is based on the neural science course our fac-

---

This paper is a slightly modified version of the graduation address given on May 16, 2001, at the Columbia University College of Physicians and Surgeons.

ulty teaches here at this college, a course for which I was privileged to serve
as first course director.

So when I am asked to what do I aspire after receiving the Nobel Prize in
the year 2000, my answer is clear: to be selected, by the graduating class of
the year 2001, to give the convocation address at the College of Physicians
and Surgeons of Columbia University! What more meaningful and satisfying
recognition can one ever imagine?

For no celebration is more satisfying for this college or more inspiring to
the intellectual community throughout the world than an academic commencement. For each commencement celebrates the entry into academic
ranks of another class of scholars. Since the task of a great university is not
to simply replicate its own image in scholarship but to create a new knowledge, it is implicit in the charge to a faculty to develop scholars who are better than we are, more knowledgeable, more thoughtful, more moral, finer
human beings.

Given that we think you are all of these things, what is there left to tell
you as you now progress from being our students to being our peers? What
are you likely to confront as you move into the next stage of your life? And,
in turn, what can we expect of you in that confrontation? Let me put these
questions, and your past 4 years in medical school, into a bit of historical
perspective.

The years you have spent in medical school—the remarkable 4 years that
spanned the transition from the twentieth century to the twenty-first century—have produced both the elucidation of the human genome and an increased understanding of the biology of the human brain. We have every
reason to expect that the revolution in genomics and in brain science will
radically change the way we practice medicine. And it will do so in two ways.
First, medicine will be transformed from a population-based to an individ-
ual-based medical science; it will become more focused on the individual
and his or her predisposition to health and disease. Second, *we* will, for the
first time, have a meaningful and nuanced biology of human mental processes and human mental disorders. If we are fortunate, your generation will
help join these two intellectual streams—that of the human genome and that
of brain science—to realize biology's aspiration for a new humanism, a humanism based in part on insights into our biology. If we are successful in advancing this new humanistic agenda, the genomic revolution and the new
insights into the biological nature of mind will not only enhance medical
care but will also change fundamentally the way we view ourselves and one
another.

The influence of biology on the way informed people think about each
other and about the world in which they live is, of course, not new. In modern times, this influence first became evident in 1859 with Darwin's insight

into the evolution of species. Darwin first argued that human beings and other animals evolved gradually from animal ancestors quite unlike *themselves*. He also emphasized the even more daring idea, that the driving force for evolutionary change stems not from the heavens, not from a conscious purpose, but from natural selection, a completely mechanistic, sorting process based on hereditary variations.

This radical idea split the bond between religion and biology, a bond based on the idea that an important function of biology was to explain divine purpose—to account for the overall design of nature. Indeed, natural selection even caused difficulty for nonbelievers because it was vague as a scientific idea.

To understand hereditary variations, scientists *first* needed to know: how is information about biological structure passed from one generation to another? This question was answered only in the first decades of the twentieth century. We owe first to Gregor Mendel and then to Thomas Hunt Morgan (of our *own* Columbia University) the remarkable discovery that hereditary information necessary to specify the construction of the organism is passed from one generation to the next by means of discrete biological structures we now call genes. Forty years later, first Avery, McCarthy, and McCloud and subsequently Watson and Crick gave us the seminal insight that the genes of all living organisms are embodied in the physics and chemistry of a single large molecule, DNA. Nature, in all its beauty and variety, results from variations in *the sequence of bases in DNA*.

In the 1960s and 1970s, our understanding of genes was further enhanced by the cracking of the genetic code, the three-letter alphabet whereby *the sequence of bases in DNA* is translated into the amino acids of a protein. This breakthrough was followed by DNA sequencing, which allowed us to read directly the nucleotide sequences that form the instructions of each gene. Creative application of these and other molecular insights made possible genetic engineering and more recently the sequencing of the human genome.

The current generation of physicians will be the first to reap the benefits of the human genome and use its insights not only to provide better care to patients—better diagnoses, better treatment—but, also, I would hope, *more individualized care,* more individually tailored diagnoses, and more individualized treatment. Indeed, one would hope that this generation will move us away from the impersonality of managed health care into a new, biologically inspired personalized medicine.

What reason do we have to believe that this will come to pass? What will we learn from the genome that might *orient* us *more* to see the patient as a person rather than as a disease state? The genome of course provides us with a periodic table of life. It contains the complete list and structure of all genes.

But it provides us not simply with an average-expectable genome. It provides each of us with *our own unique genome*. In time, our genome will be a part of our private medical record. As a result, we in academic medicine will collectively have a catalog of *all* the human genetic variations that *account for all the heritable differences between individuals.*

We now know that any two individuals share an amazing 99.9% DNA sequence similarity. This means that all the heritable differences among individuals of a species can be attributed to a mere 0.1% of the sequence. Most differences between the genomes of any two individuals take the form of very small changes, where one single base is substituted for another in the sequence of nucleotides that form a gene. These changes are called *single base changes* or *single nucleotide polymorphisms.*

We already know of about 3 million such polymorphisms, and more will be identified with time. They are spread throughout the genome and at least 93% of all genes contain at least one such polymorphism. Thus, for the first time, we will have for every gene all the polymorphic sequence variations that exist. Many of these will prove unimportant, but some of them will be fundamental to understanding disease.

These common, polymorphic variations differ fundamentally from the rare mutations that lead invariably to inherited disease, and that have been the focus of medical genetics up to now. The common polymorphisms that we now will have full access to for the first time do not cause disease per se; rather, they *influence the expression of disease; they predict our predisposition to, and our protection from, disease in all of its manifestations.*

To give but one example, there are rare genetic mutations on chromosome 21 that invariably cause an early-onset form of Alzheimer's disease in the rare person who carries the mutation. By contrast, there is a fairly common polymorphism that *does not* produce Alzheimer's disease directly. But the 17% of the population that carry this single base change polymorphism have a 10-times greater risk of developing a late onset form of Alzheimer's disease than those individuals who do not carry this polymorphism. Other genetic polymorphisms similarly predispose people to various forms of diabetes, hypertension, cancer, and mental disorders. Indeed, every disease to which we are prone—including our response to infection, to the consequences of aging, and even our very longevity itself—will be shown to be influenced by polymorphisms in our genes. As a corollary, the polymorphisms also will help reveal that complex diseases such as hypertension, depression, and Alzheimer's disease are likely not to be unitary but to be made up of a number of different, intricately related subtypes, each requiring its own distinctive medical management.

What will knowledge of these predispositions and subtypes mean for the practice of clinical medicine? This knowledge will serve to decrease the uncertainty in the management of disease. It is likely that clinical DNA testing—the search for genetic polymorphisms in ourselves and in our patients—will reveal our individual risk for all major diseases and therefore allow us to intervene prophylactically in these diseases through diet, surgery, exercise, or drugs, years before the disease becomes manifest. Indeed, genetic polymorphisms will be found to underlie the way our patients respond to these interventions, so that DNA testing will also allow us to predict individual responses to drugs and to determine the degree to which individuals are susceptible to particular side effects. This will allow the pharmaceutical industry to develop *new* targets and *new* tools to sharpen the specificity of the drugs they deliver to meet the needs of the *individual* patient.

This knowledge of *the biological uniqueness of our patients will alter all aspects of medicine.* Currently, newborn babies are only screened for treatable genetic diseases, such as phenylketonuria. Perhaps in the not too distant future, children at high risk for coronary artery disease, Alzheimer's, or multiple sclerosis will be identified and treated to prevent changes occurring later in life. For middle-aged and older people, you will be able to determine the risk profiles for numerous late-onset diseases; ideally, people at risk will know of their risk before the appearance of symptoms, so that their disease might, at least, be partially prevented through medical intervention.

The Biological Basis of Uniqueness

This new emphasis on the biological basis of uniqueness, encouraged by the human genome, brings me to my second point. Our uniqueness is reflected, in its highest form, in the uniqueness of our mind, a uniqueness that emerges from the uniqueness of our brain. Now that we understand natural selection and the molecular basis of heredity, it has become clear that the *last great mystery* that confronts biology is the nature of the human mind. This is the ultimate challenge, not just for biology but for all of science. It is for this reason that many of us believe that the biology of the mind will be for the twenty-first century what the biology of the gene was for the twentieth century.

The biology of mind represents the final step in the philosophical progression that began in 1859 with Darwin's insights into evolution of bodily form. Here, with the biology of mind, we are confronted with the even more radical and profound realization that the mental processes of humans also have evolved from animal ancestors and that the mind is not ethereal but can be explained in terms of nerve cells and their interconnections.

One reason that people have difficulty altering their view of the mind is that the science of the brain, like all experimental science, is at once mechanistic in thought and reductionist in method. We have become comfortable with the knowledge that the heart is not the seat of emotions but a muscular organ that pumps blood through the circulation. Yet some of us still find it difficult to accept that what we call mind is a set of functions carried out by the brain, a computational organ made marvelously powerful not by its mystery but by its complexity, by the enormous number, variety, and interactions of its building blocks, its nerve cells. We find it difficult to accept that every mental process, from our most public action to our most private thought, is a reflection of biological processes in the brain.

With modern imaging and cell-biological studies of brain, we are now beginning to understand aspects of both our public actions and our private thoughts: we are beginning to understand how we perceive, act, feel, learn, and remember. And the insights we so far have obtained are truly remarkable! For example, these studies show that the brain does not simply perceive the external world by replicating it, like a three-dimensional photograph. Rather, the brain reconstructs reality only after first analyzing it into component parts. In scanning a visual scene, for example, the brain analyzes the form of objects separately from their movement, and both separately from the color of the objects, all before reconstituting the full image again, according to the brain's own rules. Thus, the belief that our perceptions are precise and direct is an illusion. We re-create in our brain the external world in which we live.

We now appreciate that simply to see—merely to look out into the world to recognize a face or to enjoy a landscape—entails an amazing computational achievement on the part of the brain that no current computer can even begin to approach. All of our perceptions and actions—seeing, hearing, smelling, touching, or reaching for a glass of water—are analytic triumphs.

In addition to creating our perceptions and actions, our brain provides us with a sense of awareness, it creates for us a historical record, a consciousness not only of ourselves but of the world around us. Within the family of deep problems that confront the study of mind, the biology of consciousness must surely rank at the very top.

The brain can achieve consciousness of self, and can perform remarkable computational feats because its many components, its nerve cells, are wired together in very precise ways. Equally remarkable, we now know that the connections between cells are not fixed but can be altered by experience, by learning. The ability of experience to change connections in our brain means that the brain of each person in this audience is slightly different from the brain of every other person in this audience because of distinctive differences in our life history. Even identical twins, with identical genomes, will have slightly different brains because they will invariably have been exposed to somewhat different life histories.

The Individuality of Mental Life

It is very likely that during your careers, brain imaging will succeed in resolving these unique differences of our brain. We will then have, for the first time, a biological foundation for the individuality of our mental life. If that is so, we will have a powerful new way of diagnosing behavioral disorders and evaluating the outcome of treatment including the outcome of psychotherapy.

Seen in this light, the biology of mind represents not only a scientific and clinical goal of great promise but one of the ultimate aspirations of humanistic scholarship. It is part of the continuous attempt of each generation of scholars to understand human thought and human action in new terms.

Personalized Medicine

Your generation—the first postgenomic generation—will have adequate information from both the human genome and from brain sciences to explore, more meaningfully than ever before, the genetic contribution to mental processes. Indeed, we already know that not only psychiatric disorders but almost all long-standing patterns of behavior—from wearing bow ties to being socially gregarious—show moderate to high degrees of heritability. The human genome will thus not only aid in revolutionizing psychiatry and neurology, but it also will allow us a better understanding of normal behavior—of how you and I function.

For example, the analysis of genetic polymorphisms may at last uncover how genetic factors interact with the environment to encourage our various intellectual capabilities, our mathematical and musical talents and perhaps even our differing capabilities for creativity, for empathy, and for self-understanding. Whatever the details, we can expect that the genome will reveal new links between genetics and environment that our society will eventually have to confront.

As these and other questions are addressed, biology and medicine will help transform our society as they transform our understanding of the individuals in society. You will therefore be creating a world in which it is imperative for each individual to have sufficient understanding of this new knowledge so that we, as a society, can apply it wisely.

But like all knowledge, biological knowledge is a double-edged sword. It can be used for ill as well as for good, for private profit as well as for public benefit. In the hands of the misinformed or the malevolent, natural selection was distorted into social Darwinism, genetics was corrupted into eugenics. Brain sciences have also been, and can again be, misused for social control and manipulation.

This brings me to one final point. We are entering a world that is being changed because of advances in science and in technology *and by the social ramifications of these advances.* It will be *our obligation* to reach out to understand these advances, to evaluate them, to encourage some and restrict others. By extension, beyond our own education, *we will need to assume the leadership roles for which you have been trained to ensure the scientific literacy of the general public,* especially the scientific literacy of the patients that you will be treating.

Kandel's Laws

Let me then conclude my comments about medicine's aspirations for a new humanism by enunciating three principles that I now, in my seniority, invoke with some frequency. These principles, which I believe reflect some of my best thinking, are of such importance that I have come to refer to them, in the modesty and privacy of my own study, as Kandel's three laws.

Kandel's first law states that belonging to a university community is one of the deepest intellectual pleasures of one's life. Universities are the institutions that make society great. People from all over the world come to the United States to study in our universities because the rest of the world sees the American university as our most extraordinary national product. I will go further and say that I fully believe there is nothing more important for our society, and indeed for the world at large, than the two great missions of the university: to produce new ideas and to train young people to assume responsible roles in their society.

Belonging to a university assures you that you will be a scholar in perpetuity—one of the great sensual pleasures of life. Kandel's first law, you will appreciate, is not original. I want to remind you that the first medical school convocation in the American colonies was held at the College of Physicians and Surgeons, when *this* college conferred the first M.D. degree in the Americas, an honorary M.D., for his services to this college, to Samuel Bard, our first professor of medicine. In his commencement address on May 16, 1769—232 years ago to the very day—Samuel Bard said:

> Do not therefore imagine, that from this Time your Studies are to cease; *so far from it;* you are to be considered as but just entering upon them; and unless your whole Lives, are one continued Series of Applications and Improvement, you will fall short of your Duty...In a Profession then, like that you have embraced, where the Object is of so great Importance as the Life of a Man; you are accountable even for the Errors of Ignorance, unless you have embraced every Opportunity of obtaining Knowledge.

Kandel's second law is that within the university, teaching is a particularly rewarding activity. There is no better way to assure yourself that you understand an issue than to try to explain it to others. Teaching will guarantee that you understand the major scientific issues of your time. It will also give you a perspective on how your thinking and your work fit in with the rest of medicine.

Kandel's third law is that patient care is beyond question our most important responsibility. That is why we are here. Never let patient care take a secondary role to any other activity in your professional life. Patient welfare is the ultimate goal of biological science and it is the engine that drives the whole scientific enterprise. Here, I again want to recall for you Samuel Bard's comments of 1769:

> In your Behavior to the Sick, remember always that your Patient is the Object of the tenderest Affection to some one, or perhaps to many about him; it is therefore your Duty, not only to endeavour to preserve his Life; but to avoid wounding the Sensibility of a tender Parent, a distressed Wife, or an affectionate Child. Let your Carriage be humane and attentive, be interested in his Welfare, and shew your Apprehension of his Danger.

As I hope these three laws make clear, you should leave here confident that the best days of medical care and the best days of your lives are *ahead* of you. As a result of the training you have received at the College of Physicians and Surgeons of Columbia University, we are confident that you will be able to influence, through your knowledge and your actions, the emergence of a new humanism, a humanism made more rational by a deeper respect for the genome and a greater understanding of the human mind. You are entering an exciting time in your lives and in the history of medicine, a time that will afford you the opportunity to benefit your patients, your university, and your society in novel, important, and humanizing ways. So enjoy the future, and do it justice.

AFTERWORD

Psychotherapy and the Single Synapse Revisited

Where is psychiatry heading? What areas of psychiatry will benefit most from biology in the years ahead?

Perhaps the most important and most anticipated advances will come from the delineation of the genes that render people vulnerable to various mental illnesses and the characterization of those genes in experimental animals—in worms, flies, and mice. A second priority is the development of a new neuropathology of mental illness, one based on knowledge of how specific molecules in specific regions of the brain make people vulnerable to specific types of mental illness. A third priority is higher-resolution brain imaging technologies that will enable us to see anatomical changes in the brains of mental patients before and after treatment.

Advances along these lines will put us in a position to pinpoint the experiences that act on genetic and anatomical predispositions to disease. Understanding mental illness in genetic, anatomical, and experiential terms is likely to open up new therapeutic approaches. In addition to better drugs, we may expect better psychotherapies and better ways of selecting therapies that are effective for specific types of patients. This last theme has interested me for some time. In fact, the first essay in this volume, "Psychotherapy and the Single Synapse," makes the point that the effects of psychotherapy must ultimately be explained empirically on the level of individual neurons and their synapses, just as the effects of drugs are.

In view of our progress in the biological understanding of mental disorders, we can now ask, Is the attempt to evaluate psychotherapy in biological terms still a profitable endeavor? In the last three decades, we have developed drugs that are effective in the treatment of a variety of psychiatric disorders: obsessive-compulsive disorder, anxiety disorders, posttraumatic stress disorder, depression, bipolar disorders, and the positive symptoms of schizophrenia. Yet our experience has also made it clear that drugs alone are often not sufficient treatment. Some patients do better when psychotherapy is combined with drugs, while other patients do reasonably well with psychotherapy alone.

In her book *An Unquiet Mind*, Kay Jamison describes the benefits of both modes of treatment. Lithium prevented her disastrous highs, kept her out of the hospital, saved her life by preventing her from committing suicide, and made psychotherapy possible. "But, ineffably," she writes, "psychotherapy *heals*. It makes some sense of the confusion, reins in the terrifying thoughts and feelings, returns some control and hope and possibility of learning from it all. Pills cannot, do not, ease one back into reality" (Jamison 1996, p. 89).

Psychotherapy has not only contributed to the treatment of mental illness it has provided us with a tool for examining the workings of the mind by peeling back superficial layers of action and revealing the deeper motives. Until quite recently, there were few independent, compelling ways to test psychodynamic ideas or to evaluate the relative efficacy of one therapeutic approach over another. However, neuroimaging may give us just that—a method of revealing both mental dynamics and the workings of the living brain. Had imaging been available in 1894, when Freud wrote "On a Scientific Psychology," he might well have directed psychoanalysis along very different lines, keeping it in close relationship with biology, as he outlined in that essay. In this sense, combining brain imaging with psychotherapy represents top-down investigation of the mind, which continues the scientific program that Freud originally conceived for psychoanalysis.

Indeed, one might say that clinical imaging is of even greater importance to psychiatry than it is to neurology. One obstacle to understanding mental illness has been the limitations posed by animal models. Most mental illnesses affect functions that appear to be uniquely human—that is, language, abstract thought, and complex social interactions. As a result, we cannot as yet model a number of critical features of mental illness; these can only be studied successfully in people. Psychotherapy presumably works by creating an environment in which people learn to change. If those changes are maintained over time, it is reasonable to conclude that psychotherapy leads to structural changes in the brain, just as other forms of learning do. Indeed, we can already image people's brains before and after therapy and thus see the consequences of psychotherapeutic intervention in certain disorders.

Preliminary imaging studies have found that obsessive-compulsive disorder is associated with an increase in metabolism in the caudate nucleus. Schwartz and his colleagues at the University of California in Los Angeles have described how the increased metabolism can be reversed by a form of psychotherapy called exposure therapy as well as by selective serotonin uptake inhibitors (Schwartz et al. 1996). Moreover, imaging studies of depression commonly show a decrease in basal activity in the dorsolateral region of the prefrontal cortex and an increase in activity in the ventrolateral region. Psychotherapy and drugs reverse aspects of these two abnormalities, but here the different modes of treatment affect distinctly anatomical loci in the brain ones (for a review, see Etkin et al. 2005).

Thus, we may now be able to describe with some rigor the metabolic changes in the brain that result from drug therapy and those that result from psychotherapy. Indeed, since a variety of psychotherapies are now in use, it may be possible to distinguish among them on the basis of the changes they produce in the brain. It may be that all effective psychotherapies work through a common set of anatomical mechanisms. Alternatively, they may achieve their goals through distinctly different processes. Effective psychotherapies may even have adverse side effects, as drugs do. Describing psychotherapies in terms of empirical evidence could help maximize the safety and effectiveness of these important treatments, much as has been done for drugs. Empirical studies would also help predict the outcome of particular types of therapeutic interventions and would direct patients to the therapies most appropriate for them.

It is becoming increasingly apparent that a biological approach to psychiatry will enable us to reach a deeper understanding of human behavior. For instance, a number of experimental approaches can be used today to distinguish conscious from unconscious mental processes. These approaches are not limited to the implicit unconscious; they can also explore the dynamic and the preconscious unconscious. One way of doing this is to compare the brain activation patterns generated by unconscious and conscious perceptual states (as with the perception of fear) and to identify the regions of the brain that are recruited by each state.

Finally, biology can sharpen psychiatry's dual contribution to modern medicine: its ability to develop effective drug treatments based on neuroscience and its ability to listen to and learn from patients. We need to combine these two treatment modalities in ways that are at once objective and effective. If we are successful in this undertaking, we will join radical reductionism, which drives biology, with the humanistic goal of understanding the human mind, which drives psychiatry.

Thus, a half century after I left psychoanalysis because it was unconcerned with biology, science has progressed to the point where we now have a rudimentary biology of the mind. The goal for the next decade is twofold: first, we need to determine how specific combinations of genes give rise to altered brain anatomy that results in mental illness by increasing vulnerability to specific social and environmental experiences. Second, we need to see how drugs and psychotherapy can complement one another in the treatment of mental disorders. In taking on this twofold task, biology will be addressing some of the issues that first attracted so many of my generation to psychiatry and psychoanalysis. Moreover, with this agenda the new biology of mind will assume its ascribed role as the natural bridge between the humanities, which is concerned with the nature of human existence, and the natural sciences, which are concerned with the nature of the physical world. Indeed, in the next half, century great universities will be judged by how successfully they make that bridge and how much they contribute to our understanding of the human mind. One would hope that psychiatry and psychoanalysis would be central contributors to this historic effort to understand the mind in biological terms. If it does succeed in exercising this role, psychiatry and psychoanalysis will again capture the best and brightest of the next generations.

Acknowledgments

My laboratory is supported by the Howard Hughes Medical Institute, the Fred Kavli Institute for Brain Science, the Lieber Center for Schizophrenia Research, the Mathers Foundation, and the National Institute of Mental Health. I have benefited greatly from comments on the introduction and on the afterword from Amit Etkin, Marianne Goldberger, Christopher Kellendonk, Henry Numberg, Jonathan Polan, and Eleanor Simpson. I am also grateful to Howard Beckman and Blair Burns Potter for their editorial advice. Finally, I thank Dr. Robert Hales, editor-in-chief, and Julia Bozzolo, my book editor at American Psychiatric Publishing, Inc., for their thoughtful help in putting this book into its final form.

References

Etkin A, Pittenger CJ, Kandel ER: Biology in the service of psychotherapy, in The American Psychiatric Publishing Textbook of Personality Disorders. Edited by Oldham JM, Skodol AE, Bender DS. Washington, DC, American Psychiatric Publishing, Inc. 2005, pp 669–682

Jamison K: An Unquiet Mind. New York, Vintage Books, 1996

Schwartz JM, Stoessel PW, Baxter LR Jr, et al: Systematic changes in cerebral glucose metabolic rate after successful behavior modification treatment of obsessive-compulsive disorder. Arch Gen Psychiatry 53:109–113, 1996

INDEX

Page numbers printed in **boldface** type refer to tables or figures.